Microelectronics: Processing and Device Design

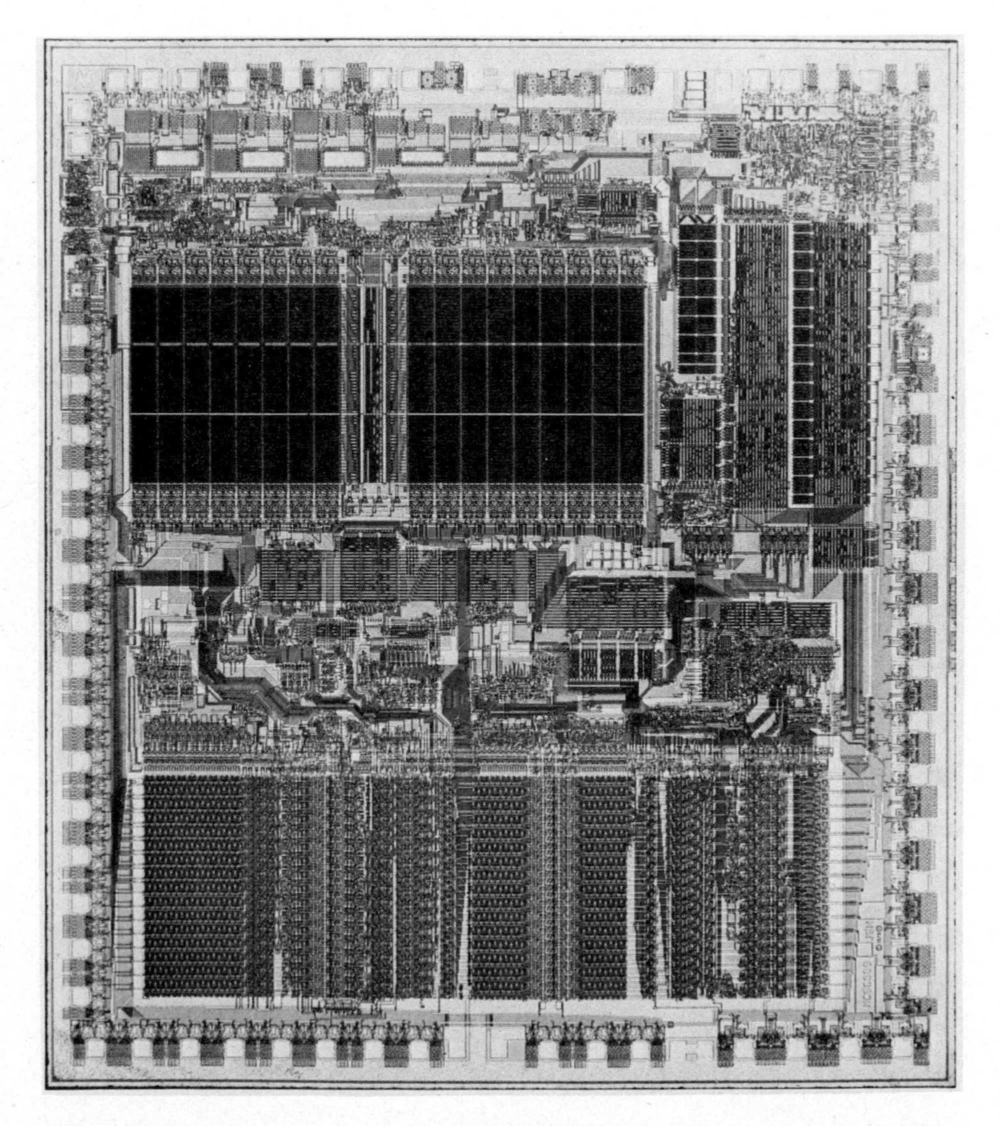

A large-scale integrated circuit, the MC68000 (Courtesy of Motorola, Inc., MOS Division).

Microelectronics: Processing and Device Design

ROY A. COLCLASER
The University of New Mexico

JOHN WILEY & SONS

New York • Chichester • Brisbane • Toronto • Singapore

Library of Congress Cataloging in Publication Data

Colclaser, Roy A
Microelectronics.

Includes index.
1. Microelectronics. I. Title.
TK7874.C63 621.381′7 79–29727
ISBN 0-471-04339-7

Printed in the United States of America

10 9 8 7 6 5

To Judi

PREFACE

The world of microelectronics is characterized by technological advances that have occurred more rapidly than in almost any other field. During the 30 years since the discovery of the transistor effect, many processing techniques have been applied to the fabrication of microelectronic circuits. As I write this, new techniques continue to be tested, for example, laser assisted doping, which may or may not have a significant impact on future products. Each process imposes certain restrictions on the operating characteristics of the devices and components fabricated in part by the process. The purpose of this book is to investigate typical microelectronics processes and their influences on device design. It is anticipated that four basic technologies, thick-film hybrid, thin-film hybrid, monolithic bipolar, and monolithic metal-oxide-semiconductor (MOS), will each continue to dominate their share of the microelectronics market, based on their particular advantages for specific applications. For this reason, all four of these technologies are discussed, but the emphasis is on monolithic bipolar and MOS integrated-circuit processing and device design.

The material in this book was developed over a period of eight years while I was teaching lecture and laboratory courses on microelectronics at the senior and beginning graduate levels at The University of New Mexico. Valuable background information was also gained while I was on sabbatical and during summer employments with the Integrated Circuit Technology Division of Sandia Laboratories, Albuquerque.

It is assumed that the user of this book has a basic understanding of electronics and semiconductor materials at a level typical of students entering the senior year in an electrical engineering undergraduate degree program. In addition, students should be familiar with device physics at a level comparable to that of *Device Electronics for Integrated Circuits* by Muller and Kamins (Wiley, 1977). The appendix provides a brief introduction to (or review of) semiconductor materials and device physics. A complete background in the latter is not a necessity for using the results of Chapters 9 and 10, but the validity of many of the design equations must then be accepted by the reader.

The book is organized essentially into two parts, with Chapters 2 to 8 focusing on processing, and Chapters 9 to 11 concentrating on device and component design. As much as possible, the individual chapters are independent, and can be read in any order. The most notable exception to this is Chapter 7, "Selective Doping Techniques," which should be read before Chapter 9, "Devices for Bipolar Integrated Circuits." Problems are included at the ends of chapters where appropriate. The problems serve several purposes. Some of them are intended to clarify the text material and to make certain of the figures more meaningful. Others are numerical exercises that familiarize students with the appropriate orders of magnitudes to be expected for certain quantities. There are also some problems that are more challenging and will take more time and effort to complete. If a fabrication laboratory course accompanies the lecture, this text will provide informative background material to help in explaining why particular processes are performed as they are. Many of the graphs in the book are adapted from the literature in a new form which makes them more meaningful and relatively easy to use.

I am indebted to my students, who have been patient through three previous versions of the manuscript. Of particular inspiration were Bill Burnett, Fred Bird, and Mark McDermott. I also thank Donald Davis, Robert Kopp, Richard Muller, Jerry Sergent, Edward Graham, and Agustin Ochoa for their helpful suggestions after reading all or parts of the manuscript. I am indebted to The University of New Mexico for assistance in preparing the manuscript and to William R. Dawes, Jr., of Sandia Laboratories, Albuquerque, for providing the opportunity to work in a modern integrated-circuits fabrication facility. Finally, I am particularly grateful to Sharon Henze Burnett for typing the manuscript, to Robert Krein for assisting in the preparation of the figures, and to Gene Davenport, editor of electrical engineering and computer science at Wiley, for his advice and guidance.

Albuquerque, New Mexico Roy A. Colclaser

CONTENTS

Microelectronics: Processing and Device Design

Chapter 1
An Overview of Microelectronics

The performance of information processing on a grand scale, either in analog or digital form, in almost zero space has been a goal, or dream, of mankind ever since the birth of electronics. Science fiction authors have casually assumed that scientists and engineers would eventually be able to produce marvelous devices like the two-way wrist television in *Dick Tracy** and the computers in Isaac Asimov's *The Last Question* (1), which increased in capability and decreased in size with each generation. Technological advances since the end of World War II have brought these dreams close to reality.

Microelectronics is that area of technology associated with the fabrication of electronic systems or subsystems using extremely small components. There are numerous ways to "shrink" electronic circuits. In this book, we will consider four technologies that are in widespread use for the manufacture of microcircuits.

Thick-Film Hybrid. Conductor, dielectric, and resistor patterns screen printed on ceramic substrates with "add-on" miniature components consisting of semiconductor devices and monolithic integrated circuits, capacitors, inductors, transformers, and so forth.

Thin-Film Hybrid. Conductor, dielectric, and resistor materials deposited by evaporation, sputtering, electro- or electroless plating on ceramic, glass, or crystalline substrates patterned by photoengraving or by depositing through a mask, with "add-on" components consisting of semiconductor devices and monolithic integrated circuits, capacitors, inductors, and so forth.

* *Dick Tracy* is a comic strip originated by Chester Gould and copyrighted by the Chicago Tribune.

Monolithic Bipolar. Bipolar junction and monopolar junction field-effect transistors, junction diodes, Schottky diodes, resistors, and capacitors formed by epitaxy, oxidation, ion-implantation, diffusion, chemical vapor deposition, and photoengraving on and within a single crystal semiconductor substrate and intraconnected by thin-film conductors.

Monolithic Metal-Oxide-Semiconductor (MOS). MOS field-effect transistors, capacitors, resistors, and conductors formed by oxidation, diffusion, ion-implantation, chemical vapor deposition, photoengraving, and thin-film deposition on and within a single crystal semiconductor or insulating substrate.

There are other possible ways for fabricating microcircuits, but most of these are based on combinations of the above technologies. Many of the terms listed above may be unfamiliar to the reader at this point, but they will be described in detail in subsequent chapters.

In this chapter, we will trace the origins of the four basic technologies and present examples demonstrating how each of the fabrication techniques could be applied to the fabrication of a microcircuit that will perform the same electronic function.

THE DEVELOPMENT OF MICROELECTRONICS TECHNOLOGY (2–6)

The first microcircuits in which components were integral with the substrates were developed by the Centralab Division of Globe-Union, Inc. in cooperation with the National Bureau of Standards. The circuits were subassemblies used in proximity fuses first employed during the Korean War. Development of these circuits started at the end of World War II. The early versions used conductors screen printed onto ceramic substrates with capacitors in chip form and miniature vacuum tubes as "add-on" components. Later developments included baked-on carbon resistors and high dielectric constant substrates which served as capacitor dielectrics. The development of resistor compositions that could be fired on the ceramic substrates was a major step in the development of thick-film hybrids.

The vacuum triode, and its refinements like the tetrode and the pentode, had served as the workhorse of the electronics industry for nearly half a century, when its worthy successor, the bipolar junction transistor was invented in 1948 by Shockley, Bardeen, and Brattain at Bell Laboratories, shortly after the invention of the point contact transistor. The vacuum tube suffered from high operating power dissipation, limited life expectancy, and fragile construction. The junction transistor had few such drawbacks, and immediately won acceptance as the best active component for many electronics applications. It should be noted that the operating characteristics of junction transistors, particularly in regard to input and output impedance levels, were significantly different from those of vacuum tubes, and brought about a change in the design philosophy necessary to accomplish specific electronic functions.

Nowhere was the transistor more welcome than in the emerging field of electronic data processing. The first large digital computers consumed as much power as a locomotive and often failed before the execution of even a short program could be completed.

The Bell Telephone System was interested in developing a highly reliable electronic switching system to replace their vast array of electromechanical relays. Vacuum tube switches were less reliable than relays, but the junction transistor outperformed both tubes and relays. As part of their new system, Bell adopted a new technology involving thin-film hybrid modules with resistors and capacitors designed to remain stable over periods in excess of 20 years. The close tolerances that could be maintained on a routine manufacturing basis using thin-film techniques resulted in a product superior to their previous efforts, at a lower cost. The small geometry components that could be obtained using these techniques resulted in circuits that could operate reliably at higher frequencies. The combination of thin-film technology and the junction transistor (and, later the monolithic integrated circuit) resulted in the improved electronic switching system so necessary for a modern telephone communications network.

The surface field-effect transistor (FET) had been proposed by Lilienfeld in 1926, but the first successful metal-oxide-semiconductor (MOS) FET was not fabricated until 1959 by Kahng and Atalla at Bell Telephone Laboratories. The major difficulties associated with the fabrication of these devices were related to the growth of ultraclean insulating layers.

Bipolar junction transistor technology had some significant developments during the 1950s. These advances set the stage for one of the most important inventions in the history of electronics, the monolithic integrated circuit. The early transistors were made from germanium, primarily by alloying or grown junction techniques. Germanium, plagued by operating temperature limitations due to its relatively small bandgap (0.7 eV compared to 1.1 eV for silicon), was gradually supplanted by silicon. One outstanding property of silicon is that it forms a stable oxide when exposed to oxidizing agents at high temperatures. This oxide provides a means of controlling the surface conditions of silicon, and acts as a protective "mask" so that impurities can be inserted by diffusion or ion-implantation into selected areas of the surface of the wafer from which the oxide has been stripped. These properties of silicon and its oxide plus developments in photolithography led to the invention of the planar bipolar transistor structure by Hoerni at Fairchild in 1958. The concept of the monolithic integrated circuit was first suggested by G. W. A. Dummer of the Royal Radar Establishment at the Electronic Components Conference in 1952. He said

With the advent of the transistor and the work in semiconductors generally, it seems now possible to envisage electronics equipment in a solid block with no connecting wires. (8)

There were a number of programs working toward this goal in the years to follow. The U.S. Army Signal Corps was sponsoring a program with RCA called the Micro-Module. The U.S. Air Force developed a program with Westinghouse called "molecular electronics," based on the concept of using the properties of materials to perform electronic functions. Wallmark and Nelson at RCA filed a patent in 1958 for a two-dimensional array of junction FET's isolated by reverse-biased *pn* junctions. Jack Kilby at Texas Instruments, who had previously worked at Centralab on thick-film hybrid circuits for hearing aids, invented what was eventually judged by the courts to be the first semiconductor integrated circuit. The circuit was an oscillator made in a single crystal germanium substrate with the components separated by mesa-etching and interconnected by wire bonding. This circuit was first fabricated successfully in 1958. Less than a year later, Robert Noyce at Fairchild brought together the developments of the previous 10 years. Using the planar process and junction isolation, the first integrated circuits, as we know them today, were fabricated.

The first commercial integrated circuits were introduced by Texas Instruments in 1960, and Fairchild's Micrologic® family appeared in 1961. These early circuits were mostly digital logic circuits which were readily accepted by the rapidly expanding computer industry.

One computer manufacturer, IBM, was planning the introduction of a new line of large computers for the mid-1960s. They did not feel that they were ready to commit themselves to the new technology for such a large project. Instead, they developed a manufacturing procedure using an automated technique for attaching special "flip-chip" transistors by means of solder bumps to thick-film substrates. This application demonstrated that thick-film hybrid technology could be used for mass production, since millions of these reliable circuits were produced. IBM called this "Solid Logic Technology." General Motors also made a high production commitment to thick-film hybrids for voltage regulators.

The success of the early digital bipolar integrated circuits provided the financial base for new, even more successful, logic circuits including the highly popular transistor-transistor logic (TTL) family. These digital circuits had almost universal appeal. The designers of analog integrated circuits were able to make use of the continued developments in technology to produce an outstanding circuit, the operational amplifier, which has become the cornerstone of the analog circuit market.

The MOSFET was not successfully fabricated until approximately the same time as the bipolar integrated circuit. It was natural for the MOS integrated circuit to develop during the same time period as that for the discrete MOSFET.

Since the introduction of the monolithic integrated circuit, there has been a continuing increase in the complexity of the circuits that could be economically fabricated on an individual chip. This has been accompanied by a gradual decrease in the minimum area occupied by typical components, due to advances in microphotolithography. Processing methods, device types, materials defects, and circuit design have been responsible for setting a limit on the die size that can be mass produced at any particular point in time. This limit is based on the ability to produce an "acceptable" yield of circuits which perform within the specifications that have been established for the particular design being implemented. Obviously, it is easier to produce saturating bipolar transistor logic circuits [which will accept a wide range of transistor common-emitter current gain (β)], and operate from a 3-V supply, than it is to produce an analog amplifier with a specific voltage gain intended to be operated from a 15-V supply. All of the devices and components within an individual circuit must function properly. A single defect, for example, a dislocation within the crystal structure occurring in an unfortunate location within a bipolar transistor, will render the entire circuit containing that transistor useless. The "acceptable" yield thus depends on the type of circuit being manufactured and the willingness of the users of these circuits to pay the premium required for a particular increment in the complexity of an electronic function. In other words, is the customer more willing to pay for a single circuit, with a smaller overall size and weight and increased reliability, than he would be for several circuits that could be interconnected to perform the same function? An increase in yield, to the point where the single monolithic circuit costs less than the individual circuits, naturally eliminates the need to even consider this question. From another viewpoint, making a single large circuit reduces the versatility available to the user of the individual circuits and introduces the concept of the custom integrated circuit as compared to the standard line or "family" of smaller individual circuits. The demand for the larger circuit must be great enough, or sufficiently "special," like a military or space application, to warrant production. It is clear that large volume products like automobile voltage regulators and hand-held calculators should be made from custom circuits, and that limited production items like implantable heart stimulators and geothermal instrumental probes should be made from standard product lines, but there are many applications in which the choice is not so obvious. The result of these varied demands from the market place has been the designation of several categories for standard monolithic integrated circuits according to size: small-scale integration (SSI), medium-scale integration (MSI), large-scale integration (LSI), and very large-scale integration (VLSI). These terms are usually restricted to digital circuits with the somewhat nebulous dividing lines associated with the number of logic gates per circuit.

The first attempts at LSI were intended to circumvent the gradual trends within the industry for increasing complexity and size based on refinements in processing techniques and materials development. This revolutionary approach was called discretionary wiring. An entire wafer (typically 2.5 cm in diameter) was to be used for a single circuit. An array of logic modules was fabricated within the wafer with test pads arranged so that each module could be individually probed and evaluated. A computer was used to determine which modules were useful, and how they could be interconnected to perform the overall task required of this gigantic circuit. Masks for multilayer interconnnection patterns were generated for each individual wafer. This was a technologically feasible method for achieving LSI, but it was an expensive technique. At the same time, MOS integrated circuit processing emerged from the development laboratory to take its place as a viable alternative to bipolar monolithic logic. Because of the higher component densities available in MOS circuitry, LSI could be achieved in circuits that were less than 1 cm on a side. LSI memory arrays, calculator chips, and digital watch circuits rapidly became commonplace. The long sought after "computer-on-a-chip," in the form of microprocessors, was soon to follow. As technology progressed, yields improved to the point that VLSI became possible in the form of 64-kilobit random access memories and charge coupled device imaging arrays capable of performing the functions of television cameras. Bipolar LSI, using a circuit design technique called integrated injection logic, has also emerged in the form of digital watch circuits and microprocessors. The development of submicrometer photolithography and new circuit concepts have enhanced the component density and circuit complexity to the point where VLSI circuitry occupies approximately the same area as previous LSI circuitry. The future of integrated circuit technology is exciting, to say the least.

Hybrid technology has also continued to develop at a rapid pace. Thin-film microwave and electro-optic circuits dominate the high frequency field. Surface-acoustic-wave devices employing thin-film patterns on piezoelectric crystals have been used to produce filter characteristics difficult to obtain by other techniques. A particularly important application of thick-film hybrids is the interconnection of LSI and VLSI chips into complex electronic systems by eliminating the large and expensive packages usually used for the individual chips.

In two decades, microelectronics has had a significant impact on the world. This far reaching impact has been due, in part, to the reduction in size and weight achieved by the shrinking of the apparatus required to perform electronic functions. Perhaps of even more importance than the changes due to size and weight have been the reduced cost, the increased performance, the reduced power consumption, and the increased reliability that have enabled microelectronics to displace numerous mechanical devices during this time period. Perhaps the most graphic example of this is in the field of digital computation. A microprocessor based digital computer system on a printed circuit board approximately 25 cm × 25 cm has more computing capacity than the first electronic computer, *ENIAC*. It is 20 times faster, has a larger memory, is thousands of times more reliable, consumes the power of a light bulb rather than that of a locomotive, occupies one-thirty thousandth of the volume, and costs one-ten thousandth as much (5). A hand-held calculator costing about one-half as much as an engineering-type slide rule is much more accurate, can perform many more functions, and even locates the decimal point for the user over a wide calculating range (typically 10^{-99} to nearly 10^{+100}). Some of the scientific programmable hand-held calculators have capabilities approaching those of the early computers. A "home" computer system, complete with video display, printer, "floppy-disk" storage, alpha-numeric keyboard, and high-level programming language costs less than an automobile. There are many other examples of the pervasive uses of microelectronics, too numerous to mention. The engineers involved with microelectronics are doing their part to make many of the authors of science fiction look like prophets.

A particularly interesting historical development and industry overview is given by Ernest Braun and Stuart MacDonald (7).

A COMPARISON OF MICROELECTRONICS TECHNOLOGIES

The four basic microelectronics technologies, thick-film hybrid, thin-film hybrid, bipolar monolithic, and MOS monolithic represent different approaches for producing circuits that will accomplish desired electronic functions. To make the optimum use of a particular technology, circuits should be designed to rely upon the unique capabilities of the fabrication technique which has been selected. To illustrate this concept, a basic electronic function, the logic inverter (NOT), has been selected to be implemented by each of the four technologies. These examples are not typical of contemporary microelectronics, but they will serve to introduce the reader to the processes and the design philosophies associated with the four technologies.

Thick-film hybrids are the largest of the microcircuits. A carefully designed thick-film hybrid microcircuit is somewhat less than one-half the size of a carefully designed printed circuit implementation of the same function. The components most readily fabricated by thick-film processes are resistors, and this technology provides the means for producing the widest range of resistor values which can be readily obtained in microelectronics, typically, any value between 0.1 Ω and 10 MΩ. Capacitors can also be fabricated using thick-film techniques, but it is common practice to use multilayer ceramic chip capacitors as add-on components, since they are smaller in size and have more reproducible characteristics. Conductor patterns and crossovers are readily produced in a thick-film production environment. Transistors, diodes, and integrated circuits are added to complete the process. These components can be attached to the circuit in standard packages, special packages, or configurations designed specifically for hybrids, or as bare chips. When necessary, miniature inductors and transformers can also be attached to the thick-film hybrid circuits. The designer of a thick-film hybrid microcircuit has essentially the same wide flexibility as that available to the designer of a discrete component electronic circuit.

The circuit selected to illustrate thick-film hybrid technology is shown in Figure 1-1. This is a diode-transistor logic inverter. The diodes and transistor will be attached in standard packages. The resistors and conductors will be printed on a ceramic substrate using thick-film inks. Resistors R_1 and R_2 will be printed using 1.0 kΩ/□ ink, and R_3 will be printed using 10.0 kΩ/□ ink. Thick-film resistors may be economically trimmed to ±1% of the specified value, if necessary.

The units associated with these inks, ohms per square (Ω/□), are units frequently encountered in microelectronics. They are derived from the following consideration. A film resistor is shown in Figure 1-2. The resistance of this structure is given by

$$R = \frac{\rho l}{tw} \tag{1-1}$$

where R is the resistance in ohms, ρ is the resistivity in ohms-centimeters, l is the length in centimeters, w is the width in centimeters, and t is the thickness in centimeters. Equation 1-1 can be rewritten in the form

$$R = R_S \frac{l}{w} \tag{1-2}$$

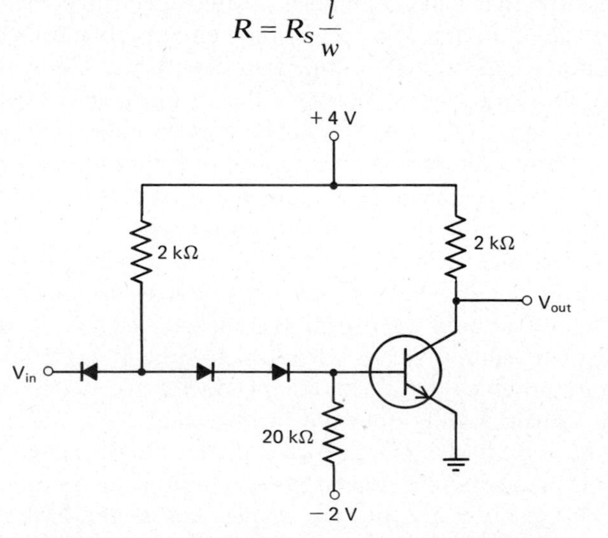

Figure 1-1. A diode-transistor logic (DTL) inverter.

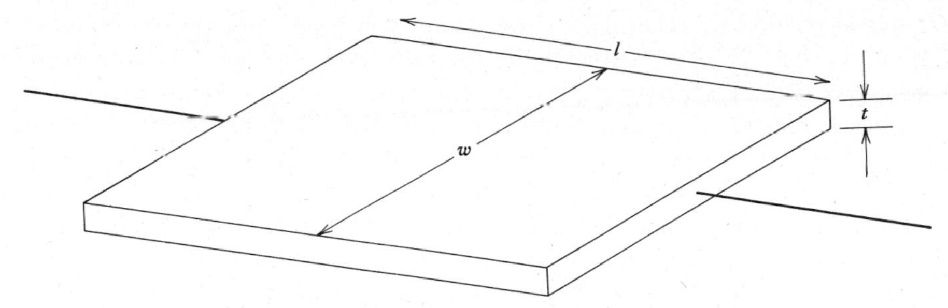

Figure 1-2. The geometry of a film resistor.

where

$$R_S = \frac{\rho}{t} \tag{1-3}$$

is called the sheet resistance and has the units of ohms.

The resistance of any square resistor, where $l = w$, will be R_S, and the resistance of any surface geometry resistor can be determined by multiplying R_S by the number of squares in the surface geometry, that is the ratio of l to w, which is called the aspect ratio. For this reason, it is common to refer to the units of R_S as ohms per square (abbreviated Ω/□).

Figure 1-3 contains a composite layout of the circuit. Resistor terminations

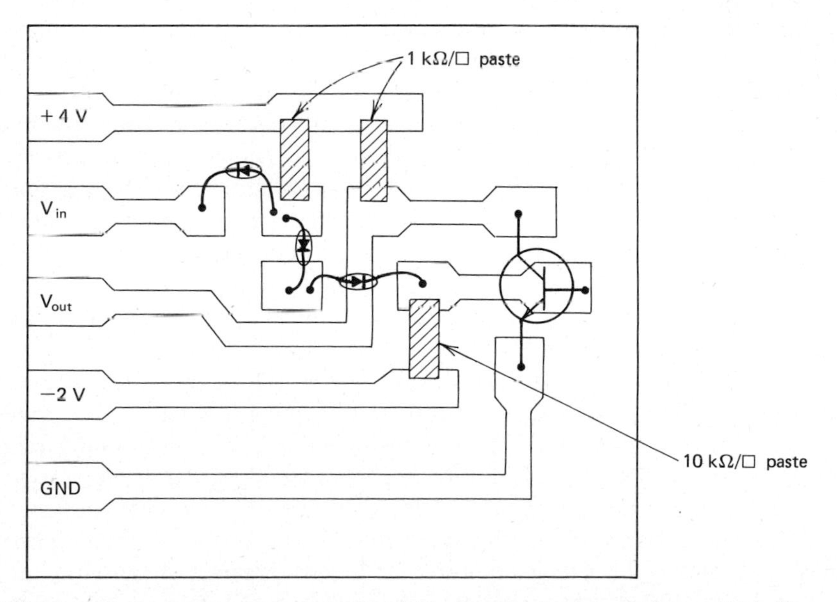

Figure 1-3. A composite layout of a DTL inverter implemented as a thick-film hybrid microcircuit.

are accomplished by overlapping the conductor print with the resistor print. The active part of the resistor is that portion between the conductor prints. Figure 1-4 shows a pictorial view of the thick-film hybrid process. In this example, "add-on" components are attached by soldering.

A thin-film hybrid microcircuit implementation of a particular electronic circuit can be as small as one-tenth the size of the same circuit fabricated by miniature printed circuit techniques, or one-fifth the size of the same circuit fabricated by thick-film hybrid techniques. This size reduction is made possible by the photoengraving process which is used to define the thin-film components and conductor patterns. The easiest components to fabricate using thin-film techniques are resistors. The range of resistor values in a single thin-film microcircuit is restricted by the common practice of depositing only one resistor film, thus constraining the designer to use only one sheet resistance. A typical thin-film resistor material is tantalum nitride, with a nominal sheet resistance of $50\ \Omega/\square$. Long meandering patterns are used to obtain large resistor values. Thin-film resistors can be trimmed by anodization or with lasers to within $\pm 0.1\%$, exhibit excellent long-term stability, and are the least sensitive to temperature variations of any resistors fabricated in microcircuits. The precise small geometries, which can be readily obtained using thin-film techniques, permit the operation of thin-film hybrid microcircuits at higher frequencies than those obtainable with thick-film hybrid microcircuits. Low valued capacitors can be fabricated by thin-film fabrication processes, and can be incorporated into thin-film hybrid microcircuits. However, "add-on" multilayer ceramic capacitors are frequently used instead of the more costly in situ thin-film capacitors. In some high-frequency thin-film hybrid microcircuits, spiral conductor patterns are used to fabricate inductors, but "add-on" components often have better characteristics and occupy less space. Bare silicon chips or special configurations like beam-lead devices and integrated circuits provide the active components for thin-film hybrid microcircuits. If bare silicon chips are used, their circuit connections are fragile wire bonds. It is, therefore, common practice to enclose thin-film hybrid microcircuits in hermetically sealed packages, which results in a significant increase in the cost, size, and weight. These increases are usually counterbalanced by the increase in reliability afforded by the package.

The circuit chosen to illustrate the thin-film hybrid process is the emitter coupled logic circuit shown in Figure 1-5, which performs the inverting function. This is a high speed nonsaturating logic circuit containing resistors and transistors. Transistors T_1 and T_2 are a matched pair of devices available on a single silicon chip. Transistors T_3 and T_4 are discrete transistors in chip form. The resistors and conductors are defined by a subtractive etching process. A glazed ceramic substrate is coated with a 500-Å layer of tantalum nitride, followed by a 200-Å layer of titanium and a 5000-Å layer of gold. The titanium layer adheres well to the tantalum nitride, and the gold adheres well

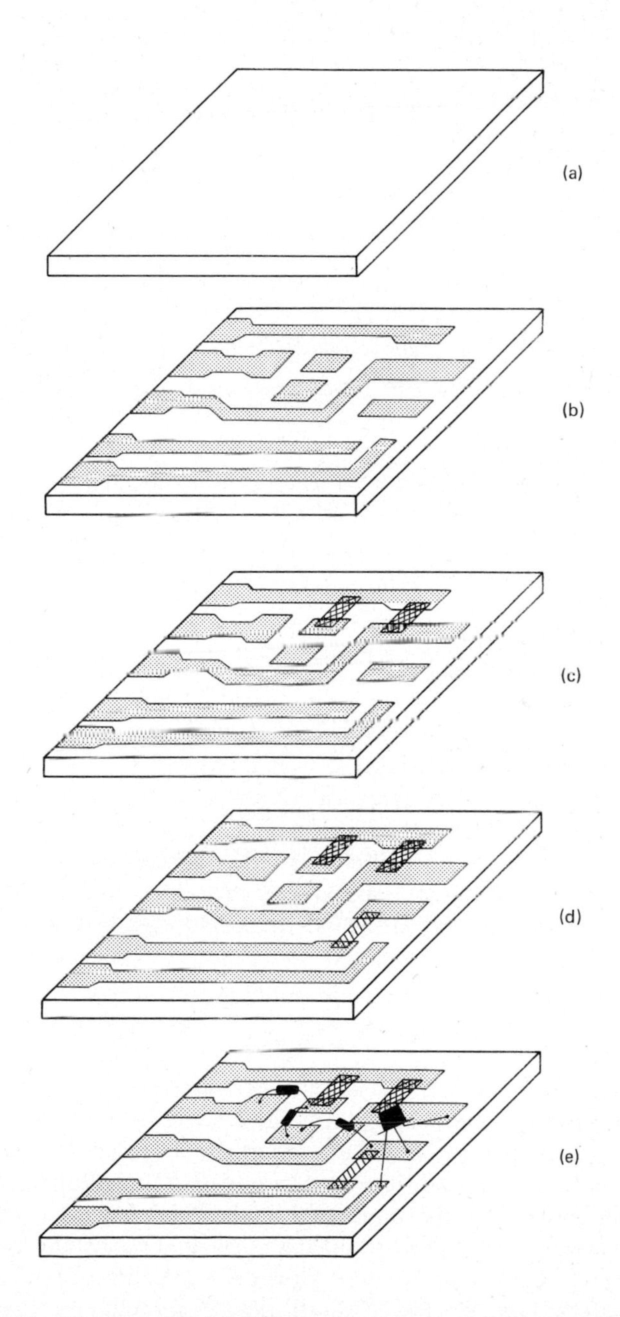

Figure 1-4. The thick-film hybrid fabrication process. (*a*) Substrate. (*b*) Conductor print, dry, and fire. (*c*) First resistor print and dry. (*d*) Second resistor print and dry. (*e*) Solder "add-on" components.

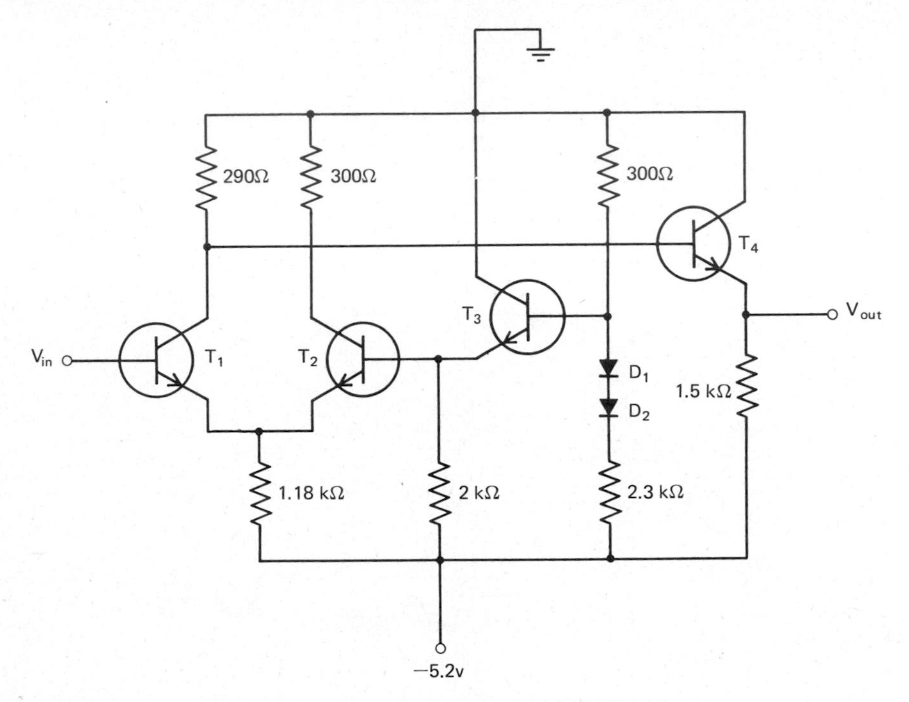

Figure 1-5. An emitter-coupled logic (ECL) inverter.

to the titanium, but does not adhere well to tantalum nitride. The conductor pattern is delineated in the gold and titanium layers by photoengraving. A second photoengraving process is used to both delineate the resistors and protect the conductor pattern which has already been defined. The layout for this hybrid is shown in Figure 1-6 and the process is illustrated in Figure 1-7 The semiconductor chips are attached to the substrate with epoxy, the substrate is attached to the package also by epoxy, and the interconnections are made using thermocompression wire bonding. The assembly is completed by attaching the lid to the package using a AuSn solder preform in an inert atmosphere.

Monolithic integrated circuits are approximately an order of magnitude smaller than thin-film hybrid microcircuits designed for the same purpose. In monolithic bipolar integrated circuit technology, as contrasted with hybrid technology, there are no "add-on" components. All of the transistors, diodes, resistors, and capacitors are fabricated within the silicon material, and intraconnected with a metal pattern on the surface. The circuit is then mounted in a package or on a hybrid and interconnected to other circuits, power supplies, displays, and so forth. Since the entire chip is subjected to the processing required to produce the most complicated device in the circuit, the bipolar transistor, it is no more difficult or expensive to make a transistor than

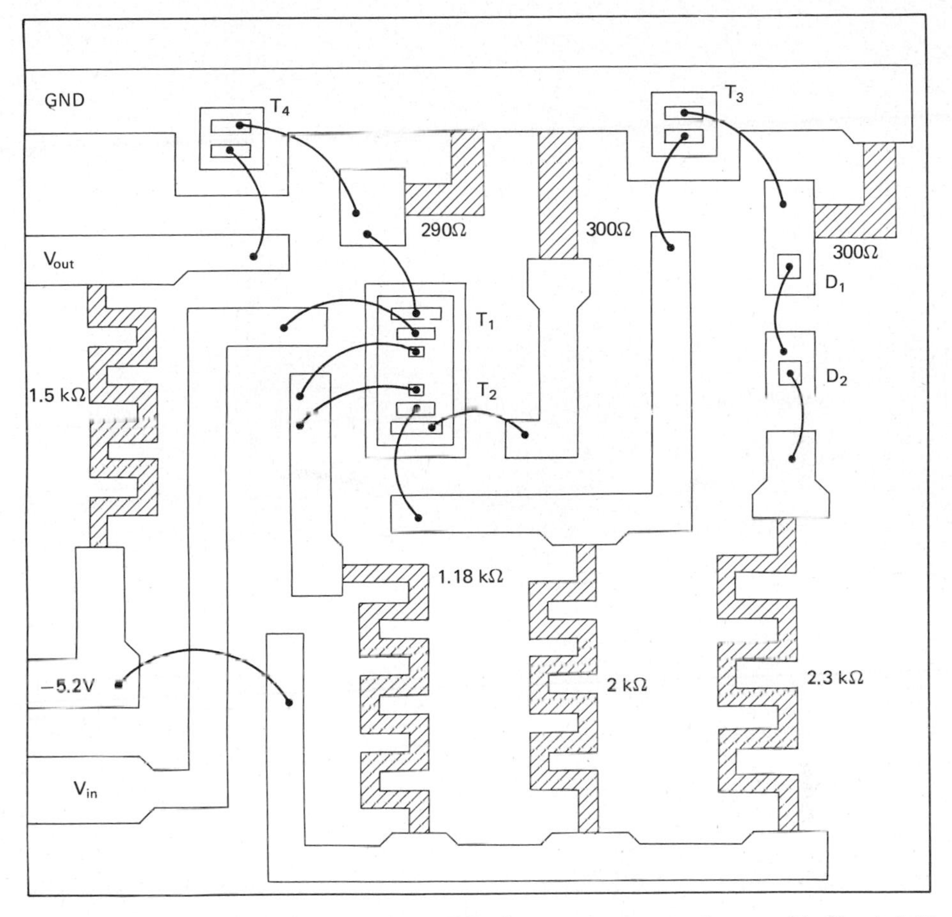

Figure 1-6. A composite layout of an ECL inverter implemented as a thin-film hybrid microcircuit.

it is to make a diode, resistor, or capacitor. In fact, transistors are usually smaller in surface area than resistors or capacitors and, thus, less expensive. The close proximity of the transistors during fabrication results in devices whose characteristics are closely matched, and changes in operating temperature produce similar effects in all of the transistors on the same chip. If circuit performance can be enhanced by using more transistors, or replacing other components by transistors, this should definitely be done. This design philosophy is much different from that which has dominated discrete component electronics design and hybrid microcircuit design. Resistors are made by the same process used to make the base portions of bipolar transistors. The sheet resistance is approximately 200 Ω/□ at room temperature, but will vary from area to area on the large wafers (75 cm in diameter or larger). This

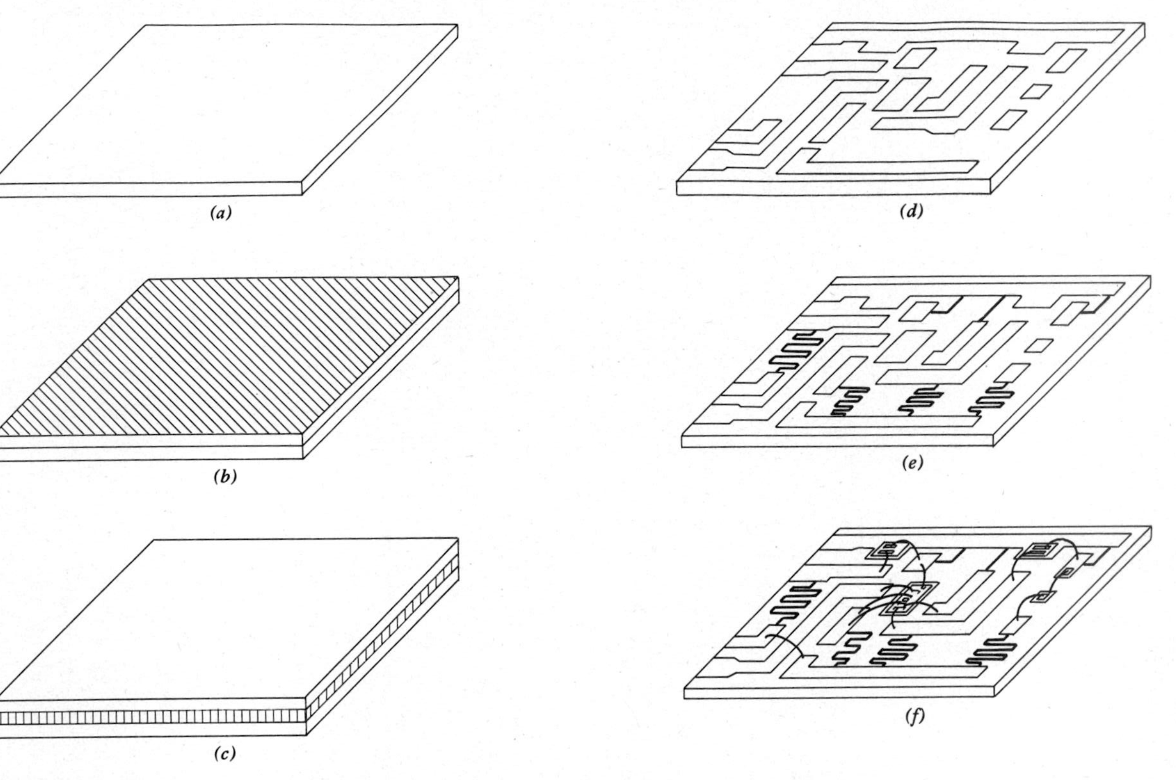

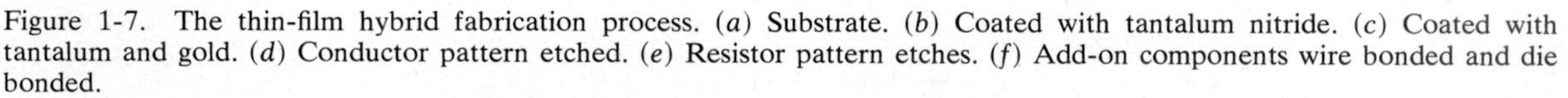

Figure 1-7. The thin-film hybrid fabrication process. (*a*) Substrate. (*b*) Coated with tantalum nitride. (*c*) Coated with tantalum and gold. (*d*) Conductor pattern etched. (*e*) Resistor pattern etches. (*f*) Add-on components wire bonded and die bonded.

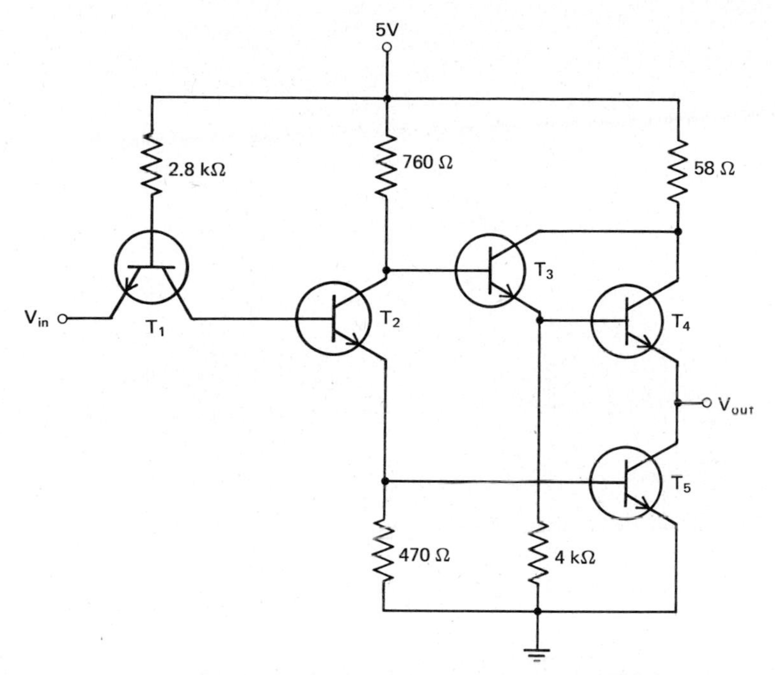

Figure 1-8. A transistor-transistor logic (TTL) inverter.

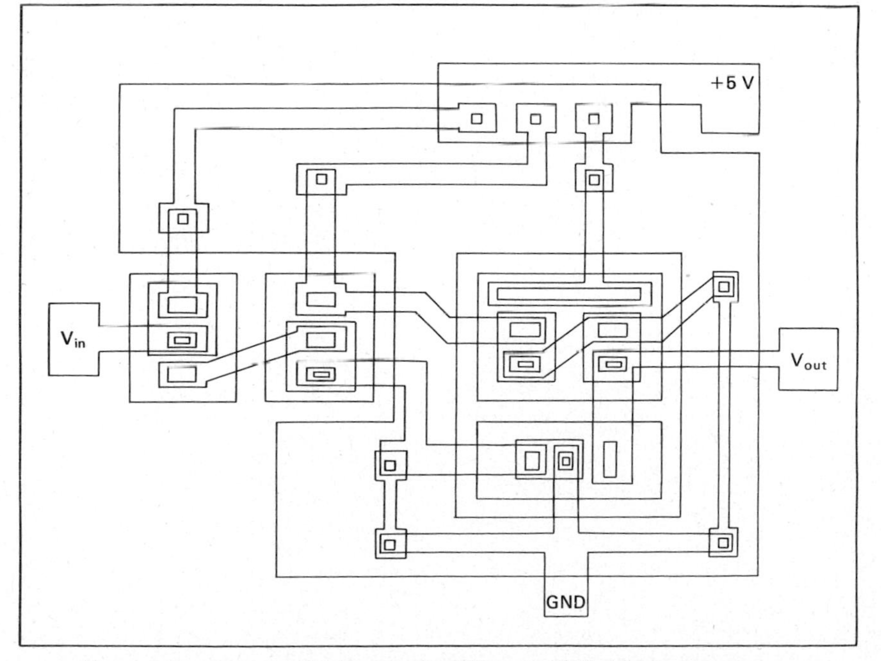

Figure 1-9. A composite layout of a TTL monolithic bipolar inverter.

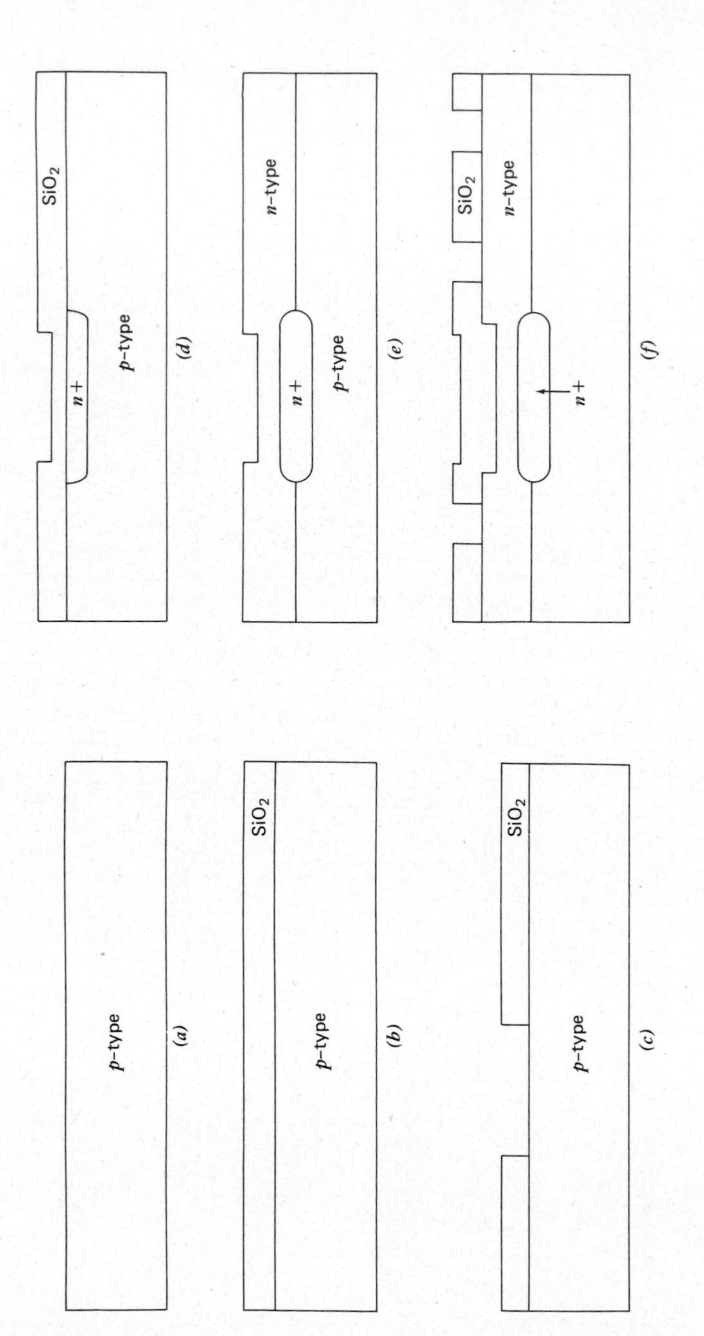
p-type
(a)
SiO2
p-type
(b)
SiO2
p-type
(c)
SiO2
n+
p-type
(d)
n-type
n+
p-type
(e)
SiO2
n-type
n+
(f)

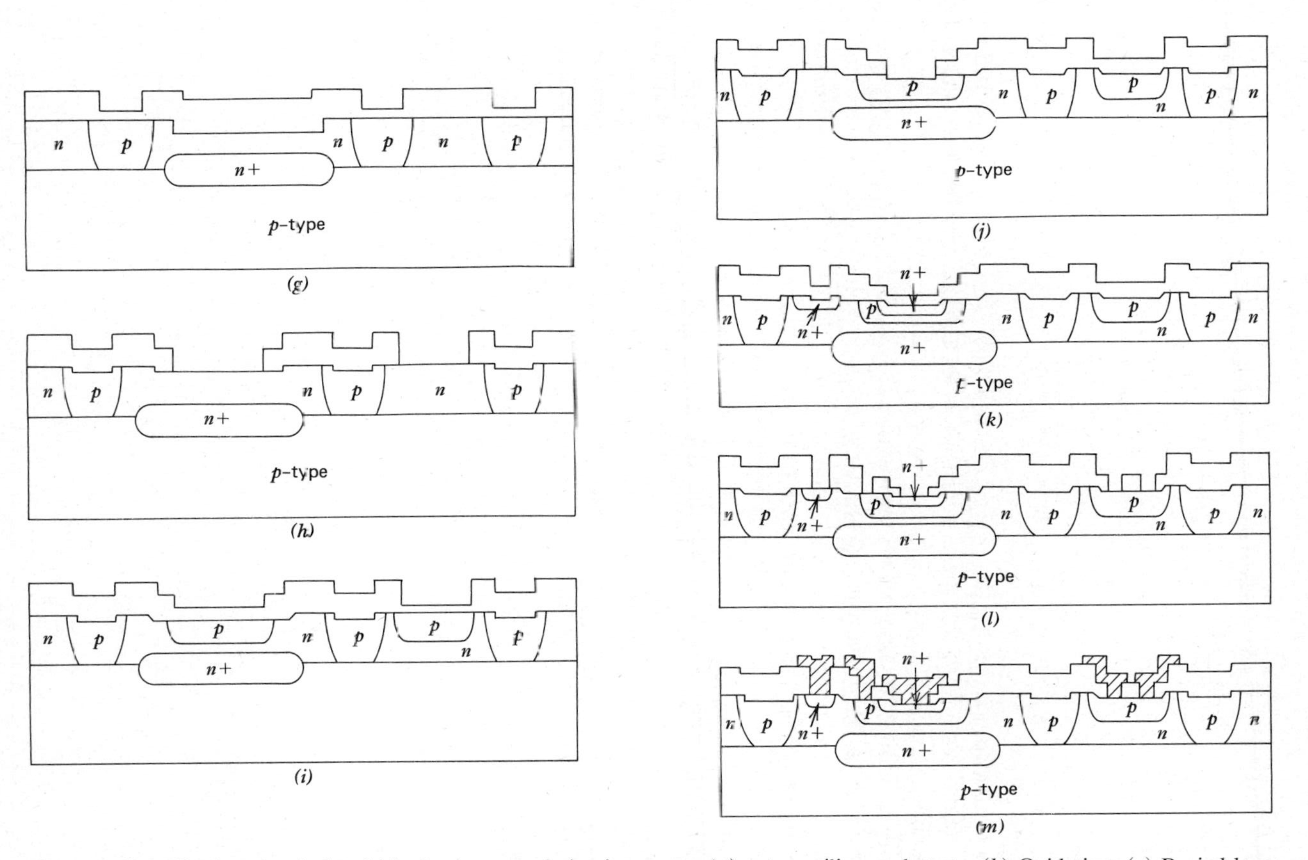

Figure 1-10. The junction isolated bipolar integrated circuit process. (*a*) *p*-type silicon substrate. (*b*) Oxidation. (*c*) Buried layer pattern. (*d*) Buried layer diffusion and oxidation. (*e*) Oxide strip and *n*-type epitaxial growth. (*f*) Oxidation and isolation pattern. (*g*) Isolation diffusion and oxidation. (*h*) Base and resistor pattern. (*i*) Base and resistor diffusion and oxidation. (*j*) Emitter and collector contact pattern. (*k*) Emitter diffusion and oxidation. (*l*) Contact pattern. (*m*) Metalization and intraconnection pattern.

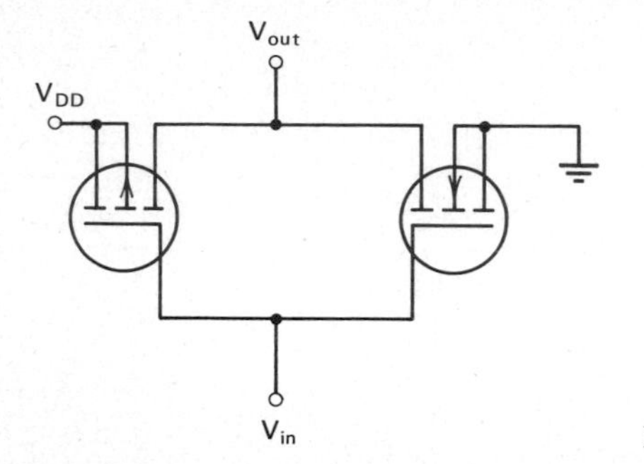

Figure 1-11. A CMOS inverter.

single sheet resistance limits the range of practical resistors in monolithic bipolar circuits. Trimming of these resistors is essentially impossible. As a result, precision resistors cannot be made by this process. This type of resistor also exhibits a significant change with temperature. Fortunately, if circuit design is based on the ratio of resistors rather than resistor values, excellent circuit characteristics can be obtained, and thermal tracking of these components alleviates the effects of temperature variations. Small valued capacitors can be fabricated in bipolar monolithic integrated circuits in two ways. Reverse biased diodes exhibit a voltage dependent capacitance, and can be used in situations where capacitance value is not important. It is also possible to make capacitors using the intraconnection metal as one electrode, the silicon dioxide layer as the dielectric, and a heavily doped silicon region, like that used for emitter regions in bipolar transistors, as the other electrode. This type of capacitor, although low in capacitance density, is voltage independent and can serve

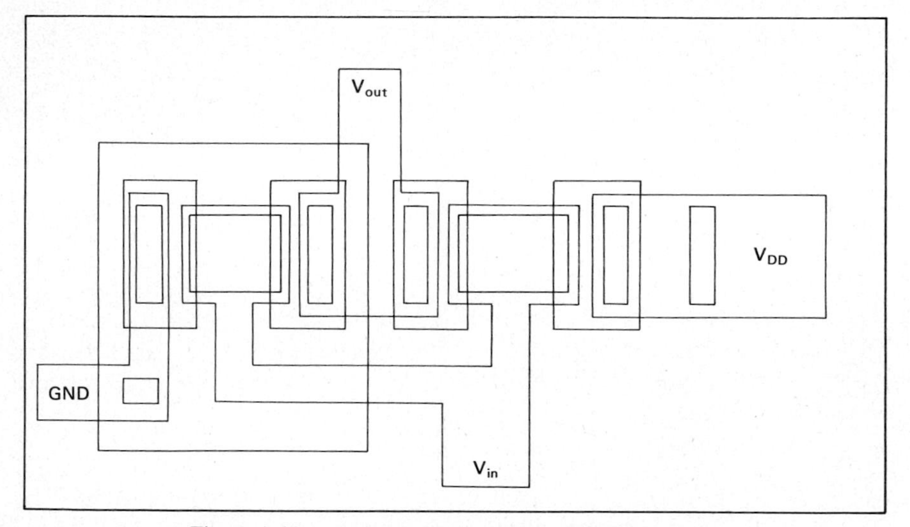

Figure 1-12. A composite layout of a CMOS inverter.

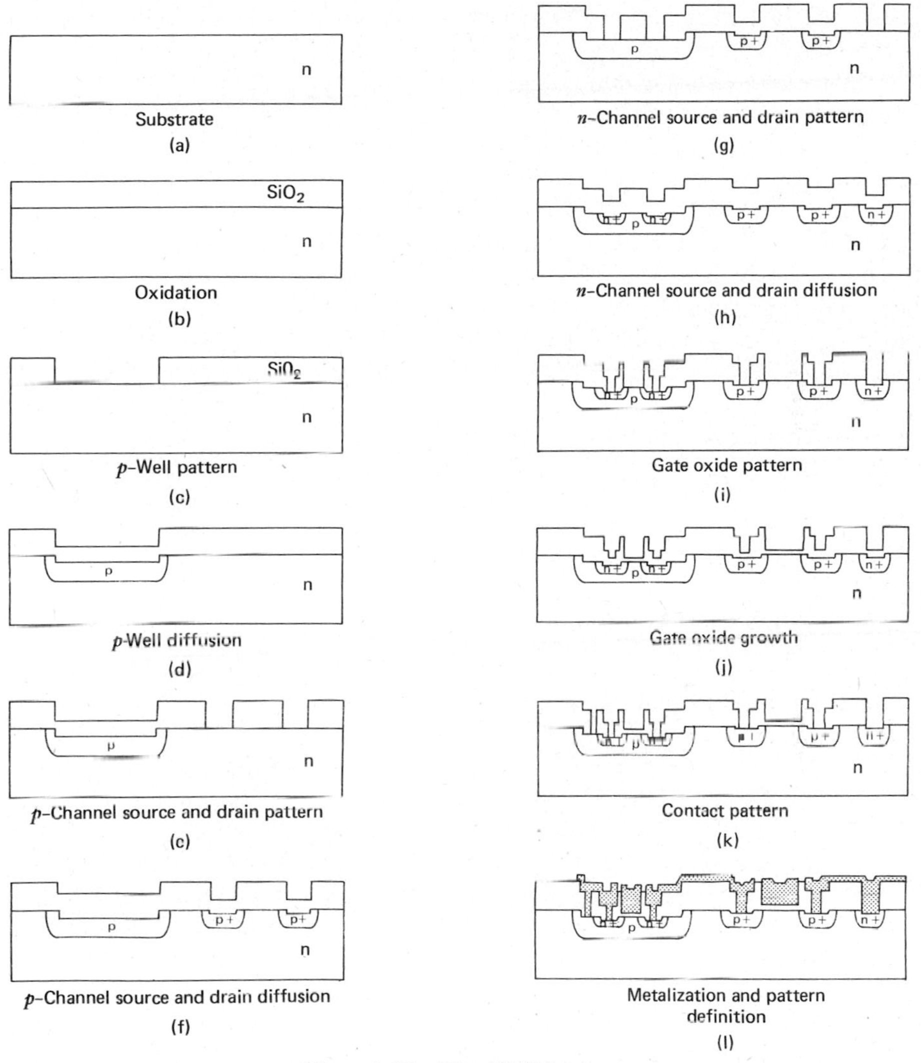

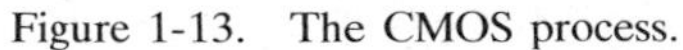
Figure 1-13. The CMOS process.

where capacitor value must remain constant. Inductors are not used in monolithic bipolar integrated circuits. Even with all these restrictions, it is possible to realize a wide variety of both analog and digital electronic functions using monolithic bipolar microcircuit techniques if the designers make careful use of the particular advantages of the technology.

To illustrate the bipolar monolithic integrated circuit technology, a basic transistor-transistor logic inverter has been selected. This circuit is shown in Figure 1-8. A layout for this circuit is shown in Figure 1-9, and the process is illustrated in Figure 1–10. The components are electrically isolated from one another by *pn* junctions.

In metal-oxide-semiconductor (MOS) monolithic integrated circuit technology, field-effect transistors (FET's) are the essential components. In fact, these devices are used as both active and passive components to the virtual exclusion of resistors and inductors. Capacitors are sometimes used in these circuits in the form of MOS structures. MOSFET's are made with *p*-type channels or *n*-type channels. Many of these devices have no channels until a threshold bias voltage is applied between the gate and substrate. These devices are called *enhancement* mode FET's. If channels exist with no applied bias, both *depletion* and *enhancement* modes of operation are possible.

The circuit selected to illustrate MOS monolithic integrated circuit technology is a complementary MOS (CMOS) inverter. This circuit makes use of an *n*-channel enhancement mode FET and a *p*-channel enhancement mode FET as shown in Figure 1-11. The layout for this circuit is shown in Figure 1–12 and the process is shown in Figure 1–13.

SUMMARY

Microelectronics has become a dominant factor in electronics in the relatively short time since its introduction in the mid-1950s. The original driving force behind the development of microelectronics was the need for very small, light electronic circuits in military applications, with an assist from "science fiction" authors. The additional benefits of low power consumption, reliable operation, low cost, and long life helped to make microelectronics important in virtually all aspects of electronics. Four major technologies in microelectronics have developed, thick-film hybrid, thin-film hybrid, bipolar monolithic, and MOS monolithic. The circuits fabricated by each of these technologies possess unique characteristics which assure the future of all four fabrication techniques.

REFERENCES

1. Isaac Asimov, "The Last Question," *Science Fiction Quarterly* (November, 1956). [Also available in Isaac Asimov, *Nine Tomorrows* (New York: Doubleday, 1959); reissued by Fawcett, 1971.]
2. D. W. Hamer and J. V. Biggers, *Thick Film Hybrid Microcircuit Technology* (New York: Wiley, 1972).
3. J. S. Kilby, *IEEE Transactions on Electron Devices* ED-23, 7(1976).
4. D. Kahng, ibid.
5. R. N. Noyce, *Scientific American* 237, 3(1977).
6. A. B. Glaser and G. E. Subak-Sharpe, *Integrated Circuit Engineering* (Reading, Mass: Addison-Wesley, 1977).

7. E. Braun and S. MacDonald, *Revolution in Miniature* (Cambridge: Cambridge University Press, 1978).

8. G. W. A. Dummer, *Proceedings of the Symposium of the IRE-AIEE-RTMA*, Washington, D.C., May 1952.

PROBLEMS

1-1. Determine the number of squares required to fabricate resistors of 22Ω, 150Ω, 680Ω, 4.7 kΩ, 33 kΩ, and 1 MΩ, if the sheet resistance is: (a) 50 Ω/□, (b) 100 Ω/□, and (c) 1000 Ω/□.

1-2. The resistivity of a semiconductor is given by $\rho = (q\mu n)^{-1}$ where q is the magnitude of the electronic charge (1.602×10^{-19}C), μ is the mobility, and n is the carrier concentration. Find the resistance of a semiconductor bar which is 400 μm long, 25 μm wide, and 5 μm thick, if n is 10^{19} cm^{-3}, and μ is 135 cm$^2 \cdot$ V$^{-1} \cdot$ s^{-1}.

1-3. The sheet resistance of a resistor structure formed by the diffusion of an impurity of the opposite impurity type into an homogeneously doped semiconductor like silicon depends on the distribution of the impurity as a function of depth from the surface and the concentration of the background impurity. Curves relating the surface concentration to the product of the junction depth and the sheet resistance with background concentration as a parameter are shown in Figures 7–11 through 7–14 for Gaussian and complementary error function impurity distributions. Using these curves, find the surface concentrations necessary to obtain a sheet resistance of 200 Ω/□ for a p-type Gaussian diffusion with a junction depth of 3 μm in n-type silicon with a background concentration of 10^{16} cm^{-3}, and 10 Ω/□ for an n-type complementary error function diffusion with a junction depth of 2 μm in p-type silicon with a background concentration of 10^{18} cm^{-3}.

Chapter 2
Pattern Generation and Photoengraving

Photographic processes are essential to microelectronics. Accurate, reproducible lines with widths as small as 4 μm are routinely produced in monolithic integrated circuits by exposing photosensitive polymers to ultraviolet light through masks which have been photographically fabricated. Submicrometer patterns can be achieved using materials that are sensitive to electron beams or soft x-rays. Thin-film hybrids are fabricated using the same basic photoprocesses as those used in monolithic integrated circuits. The relatively large patterns associated with thick-film hybrids are defined by screen printing, but the patterns are transferred to the screens photographically.

PHOTOSENSITIVE MATERIALS

There are two types of photosensitive materials used for creating photographic images for microelectronics, silver halide emulsions and photoresists. Silver halide emulsions have been used on photographic films and plates since the invention of photography. Photoresists, in their present form, are photo-cross-linked polymers which are resistant to the attack of numerous chemicals and vapors, thus making possible the photoengraving techniques that are the basis of microelectronics.

Silver Halide Emulsions

Finely divided grains of silver halides, principally silver chloride (AgCl), silver bromide (AgBr), and silver iodide (AgI), are suspended in gelatin to form the basis for a typical photographic emulsion. Other ingredients are added which influence the spectral sensitivity of the material. An important additive is allyl isothiocyanate ($CH_2:CHCH_2NCS$) (mustard oil) which results in the conversion of some of the silver halide molecules into silver sulfide (Ag_2S). In the pure state, silver halide grains are essentially insensitive to light. In the presence of Ag_2S, however, silver halide grains that have been exposed to light are readily converted to metallic silver in chemical solutions that have very little effect on unexposed grains. The discovery of the effect of mustard oil was traced back to the fact that the gelatin used for a particular batch of plates had come from a group of cows who had eaten mustard (1). The creation of a photographic image in a photosensitive emulsion which has been coated on a glass or plastic backing is performed in the following manner. When a photographic emulsion is exposed to light, there is no visible effect on the emulsion. However, the energy levels of certain grains have been increased to an excited state. The pattern of these excited grains is referred to as a latent image. The formation of the latent image is not well understood. If one accepts the hypothesis (2) that particular silver halide grains contain Ag_2S specks, these grains have higher energy levels than pure grains. When an impure grain absorbs a photon, the energy level of the grain is increased to an excited state. The silver halide grain is assumed to be composed of positive silver ions and negative halide ions. The absorption of the photon results in the formation of an electron-hole pair. The hole neutralizes a halide ion, and the electron is trapped at the surface by a Ag_2S speck. An interstitial silver ion is attracted to this point and neutralized by the electron. As this process is repeated, halogen molecules are formed which escape, leaving the grain with an imbalance of silver. When these unbalanced grains are subjected to a reducing agent, they are transformed into silver. The electrically balanced grains are not transformed. This process is called "developing," and the most popular chemical to accomplish this is alkaline pyrogallol [$C_6H_3(OH)_3$]. The amount of silver formed in this process is directly proportional to the uncombined silver that has been formed during the creation of the latent image (3). The image is "fixed" by removing the unconverted silver halide grains from the emulsion after the excited grains have been converted to silver. Sodium thiosulfate ($Na_2S_2O_3$) reacts with the silver halide to form water soluble salts. This fixing agent was first used by the English astronomer, John Herschel, who thought that it was sodium hyposulfite. This is the reason that the fixing agent is erroneously referred to as "hypo." The result of this process is a silver pattern on a glass or plastic backing which is directly related to the pattern of the intensity of light to which the photosensitive emulsion has been exposed.

The spectral sensitivity of the silver halide activated with silver sulfide is limited to the blue and violet portions of visible light. The addition of dyes to the emulsion results in an increased spectral sensitivity. The proper combination of dyes results in an emulsion with a spectral sensitivity that is nearly uniform over the frequency range of visible light. These emulsions are called "panchromatic."

Emulsions are usually coated on glass plates or plastic films. The finite thickness of the base results in an undesirable effect called "halation." Not all of the incident light is absorbed by the emulsion, and those rays that are not perpendicular to the surface can be reflected from the back side of the plate. This results in a fogged area around the image like a halo. Halation can be decreased by the use of an antihalation coating on the back side of the plate. This coating contains colored material which absorbs the colors to which the emulsion is sensitive, and thus reduces the reflections that cause halation. An additional difficulty encountered in emulsion films is due to scattering of light within the emulsion itself. This effect contributes to the undesirable fogged area around the image and is referred to as "irradiation." Both halation and irradiation degrade the quality of patterns produced in emulsion plates and films.

Emulsion Plates for Microelectronics (4–10). The only suitable silver halide emulsion photomasks used in microelectronics production make use of Lippmann-type high resolution emulsions. These are available commercially as Kodak High Resolution Plate (HRP®), Kodak High Resolution Plate (Type 2), and Agfa-Gevaert Micromask.®

The glass plates upon which these emulsions are coated are of special quality. The flatness is carefully controlled, and the plates are carefully inspected for defects. Typical plates are 1.52 mm thick and must transmit a significant portion of the ultraviolet (UV) since photoresists sensitive to UV light are conventionally exposed to light that has been transmitted through the glass plates.

Kodak HRP has a micro-fine-grain emulsion coating with a grain size less than 0.1 μm and an average size of 0.05 μm. The resolution capability is 2000 lines/mm. Before processing, the emulsion coating is 6 μm thick. After exposure and development, the emulsion in a large geometry exposed area is approximately 4 μm thick. An unexposed emulsion layer is approximately 2 μm thick. For narrow exposed lines, the emulsion thickness will vary between 2 μm and 4 μm. When narrow lines and wide lines are in close proximity in the image, it is difficult to transfer the image from one plate to another by contact printing. The noncontact of the narrow lines results in diffused light which degrades the transferred image of the lines.

Kodak HRP has an antihalation backing in the form of a colored gelatin, and is most sensitive to light with a wavelength of 5450 Å. Kodak HRP Type 2 has a different emulsion coating which requires approximately twice the exposure of HRP. It responds well to both blue and green light, and was designed to limit the effects of optical scattering without the use of an antihalation backing. This reduces the presence of unwanted extra particles and gels, thus simplifying the development procedure. Agfa-Gavaert Micromask plate has a silver halide emulsion with an average grain size of 0.068 μm. There is an antihalation agent included in the emulsion rather than coated on the back. The peak spectral sensitivity is in the range between 5250 Å to 5460 Å.

The suggested complete procedure for processing the images on emulsion plates is available in the technical information from the manufacturers of the plates. For microelectronics applications, there are special instructions, which are illustrated in the flowchart of Figure 2-1. The dehydration process and final clean make use of methanol to remove water from the emulsion. The first of these baths uses a mixture of 50% methanol and 50% deionized water. If 100% methanol were used in this bath, the emulsion would swell, changing the dimensions of the image.

Emulsion photoplates are used for master reductions, 10× reticles, final reductions with step-and-repeat, and contact and projection printing of

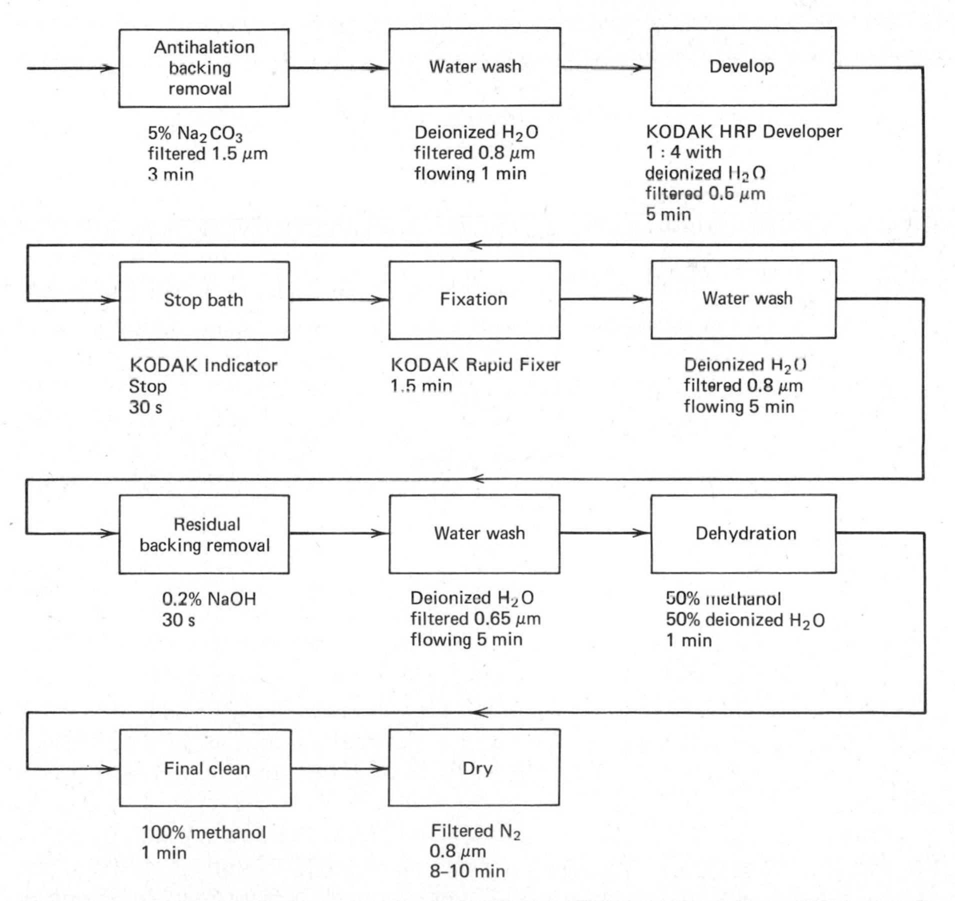

Figure 2-1. A process sequence for developing Kodak High Resolution Plate (HRP) for microelectronics applications (4).

photoresists on microelectronics wafers. Most thin-film hybrid patterns are delineated from emulsion masks. Thick-film patterns are transferred to screens from emulsion masks.

Photoresists (11–16)

Photoresists are photosensitive materials that exhibit chemical resistance, have film-forming properties, and are reasonably adherent to various surfaces. There are two basic types of photoresists, negative and positive. Negative photoresists become insoluble when exposed to light. Positive photoresists, on the other hand, become soluble when exposed to light. It is, therefore, possible to use the same mask to delineate two patterns, one in which the photoresist remains in the regions where the mask is clear, and one in which the photoresist remains in the regions where the mask is dark. A number of photoresists have been formulated which have different spectral sensitivities. The most commonly used photoresists are sensitive to ultraviolet (UV), but it has been necessary to produce special materials that are sensitive to electron beams and soft x-rays to delineate patterns with line widths smaller than 1.5 μm.

Negative UV Photoresists. Negative UV sensitive photoresists have been the most frequently used photoresists in the microelectronics industry. A number of these photoresists are in common use. Their properties are described below.

In 1953, Kodak introduced a photoresist that was designed for the printed circuit and chemical milling industries. It was called KODAK Photo Resist (KPR®). This was the first photoresist to make use of the photoactivated cross-linking of synthetic polymers to render them insoluble in a particular organic solvent. The photosensitive polymer in KPR is polyvinyl cinnamate. This polymer exhibits a low degree of photosensitivity which is due to the presence of unsaturated carbon bonds within the polymer chain. These are bonds in which two pairs of electrons are involved. When one of these molecules absorbs energy, one of the electron pairs can be shared with a carbon atom in an adjacent molecule, resulting in a cross-linked structure. This process can be substantially enhanced by the inclusion of certain organic sensitizer molecules, which absorb photons of a particular energy range (UV), and interact with the polymer molecules to activate the cross-linking process. It is also possible to produce cross-linking by thermal excitation, which makes it necessary to carefully control the temperature cycle of these materials prior to exposure.

With the development of the microelectronics industry came a demand for photoresists capable of adhering to different surfaces and permitting the etching of very fine lines. The second generation of synthetic polymer photoresists was formulated to meet these demands.

The resin system for these photoresists is cyclized polyiosoprene rubber. This is a tough, resinous material that is resistant to both acids and alkalies and adheres well to metal surfaces. This resin is not photosensitive. The average molecular weight of the polymer is 60,000 AMU. The resists are made photosensitive by the addition of photoinitiators in the form of azides, compounds containing one or more $-N_3$ groups. Typically, the resins outnumber the photoinitiators 50–70 to 1. Most of the useful azides also contain double bonds and are very active toward cross-linking. The compounds are activated by exposure to light in the 2600- to 4700-Å wavelengths.

The solvent system for these photoresists is usually xylene. Development is performed using a mixture of 50% xylene and 50% Stoddard's solvent (a petroleum distillate that contains both aromatic and aliphatic solvents). Developers and rinses compatible with photoresists are available from the manufacturers.

The first products in the second generation of photoresists were KODAK Metal Etch Resist (KMER®) and KODAK Thin Film Resist (KTFR®). KMER showed great promise for use in the semiconductor industry because it adhered well to both silicon dioxide and metals. Unfortunately, in order to obtain reproducible results with KMER, it is necessary to precondition the photoresist by centrifuging and filtering. KTFR uses an improved polymer and the same photoinitiator as KMER. Less preconditioning is required, due to improved control over the raw materials, but KTFR does not adhere well to silicon dioxide. Adherance can be improved by priming the oxide surface with a conditioning agent like hexamethyldisilizane (HMDS). Some microelectronics manufacturers use mixtures of KMER and KTFR in preconditioned forms to achieve results that are superior to those obtained with either photoresist used separately.

More recently, photoresists have been formulated specifically for the integrated circuits industry. The chemistry for these materials is similar to that for KMER and KTFR, but they are in a form that is ready to be applied to the wafers. Two of these products are KODAK Micro Resist 747® and HUNT Waycoat IC Resist®.

All of these photoresists based on cyclized rubber share a common property. The presence of oxygen during exposure significantly reduces the ability of the photoinitiator to promote cross-linking. This can reduce the thickness of a developed photoresist coating by 20% (14). To avoid this problem, exposure is performed in nitrogen or vacuum, or the photoresist coated wafer is forced into contact with the mask by air pressure. Hunt has introduced a photoresist (Waycoat HR Resist®) in which the oxygen sensitivity is significantly reduced. This photoresist also reduces effects similar to halation when the material to be etched is highly reflective, like aluminum or chromium.

Negative UV sensitive photoresists are used in most of the integrated circuit applications except those requiring patterns smaller than 2.5 μm. In order to form an effective protective layer, negative photoresist coatings are typically 0.8 to 1 μm thick. As a general rule, the coating thickness should not exceed one-third of the minimum pattern size if sharp definition is to be obtained using negative photoresists.

Positive UV Photoresists. Positive UV sensitive photoresists are more expensive, and more expensive to use than negative photoresists. The energy required to expose them is greater, resulting in a longer exposure time, and, thus, a reduction in throughput. A distinct advantage of positive photoresists is that patterns can be delineated as small as the thickness of the photoresist layer, due to the difference in the photoactivation mechanism. It is, therefore, possible to define 1-μm lines with a 1-μm layer of positive photoresist, a thickness which is virtually pin-hole free. Positive photoresists are used for most patterns smaller than 2.5 μm.

Positive photoresists also contain resins and sensitizers. Unlike negative photoresists in which the polymers outnumber the photoinitiators by 50 or more to 1, the ratio of resins to sensitizers in a positive photoresist is typically between 2 and 4 to 1. In unexposed positive photoresists, the resin-activator mixture is soluble in an organic solvent but insoluble in water. The activators are diazo compounds that are decomposed when exposed to UV radiation ($\lambda = 3850$ Å) and become deactivated. The combination of the photodecomposed molecules and the resin is soluble in water. An alkaline-aqueous developer is used to remove the exposed positive photoresist. The unexposed diazo molecules promote coupling between the resin molecules in the presence of an alkaline solution. Positive photoresists permit the clean definition of extremely fine lines.

Commercially available positive photoresists include SHIPLEY AZ-111®, AZ-1350®, and AZ-2400®, KODAK Micro Positive Resist 809®, GAF Microline 102®, and HUNT Positive HR Resist®. These are based on low molecular weight phenolformaldehyde resins with diazo-ketone sensitizers. Developers for these photoresists are available from the manufacturers.

Positive photoresists are insensitive to oxygen during exposure. They are, however, sensitive to humidity. It is recommended that the relative humidity be controlled to approximately 50% for best results.

Electron Beam and x-ray Photoresists. The negative and positive UV sensitive photoresists described in the previous sections are also sensitive to electron beam (*e*-beam) and soft x-ray (4 to 9 Å) exposure. However, the sensitivity of these photoresists to this type of radiation is relatively low, requiring long exposure times and low throughput. Special photoresists have been developed which exhibit significantly higher sensitivity to *e*-beam and x-ray exposure.

As indicated previously, reliable resolution of negative UV photoresists in a production situation is limited to 2.5 μm, and that of positive UV photoresists is limited to 1 μm. In order to understand the limits of defining small geometries in a reproducible fashion, the method of exposure must be carefully examined. If contact printing is to be used, it should be recognized that some of the photoresist will adhere to the mask after the mask is separated from the wafer, requiring frequent cleaning of the mask. Small variations in the wafer surface geometry due to bowing of the wafer after high temperature processes and previous etching processes make intimate contact between the wafer and the mask difficult. Flexible mask materials permit conformal contact printing, but this technique is not a viable production method. Proximity printing, in which the mask is separated from the wafer by several micrometers yields a longer mask life but introduces the possibility of standing wave patterns occurring in the photoresist, particularly when the wavelength of the exposing light is of the same order of magnitude as the width of the patterns being exposed. A similar problem is encountered if the image is projected onto the photoresist using a 1:1 lens. The edge profile of a 1-μm line defined in AZ 1350 positive photoresist is shown in Figure 2-2. Diffraction problems can be essentially eliminated by exposing with an *e*-beam or soft x-rays which have two orders of magnitude shorter wavelengths than those of the UV radiation used in optical photolithography requires photoresists that are sensitive to *e*-beam or x-ray exposure.

The photoresists that have been developed for *e*-beam exposure are also sensitive to soft x-rays. The absorption of x-rays results in the liberation of 1- to 3-keV electrons, which produce essentially the same results as bombarding the photoresist with 7.5- to 14-keV electrons during *e*-beam exposure.

Negative *e*-beam sensitive photoresists are based on the copolymer poly(glycidal methacrylate-co-ethyl acrylate), which is designated COP. Exposure to 10-keV *e*-beams results in the cross-linking of these molecules, preventing dissolution during development. This photoresist is at least an order of

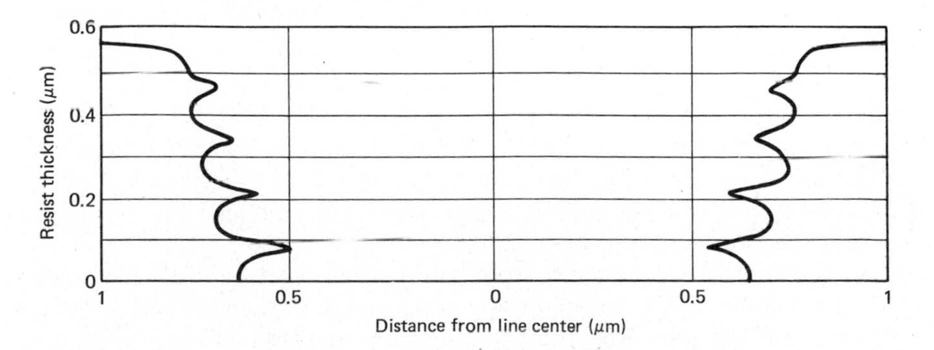

Figure 2-2. A typical edge profile of positive photoresist used to define 1-μm lines (25). © 1975 IEEE. Reprinted from *IEEE Trans. on Electron Devices*. July 1975, 462.

magnitude more sensitive to e-beam exposure than typical UV sensitive negative resists.

Positive e-beam sensitive photoresists are based on poly(methyl methacrylate), which is designated PMMA. Exposure to 7.5- to 14-keV e-beams causes a splitting of the large molecules into smaller molecules which have a faster dissolving rate in the developer than the unexposed molecules.

Neither of these e-beam photoresists is completely satisfactory. There is a continuing effort to develop e-beam/x-ray photoresists that have high sensitivity, high resolution, and resistance to chemical, ion, and plasma etching.

Dry Film Photoresists. Negative UV sensitive photoresists are also available in the form of a 15- to 75-μm-thick dry film sandwiched between a polyester film and a polyolefin film. The resolution of this film is limited to 25 μm. The photoresist is applied by removing the polyolefin film and using heat and pressure to cause it to adhere to the substrate. Exposure takes place through the polyester film. The principal use of this type of photoresist in microelectronics is to define patterns for the selective electroplating of thick "bumps" on wafers to facilitate beam tape or flip-chip bonding.

PATTERN LAYOUT AND GENERATION

The transformation of an electronic circuit from a schematic diagram to a microcircuit is a complicated procedure. For hybrid circuits, the line widths are large compared to those in monolithic integrated circuits. The layouts for hybrid microcircuits are typically 10× to 50× final size. They are reduced with a single reduction onto emulsion plates and then transferred to the substrates. For monolithic integrated circuits, the final size master masks are usually made by a step-and-repeat reduction from 10× reticles. Several methods for the generation of 10× reticles are in current use. It is also possible to generate patterns at final size by using an e-beam pattern generator.

Layout

In many cases, the first step in the generation of patterns for microcircuits is to draw a large scale composite layout of the mask set. This is typically 100× to 2000× final size. As the circuit complexity increases and the line widths decrease, the magnification must increase.

The composite layout can be transformed into a set of oversized artwork by several techniques. A simple, inexpensive, and flexible method is to inscribe the pattern on ULANO Rubylith® (a red strippable plastic film on a polyester base) using a manual precision drafting table called a coordinatograph. The red plastic is stripped from the artwork by hand in the regions where the pattern is to transmit light, and remains where the pattern is to be opaque. Red is used because the emulsion used in high resolution plate (HRP) is insensitive to red light. The inscribing of the Rubylith can be automated by using a digitizer to convert the information in the drawing into digital form, and using the results to control an automatic cutting machine. Both of these methods require manual stripping of the Rubylith, which, on a complicated layout, can result in errors. An alternate method is to use the output of the digitizer to control a variable aperture photoplotter, which exposes the pattern on an emulsion coated film or plate. This technique is more accurate than the cutting of Rubylith, and is usually performed at 100× final size.

The next step after the generation of the oversized artwork is a photographic reduction to a 10× reticle. The camera for this reduction must satisfy several stringent requirements. The lens must be corrected for aberrations. There are seven elementary aberrations, five of which are independent of wavelength, and two of which depend on the color of the light used (28). In general, a lens can be fully corrected for any five of the seven aberrations. Since a monochromatic light source can be used for the reduction of microelectronics artwork, the chromatic aberrations can be ignored and a nearly perfect lens can be made for this purpose. The ability of a lens to resolve the differences between fine lines is limited by diffraction patterns. The lens opening, which is inversely proportional to the f-number of the lens (29), must be as large as possible to increase the resolving power. Unfortunately, decreasing the f-number increases the aberration effects, requiring a compromise in lens design. The resolution limit in lines per millimeter is given by

$$\text{resolution limit} = \frac{10^6}{\lambda \times f\text{-number}} \tag{2-1}$$

where λ is the wavelength in nanometers. Another important lens criterion is the useful field. Due to the curvature of the lens, the sharpest image is restricted to a field whose diameter does not exceed one-fifth of the focal length of the lens. If the artwork is assumed to be square with dimensions $A \times A$, and is to be reduced by a factor of R, the required focal length, F, is given by

$$F = \frac{5A\sqrt{2}}{R} \tag{2-2}$$

with F and A having the same dimensions. for example, if A is 250 cm, and the reduction required is 50, the minimum focal length is 35.35 cm. The camera dimensions necessary to perform this reduction can be determined in the following manner. The lens-to-image distance, L_1, is given by (25)

$$L_1 = F\left(\frac{1}{R} + 1\right) \tag{2-3}$$

which, for this example is 36.06 cm, and the lens-to-artwork distance, L_2, is given by

$$L_2 = F(R+1) \tag{2-4}$$

or 1802.85 cm for this example. It is obvious that this camera is huge. This type of camera must be rigidly mounted on a vibration-free slab. The artwork in these cameras is back-lighted, using a diffuse light source that is usually filtered to provide a green monochromatic light for exposing Kodak HRP. The environment for the camera must be carefully controlled, and the artwork is usually produced and retained in the same area to prevent distortion due to the effects of temperature and/or humidity on the polyester backing material. For large scale integrated circuits, it may be necessary to perform two reductions to obtain a 10× reticle, or to perform the reduction in sections, and photocompose the reticle.

This approach to pattern generation, using oversized artwork, will continue to be a viable technique, primarily because of the simplicity of the process and the significant capital invested in the equipment used to implement the approach. There are difficulties associated with this technique, particularly for dense, large-scale integrated (LSI) circuits and very large-scale integrated (VLSI) circuits. These difficulties have spurred the development of different approaches for the creation of microelectronics artwork.

Computer Aided Design

The availability of computers has made possible a new medium for the layout of microelectronic circuits. A significant effort has been applied to the development of software for interactive graphics systems enabling the operator to completely describe circuit layout electronically. The benefits derived from this type of system more than make up for the impressive capital investment.

Computer aided design (CAD) systems are available in several forms. A typical system might include several different levels at which the operator may enter the system. The basic level contains the capability for specifying rectangular patterns on a cathode ray tube (CRT) which can be positioned using a light pen or joystick arrangement. This system is used for the generation of individual devices or circuit modules. It is possible to draw and display all of the different mask levels either individually or in composite form. There is also a zoom capability with this type of system so that specific areas can be examined in detail. A second level of entry into a CAD system permits the operator to make use of an established library of devices and circuit elements. With this type of system, the operator calls out the components necessary for the circuit, positions the circuit elements, and indicates the points which must be interconnected to provide the desired function. A third level of interaction makes use of predesigned circuit modules which are to be interconnected to form a large circuit. A typical example of this would be the interconnection of standard logic modules to form an LSI logic array. The operator can place the modules and specify the interconnections manually, or make use of a more sophisticated software package which accepts the logic diagram as an input and automatically places the modules and performs the desired interconnections. Other features of advanced CAD systems include circuit simulation at the static logic level, transient analysis and propagation delay simulation, and the specification of a complete testing procedure for the circuit. With an increase in the sophistication of CAD, information in addition to the layout is made available to the circuit designer about the operation of the microcircuit.

The output of the CAD system is usually in the form of digital information stored on a magnetic tape. The tape can be used to control a photoplotter for a 100× final-size set of patterns, a precision pattern generator for a 10× final-size set of patterns, or *e*-beam lithography system for a 1× final-size set of patterns.

Pattern Generators

The photoplotter described above is the forerunner of the precision reticle generator in common use in the microelectronics industry. The reticle generator is designed to work at final size for hybrids and 10× final size for monolithic microcircuits.

The patterns are usually generated in Kodak HRP using a variable rectangular aperture and a flash lamp. Curves can be generated by multiple exposures of successively rotated rectangles. The positional accuracy of the stage is controlled by laser interferometers, with resolutions of ±0.6 μm. The address resolution is typically 0.1 μm, aperture size can be controlled in 1-μm increments from 4 μm to 3000 μm, and the area that can be exposed is 1.5×1.5 cm. The accuracy of 10× reticles generated on these pattern generators is superior to that obtainable with patterns created at 200× final size using Rubylith.

Pattern generators designed to work at final size have been developed using *e*-beam sources. They work on photoresist either for making masks or for direct exposure on the wafer. Direct exposure on the wafer is a time consuming process and is only used when the accuracy and line widths require the precision obtainable with an *e*-beam lithography system. There are two basic types of *e*-beam exposure system, vector scan and raster scan. In a vector scan system, the beam is directed to each individual location that is to be exposed. Resolutions down to 0.1 μm have been obtained using this type of system, but exposure times of several hours are necessary for a complex pattern on a wafer with a diameter of 7.5 cm. It is possible to form a rectangular beam shape by a variable aperture technique which reduces the exposure time required for a vector scan system by permitting the exposure of an area instead of a point. In a raster scan system, the $x-y$ table is in continuous motion and the e-beam is swept over a 256-μm distance at right angles to the motion of the table. The beam is selectively blanked and unblanked under computer control to expose the desired pattern. The time required to expose a mask or wafer is independent of the complexity of the pattern, since the same raster is used. In early systems, it took 40 min to expose a 7.5-cm wafer. The beam diameter is typically 0.25 μm and four beam spots are required to define a minimum line width, resulting in a resolution of 1 μm. Pattern generators using e-beams provide a viable alternative to reticle generators for the production of precision masks.

Step-and-Repeat Systems

The final-size master mask sets are made from 10× reticles by a system in which the image is multiple exposed on a plate that is mechanically stepped, except in the case of an e-beam final-size lithography system. The 10× reticles are generated either by reduction from an oversized set of artwork, or by a reticle generator.

The master mask set can be made in Kodak HRP or a wear-resistant hard surface material. If hard surface masks are used, photoresists must be exposed in the step-and-repeat system, requiring a UV light source, lenses corrected for UV, and longer exposure times than those used for HRP exposure.

In order to assure that each individual circuit will be aligned on each mask level, it is desirable to fabricate an entire mask set simultaneously. This is accomplished by using a multiple-barrel step-and-repeat camera. In this system, a reticle is placed in each barrel, one for each mask level. The plates to be exposed are placed on a single table, so that they are all stepped and exposed simultaneously. If the equipment is designed for emulsion plate, the table is moved continuously and the exposure lamps are strobed. If photoresist is to be exposed, the table must pause for each exposure. Test patterns are inserted at symmetrical locations on the mask for the evaluation of the process. The test patterns also assure that the circuit patterns on each mask level are aligned to the corresponding patterns that were exposed simultaneously on the step-and-repeat system. One drawback associated with a multibarrel step-and-repeat system is that, if a defect occurs in one of the masks in a complicated set, it is necessary to remake the entire mask set.

Another approach to step-and-repeat processing is to perform a direct wafer exposure by a reduction of a 10× reticle. Line widths of 1.25 μm have been defined in positive photoresist using this type of system. This is a relatively inexpensive and rapid alternative to direct exposure e-beam lithography.

Once the master mask set has been generated, the working mask sets are produced. The working masks can be either emulsion plates or hard-surface masks. They can be defined by contact or 1:1 projection printing from the master mask set. If both the master masks and the working masks are emulsion plates, a reversal development process is often used in the production of the working masks to provide a positive rather than negative image. Positive photoresist is used to produce this result on hard-surface working masks.

MASK MATERIALS

In this section, the materials used for the transfer of patterns to microelectronic substrates are described. These include emulsion plates and hard-surface plates for optical photolithography, masks for x-ray photolithography, etched metal masks for thin-film depositions and fine line thick-film printing, and emulsion screens for thick-film printing. The selection of the materials for masking is based on cost, minimum line width, the process to be employed, and the number of circuits that must be produced.

Emulsion Masks

Emulsion masks like Kodak High Resolution Plate (HRP) are in widespread use for contact printing to photoresist coated substrates for monolithic integrated circuits and thin-film hybrid microcircuits. They are the lowest cost masks and can be profitably employed for line widths down to 5 μm. The major difficulty experienced with HRP masks is the damage that they incur during the contact printing process. Photoresist particles become imbedded in the emulsion, giving rise to an increasing number of defects with repeated usage. It is difficult to clean emulsion masks without producing additional defects. For typical production runs, emulsion masks are discarded after 10 or fewer printings. Off contact printing or projection printing substantially increases the usable life of an emulsion mask.

Hard-Surface Masks (32–34)

In order to produce a mask that could be used for contact printing and not deteriorate with repeated use, a number of mask materials have been developed which can be readily exposed to the strong solvents and acids commonly used for the removal of photoresists. These hard-surface mask materials are deposited on precision glass plates by thin-film techniques and patterned using the photolithographic process. The deposited films are typically 1000 to 2000 Å thick, which are one-fortieth to one-twentieth the thickness of a typical emulsion film. It is possible to define more precise patterns using hard-surface materials than emulsion films. All *e*-beam generated masks are generated in hard-surface materials. Since small line widths can be generated in hard-surface materials, masks made from these materials are also used for proximity and projection printing applications.

The most popular hard-surface mask materials are chromium, chromium oxide, iron oxide, and silicon. Chromium is an excellent hard-surface mask material with a higher optical density than developed emulsions. Unfortunately, chromium is highly reflective, which presents problems for mask alignment, particularly for dark field masks, and for noncollimated exposure light which reflects from the substrate and then from the mask, resulting in poor edge definition. In the latter case, antireflection coatings of chromium oxide or gold have been used to alleviate the problem. Iron oxide is very popular for hard-surface masks. In order to obtain the same optical density at UV as a 1000-Å chromium layer, it is necessary to use a 2000-Å layer of iron oxide. The iron oxide is deposited in amorphous form, etched in hydroiodic acid, and subjected to a time-temperature cycle (Typically 10 minutes at 300°C) to crystallize and make it impervious to attack. Iron oxide is the most abrasion-resistant mask material and has the additional advantage of being translucent to visible light. This results in a see-through mask that is more readily aligned if a dark field mask is required. Polycrystalline silicon and chromium oxide also result in see-through hard-surface masks.

Hard-surface masks are considerably more expensive than emulsion masks, and, since they use photolithography for pattern definition, require longer times for fabrication. If working masks are hard surface, the master masks should also be hard surface, to prevent degradation of the master masks during contact printing to produce the working masks. For these reasons, emulsion masks will continue to be the dominant masks for the production of small-scale and medium-scale integrated circuits, as well as for thin-film hybrid microcircuits.

x-Ray Masks (35–38)

Masks for x-ray lithography must absorb or transmit x-rays, rather than UV light, and are radically different in construction than those used for conventional photolithography. Gold is an excellent absorber of soft x-rays (wavelengths between 2 and 15 Å), and is the most popular material for this purpose.

Unfortunately, it is difficult to find a suitable material to act as a support for the pattern. Beryllium offers the least attenuation for soft x-rays, but a thickness in excess of 25 μm causes too much attenuation. Several mask substrate materials have been successfully demonstrated. One approach is to use silicon as the substrate. A gold-on-silicon x-ray mask is fabricated in the following manner. A 3-μm n-type epitaxial layer is deposited on a 200-μm-thick heavily doped n-type silicon substrate. A 0.1-μm aluminum oxide layer is sputtered on top of the epitaxial layer, followed by a 300-Å layer of chromium, for adhesion, and a 3000-Å layer of gold. The pattern is defined by e-beam lithography and etched. The heavily doped substrate is selectively removed by an electrochemical process in a dilute HF solution. The result is a gold pattern supported by a 3-μm silicon membrane, which has a 200-μm thick silicon boundary for mechanical strength. The etch rate for the heavily doped silicon is much greater than that of the lightly doped silicon, permitting reasonable control over the process. A similar process is used to create a silicon nitride membrane supported by silicon as a mask substrate. A third approach makes use of a polyester film supported on an aluminum ring as the substrate for the gold pattern. In some cases a 300-Å layer of gold is deposited and used as the basis for electroplating a 10,000-Å pattern as defined by e-beam lithography. This results in the situation where the thin layer of gold, which was protected by the photoresist, acts as the transmission windows for the x-rays, and the thick layer of gold acts as the absorption medium for the x-rays. Since x-ray mask substrates are extremely thin, the application of a low pressure (typically, 0.1 atm) to the back of the mask permits it to conform to the curvature of the substrate. As x-ray lithography becomes a production technique, other mask materials will be investigated to satisfy the stringent requirements of this process.

Etched Metal Masks (39)

Etched metal masks are sometimes used for defining relatively large patterns for thin-film depositions. For these situations, the mask is usually clamped to the substrate and the material is deposited on both the mask and the areas of the substrate exposed through the openings in the mask. The alignment of patterns from one level to another is difficult unless the patterns are extremely simple. A much more important application of etched metal masks is their use as stencils for fine-line (200 μm) printing in thick-film hybrid microcircuits. Etched metal stencils are considerably more expensive than conventional screen printing and their use must be justified by the fine-line requirements.

For simple patterns with large openings, a beryllium copper alloy foil, typically 125 μm thick, is etched with ferric chloride using photolithography. For more complex patterns, a composite metal structure is used. This consists of a 12.5 μm electroplated nickel layer on a 100-μm beryllium copper foil. The beryllium copper provides mechanical support for the nickel. The pattern is photoengraved in the nickel and part of the way through the beryllium copper with ferric chloride, while the backside of the beryllium copper is protected with acid resistant lacquer. The etchant is changed to chromic acid, which attacks the beryllium copper but not the nickel. When the beryllium copper has etched through, about 75 μm of undercutting has occurred, leaving

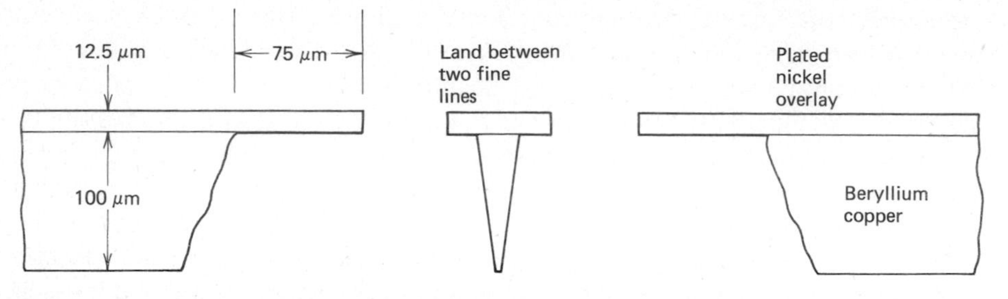

Figure 2-3. A bimetal etched metal mask.

a sharply defined pattern in the thin nickel layer, supported except at the edges of the cuts with beryllium copper, as indicated in Figure 2-3.

If a thick-film stencil is to be fabricated, the same procedure is used for defining the pattern, followed by bonding the foil to a stainless steel screen. This can be accomplished by welding, brazing, or the use of adhesives. It is also possible to electroplate nickel directly on the screen and define the pattern by photoengraving. For fine-line stencils, it may be necessary to use an etched solid metal pattern. As line widths approach the screen mesh spacings, alignment between the pattern and the mesh becomes important. A solid metal stencil is photoengraved from both sides. The pattern on one side is the pattern to be printed, and the pattern on the other side is a custom set of mesh openings which aligns perfectly with the pattern to be printed. Molybdenum is the most popular material for solid metal stencils. In some cases, solid metal stencils without mesh patterns have been used for fine-line printing.

Screens for Thick-Film Hybrid Microcircuits (40, 41)

Patterns for thick-film microcircuits are screen printed on ceramic substrates. The patterns are usually transferred to the screens from emulsion plates by contact printing to an emulsion that is uniformly coated on the screen. As discussed in the previous section, etched metal masks are bonded to screens for high volume production, and etched metal stencils are used for fine-line printing.

There are several forms of the emulsion-screen combination. The most frequently used method consists of coating the screen with an emulsion that is either photosensitive or can be sensitized after application. This emulsion is in powder form, mixed with a suitable vehicle, or as a uniform layer on a plastic backing that is rendered gel-like by soaking and pressed into the screen. The pattern is transferred to the emulsion by a contact printing process. The result is a tough, durable pattern with a reasonable line definition. Patterns are replaced after 1000 to 10,000 prints. Another approach, used for prototype development, makes use of a photosensitive emulsion on a polyester backing.

The emulsion is exposed through the backing film, so that the energy of the exposure determines the thickness of the emulsion after development. The development process includes a warm water rinse, which leaves the emulsion in a soft, gel-like state. The screen is then forced into the soft emulsion on a flat surface. After the emulsion drys, the backing sheet is peeled off, leaving a smooth surface on the bottom of the screen. This process offers quick turn-around and good pattern definition, but is limited to less than 100 prints.

The orientation of the pattern relative to the screen mesh becomes significant for fine-line printing. It is common practice to use a microscope to align the pattern either in parallel with the mesh or at 45° to the mesh, for optimum line definition.

PHOTOLITHOGRAPHY

Photolithography, or photoengraving, is the process of transferring a pattern from a mask to a substrate. It is an essential technique for monolithic integrated circuits and most thin-film hybrid microcircuits. The process has been automated in semiconductor device fabrication facilities.

The steps in a photolithography process are

1. Substrate preparation.
2. Photoresist coating.
3. Pre- or soft baking.
4. Mask alignment.
5. Exposure.
6. Development.
7. Post- or hard baking.
8. Etching.
9. Stripping of the photoresist.

Substrate Preparation (42)

In order for photoresist to adhere well to a surface, it is necessary that the surface be free of organic contaminants and water. Most photoresists are hydrophobic, while one of the most often encountered surfaces, thermally oxidized silicon, tends to be hydrophyllic. Special surface pretreatment is essential for successful photoengraving of silicon dioxide.

The most effective pretreatment is a high temperature (300 to 1000°C) bake in an oxidizing atmosphere for 30 minutes. Since this is a typical situation encountered in the diffusion and oxidation processes associated with integrated circuit fabrication, silicon wafers are frequently taken directly from the diffusion furnace to the photoresist application station. If this is not the case, the substrates are usually subjected to a mechanical scrubbing operation, either a pressurized spray of deionized water or a rotating nylon brush with deionized water, on a spinning chuck, followed by a nitrogen blow-off. The substrates are then baked in an infrared conveyer oven to dehydrate the surface.

Photoresist adhesion can also be promoted by the use of a primer. The most common materials for this purpose are solutions of hexamethydisilizane (HMDS), trichlorophenysilane (TCPS), or bistrimethylsilyacetamide (BAS) in xylene. The priming solution removes any residual moisture on the substrate surface. The primer is applied by pumping several drops on the substrate, and spinning at 3000 to 5000 r/min for 15 to 30 s. Most of the solvent evaporates during the spinning cycle. The primer is applied immediately before the photoresist is applied.

Photoresist Coating (43–46)

The most popular technique for obtaining a thin, uniform coating of photoresist on a microelectronics substrate is by spinning. The substrate is held on a vacuum chuck, while several drops of photoresist are dispensed automatically and allowed to spread over the surface. The substrate is accelerated at a high rate to a constant speed, which is maintained for a specified time, depending on the viscosity of the photoresist and the desired thickness of the coating.

The creation of a uniform photoresist coating is a complex process. During the first few revolutions, the centrifugal force acting on the photoresist causes it to bulge out over the edge of the substrate. When this force exceeds that due to surface tension, most of the photoresist is expelled from the wafer (43). The total time from start-up until this event is typically less than 0.20 s, depending on the acceleration characteristics of the spinner. In order to prevent the redeposition of expelled photoresist, the equipment should be designed with a downdraft at the chuck. The photoresist remaining on the wafer after this initial expulsion, except for solvent evaporation, stays there for the remainder of the spin cycle. This photoresist redistributes itself in a series of waves which are easily observed, due to interference patterns, as a series of colors while the substrate is spinning. The redistribution of the photoresist results in an edge buildup on a round substrate. This buildup can be described as a squeezing of the photoresist against an invisible wall due to the surface tension at the edge of the substrate, with the photoresist rim becoming higher and thinner with time. The resultant rim is so thin that it is invisible except by means of an interferometer, and is typically 10 times as high as the thickness of the photoresist in the center of the wafer. The profiles expected from spin-coating of photoresists on round substrates range from an up-dish at low speeds, with a gradual buildup near the edges, to a down-dish at high speeds, with the center thickness greater than the edge thickness, except for the rim. High acceleration rates minimize dish effects. Nonround substrates cannot receive uniform coatings, due to turbulence in the photoresist caused by the corners.

The thickness of the final layer depends upon the spinning speed and the viscosity of the photoresist, which is a function of the solids content. The solids content can be varied by adding thinner to the photoresist. Empirical results indicate that the thickness of the photoresist layer can be predicted from the following relationship (43).

$$t = \frac{kp^2}{\sqrt{w}} \tag{2-5}$$

where t is the thickness in angstroms, k is a constant associated with a particular spinner (usually between 90 and 110), p is the percent solids in the photoresist, and w is the spinner speed in revolutions per minute divided by 1000.

The total spinning time has little effect on the coating thickness, once the redistribution is complete. Since the redistribution usually takes 15 to 20 s, spinning times of 30 to 60 s are typical. Photoresist thicknesses of 8000 to 15,000 Å are common in the integrated circuits industry.

Photoresist application is one of the most critical steps in microelectronics technology. The environment in which it is performed must be carefully controlled. Dust particles represent an important hazard to successful photolithography. Particulate contamination on the substrate prior to photoresist application results in nonuniform coating. Dust particles falling on the substrate during and after spinning are the major causes of pinholes in the photoresists. For these reasons, photoresist application takes place in a vertical laminar flow *Class 100* (less than 100 particles greater than 0.5 μm in diameter per cubic foot) clean area. It may also be necessary to control the relative humidity in the photoresist application area, since some photoresists do not perform well at either extreme of humidity level.

Pre- or Soft Baking

The purpose of prebaking is to remove the solvent from the photoresist coating. This can be performed in an oven with an air or nitrogen atmosphere, under vacuum, or in an infrared conveyer arrangement. The temperature-time cycle for prebaking must be sufficient to drive off the solvent without thermally cross-linking the photoresist. Consult the manufacturer's data sheets for the optimum prebake cycle for the photoresist being used.

Mask Alignment and Exposure (47–55)

The alignment of masks to patterns on substrates, and the subsequent exposure of photoresist through the masks, requires some of the most intricate equipment used in the fabrication of microcircuits. The design rules for monolithic integrated circuits often require alignment precision of 0.5 μm. Exposure systems have been developed for contact, proximity, and projection printing. Special alignment techniques are necessary for the *e*-beam and x-ray exposure systems associated with submicrometer line widths.

Most of the mask-alignment equipment in the semiconductor industry is manual. The substrates are held on a vacuum chuck, which is part of a precision x-y table with rotational adjustment capability. In a contact printing system, the substrate is separated from the mask by a distance of 25 to 125 μm during the alignment procedure. Split-field optics are used so that the operator can observe two areas of the mask simultaneously. After the mask has been aligned to the pattern on the substrate, the substrate is clamped to the mask either by vacuum or air pressure, the microscope is moved away and replaced by a collimated UV source, and a shutter is opened for a specified time. When the shutter is closed, the substrate is separated from the mask, and then removed from the vacuum chuck. The substrate is now ready for the development process. The light source used for illumination during the alignment phase is filtered to prevent premature exposure of the UV sensitive photoresist. Mercury vapor lamps are popular sources of UV for photoresist exposure, usually collimated through a quartz optical system. In some cases, a light integration is performed to control the time for optimum exposure of the photoresist. Minor nonoptimum exposure results in slight differences between line widths on the mask and the corresponding photoresist patterns.

Proximity printing has been used as a technique for the extension of mask life. The optimum separation between the mask and the substrate is 10 to 30 μm depending on the minimum line width to be reproduced. Diffraction effects can be important for fine line widths. The basic equipment is similar to that used for contact printing.

In a projection system, a lens is interposed between the mask and the substrate. Since there is no possibility of contact between the mask and the substrate, the mask life is essentially indefinite. There are significant constraints on the projection lens, particularly if large diameter substrates are to be exposed. Diffraction effects are also important in projection printing when line widths approach 1 μm.

Automatic mask alignment is available from several equipment manufacturers. By the use of special alignment targets on the substrate, a closed loop system is employed to produce 0.5-μm alignment on a repeatable basis. The important consideration in this type of system is a mechanical prealignment process which must position the targets within 1000 μm of the corresponding targets on the mask. The rotational error after prealignment must be within 3° on a 50-mm-diameter substrate, 1.8° on a 75-mm-diameter substrate, and 0.8° on a 100-mm-diameter substrate. Automatic mask alignment has had difficulty keeping pace with other substrate processing techniques.

The alignment of masks for x-ray exposure of submicrometer line widths presents a challenge for optical equipment. The limited depth of focus of high-magnification lenses makes it difficult to simultaneously view the mask pattern and the pattern on the substrate. The ability of the operator to determine correct alignment is seriously hampered under these conditions. It will probably be necessary to use an automatic system responding to hard x-rays from special alignment targets to achieve the required 0.1-μm positioning accuracy required for high density integrated circuits.

In some cases, it is necessary to align a mask to patterns that are not on the surface of the substrate. Examples of this include diffused regions underneath epitaxial layers and situations when it is necessary to define a pattern on the back of an integrated circuit substrate. The optical alignment system is replaced with an infrared imaging system on a conventional mask aligner to perform this operation.

An entirely different approach to photolithography is represented by the electron image projector (55). The mask is an e-beam generated chromium on quartz-glass pattern that is coated with a cesium iodide photocathode. The mask is illuminated from the back by a diffuse source of UV, resulting in the emission of electrons from the areas of the photocathode not covering the chromium pattern. The electrons are accelerated by a potential difference, typically 20 kV, between the photocathode and the substrate to be exposed. Focusing is accomplished by means of a magnetic field parallel to the path between the photocathode and the substrate. Alignment is performed by an automatic system during the first second of a 10-s exposure. Tantalum oxide targets are defined on the substrate before the first pattern is defined. These targets are more efficient emitters of x-rays than silicon when they are bombarded with electrons. Transverse magnetic fields are used to reposition the electron image until the x-ray generation is a maximum, thus ensuring proper alignment. This type of electron exposure takes much less time than raster or vector scan e-beam pattern generation.

Development

Photoresist development is usually performed by spraying a proprietary solution provided by the photoresist manufacturer. This is followed by a spray rinse of a different solvent, and a nitrogen blow-off to remove the solvents.

Some microcircuit manufacturers perform an inspection at this point in the process to assure that the development is clean. If residue remains in the pattern, the substrate can be cycled through the development process a second time, or the photoresist can be removed and the photolithography process repeated.

Post- or Hard Baking

After development, the photoresist has the consistency of rubber. If it is baked at a temperature between 120 and 180°C for periods from 20 to 30 min, the photoresist will harden and adhesion to the substrate will be significantly improved. The postbake time-temperature cycle depends on the type of photoresist and the material on the surface of the substrate. In general, increasing the temperature of the postbake increases the photoresist adhesion. Care should be exercised to limit the postbake temperature at least 20°C below the dehydration bake temperature to assure that gases adsorbed at the substrate surface do not lift off the photoresist. In addition, if a high postbake temperature is used, it may be difficult to remove the photoresist in the subsequent stripping process. Postbakes are usually performed in an air atmosphere in an oven or infrared conveyer system.

Etching (56–69)

The etching process depends upon the particular material to be removed. Wet chemical etches have been the dominant techniques employed in microelectronics fabrication, but dry processes like sputter etching, plasma etching, and ion-beam etching are becoming popular for many applications.

Wet chemical etching can be performed by simple immersion, or with agitation or spraying. A common method for agitation is to use nitrogen jets in the etching bath. Most etch rates are temperature dependent, requiring careful control for optimum results. Plasma etching equipment comes in two forms, barrel reactors and planar reactors. In a barrel reactor, the substrates are mounted vertically in a fused silica carrier. The plasma is created by passing reactant gases through an RF field (13.56 MHz) created by a coil outside of the reactor chamber. In most cases, a perforated aluminum cylinder surrounds the substrates which shunts the RF field and confines the plasma between the reactor wall and the cylinder. The reacting species passes through the perforations in the cylinder and etches the substrates. In a planar plasma reactor, the plasma is generated between two parallel electrodes spaced, typically, 4 cm apart within the reactor chamber. The electrodes are excited by a 400-kHz generator. The substrates are immersed in the plasma in a planar reactor.

Sputter etching is a form of sputtering (see Chapter 6) where the substrate is the target. The rate of removal depends on the material being sputtered. In many cases, photoresist is removed at a similar rate to that of the material to be etched. In these situations, a deposited metal is used to protect the substrate where etching is not desired.

Ion-beam etching is a form of sputtering that makes use of a collimated ion beam with energies in the range between 500 eV and 1 keV. The selectivity of ion-beam etching, like that of sputter etching, is poor. For example, the etch rate for gold is less than two times that for PMMA electron resist, and the etch rates of SiO_2 and Si are approximately equal (69). Submicrometer patterns with essentially vertical sidewalls have been produced using ion-beam etching.

In the following sections, some etching processes common in the microelectronics industry are described. The reader is referred to the literature for more details on specific etching processes.

Etching of Thermal SiO_2 on Silicon. One of the most frequently encountered photolithographic processes in monolithic integrated-circuit fabrication is the selective etching of thermally grown SiO_2 on silicon substrates. Wet chemical etching using HF or buffered HF is the most popular means for accomplishing this process. A special technique using plasma etching equipment has been developed for the selective etching of SiO_2 on silicon.

SiO_2 is readily attacked by room temperature HF, while silicon is not. It is, therefore, possible to selectively remove SiO_2 with HF solutions. As supplied by the manufacturers of electronic grade chemicals, the HF concentration is 49% in water. This is too strong a concentration for controlled pattern definition. The HF can be diluted to 5 to 10% with deionized water for a more controlled etch. Dilute HF loses its acidity during the etching process and must be replaced frequently. A more common etchant contains buffering agents like ammonium fluoride (NH_4F) which helps to maintain a constant acid-base ratio. A typical etchant mixture contains 6 parts of 40% NH_4F to 1 part of 49% HF, which etches 1600 Å of SiO_2 per minute at 30°C. Premixed buffered oxide etchants are available from a number of suppliers of electronic grade chemicals. Since neither HF nor water will wet clean silicon but both will wet SiO_2, it is relatively easy to determine when the etch is complete. It is common to simultaneously etch both control substrates (which have been through the same process steps with the exception of pattern definition) and the substrates that are to be patterned. After the control substrates become hydrophobic, the etch is continued for approximately 30 s. The etching is stopped by immersion in water. To ensure uniform etching, the substrates are dipped in a wetting agent like Triton-X 100® (Dow Chemical Company) prior to immersion in the etching solution.

The etching rate depends on the conditions under which the oxide is grown. Thermal oxides grown in steam etch slightly faster than those grown in dry oxygen. The presence of impurities in the oxide can significantly alter the etch rate. A high concentration of boron, which converts the oxide into a borosilicate glass, results in a much lower etch rate. Pretreatment by immersion in fluoboric acid (HBF_4) changes the structure of the glass, and enhances the etch rate. A high concentration of phosphorus, which converts the oxide into phosphosilicate glass, results in a very high etch rate.

Conventional plasma etching, using a mixture of Freon 14 (CF_4) and oxygen can be used to etch SiO_2 at a rate of 100 Å/min. Unfortunately, the same etchant also etches silicon at a higher rate. It is very difficult to control this process. If anhydrous HF is used in plasma etching equipment, at temperatures between 150 and 190°C and pressures between 0.1 and 30 torr, etch rates between 300 and 1500 Å/min for SiO_2 are observed with essentially no etching of silicon. The direct etch rate of SiO_2 depends on the amount of water present on the surface of the oxide. Most thermal oxides are very dry, resisting attack from the etchant. Another approach, called permeation etching (62), makes use of the rapid etching of SiO_2 by HF in the presence of compounds containing carbon and hydrogen, but no oxygen. Negative photoresist is an excellent source of this type of hydrocarbon. The process consists of defining the pattern with negative photoresist in the areas that are to be etched, rather than the areas that are to be protected in a direct etching process. The exposed SiO_2 is treated in a plasma of CF_4 and O_2, which renders the oxide impervious to attack by HF. The subsequent exposure of the substrates to HF results in a rapid etch of the SiO_2 under the photoresist. The process is concluded by stripping the photoresist in an oxygen plasma. Planar plasma etching of SiO_2 with C_2F_6 has been demonstrated with an etch rate 15 times that of Si.

Etching of Aluminum Thin Films. The most popular metal material used for the intraconnection of elements within monolithic integrated circuits is aluminum. In some cases, the aluminum contains 1 to 2% silicon and 4 to 5% copper to improve its electrical characteristics. The aluminum is deposited on a variety of materials, like oxides and glasses, silicon nitride, and single-crystal and polycrystalline silicon, and frequently encounters significant vertical steps on the substrate. It is difficult to define fine lines in aluminum over steps without significant necking of the material. A further complication is the native protective oxide that forms on aluminum when it is exposed to air. Wet chemical etching of aluminum is popular, and some success has been obtained using planar plasma etching.

Aluminum can be etched with KOH, NaOH, HCl, and H_3PO_4, none of which attack most of the common substrate materials. Potassium and sodium ions have unfortunate properties in MOS devices and are avoided in this type of processing. Most aluminum etchants are based on H_3PO_4 (73%), with the addition of HNO_3 (4%), acetic acid (CH_3COOH) (3.5%), and deionized water (19.5%). This is used at temperatures between 30 and 80°C, with an etch rate of approximately 1 μm/min at 50°C. The etching is performed either by immersion or spraying. Gas bubbles occur during etching, necessitating agitation for uniform results. The etching process is terminated by a deionized water rinse.

Aluminum can be etched in a CCl_4/He plasma at 300-μm pressure in a planar plasma reactor. The active reactants are short-lived and the substrates must be immersed in the plasma discharge. The etching process is highly anisotropic, with the ratio of lateral to vertical etching being less than 1:10. Necking at vertical steps is virtually eliminated with this type of process. After the native oxide is removed, which may take several minutes, the aluminum is etched at 1800 Å/min.

Silicon Nitride Etching. Si_3N_4 is used as a protective layer in many integrated circuits, and is also used as a gate dielectric in some MOS memory devices. It can be wet-chemically etched in HF or orthophosphoric acid, but plasma etching is very effective. In a barrel reactor, with a SiF_4/O_2 plasma, the etch rate for Si_3N_4 is 5 times that of silicon and 50 times that of SiO_2.

Silicon Etching. Single-crystal silicon and polycrystalline silicon are etched in a number of situations associated with the fabrication of microcircuits. Silicon is etched by several solutions, some of which attack the material isotropically, while others attack particular crystal planes more rapidly than others. Plasma techniques are also effective for the etching of silicon.

The most common isotropic etches for silicon are those using various proportions of HNO_3, HF, CH_3COOH, and/or water. The effect of the HNO_3 is to oxidize the silicon, while the HF aids in the formation of complex soluble ions from the oxidized silicon. The CH_3COOH acts as a diluent. HF is supplied at 49% in water, and HNO_3 is supplied at 70% in water. The maximum etch rate is 800 μm/min at room temperature and corresponds to a ratio of 7 parts HF to 3 parts of HNO_3. Providing an excess of either active acid or adding a diluent reduces the etch rate to a more controllable value.

Anisotropic etching of silicon is used for certain processes in monolithic integrated circuits. These include dielectric isolation, silicon-on-sapphire circuits, vertical MOS circuits, and the separation of beam-lead circuits. Etchants that attack silicon preferentially include KOH (64, 65) and hydrazine hydrate (66). These etchants remove (100) crystal planes (see Chapter 3) more rapidly than (110) crystal planes, and much more rapidly than (111) crystal planes. The etching ratios are typically 40:30:1. A preferential etch can be used for the photoengraving of *V*-shaped self-stopping grooves in (100) oriented silicon. Since the (111) planes make an angle of 54.7° with the (100) planes, the width of the surface pattern essentially determines the depth of the groove. The depth, d, is given by

$$d = \frac{w}{2} \tan 54.7° = 0.706w \tag{2-6}$$

where w is the width of the surface pattern.

Polycrystalline silicon is used as a gate material in certain types of MOS integrated circuits (see Chapter 10). It can be etched using an isotropic silicon etch like 10 parts HNO_3, 1 part HF, and 10 parts water, or by a plasma etching technique. Polycrystalline silicon is selectively etched in a barrel reactor in a CF_4/O_2 plasma at a rate 25 times that of SiO_2. Single-crystal (111) oriented silicon etches at a slightly lower rate in the same system, with an etching rate 17 times that of SiO_2.

Chromium Etching. Chromium thin films on glass, used for working masks in the integrated circuits industry, can be etched in either alkaline or acid solutions (32), or by plasma techniques.

A typical alkaline etchant for chromium is prepared by mixing 1 part of a solution of 500 g of NaOH in a liter of deionized water with 3 parts of a solution of 333 g of $K_3Fe(CN)_6$ in a liter of deionized water. When freshly mixed, this solution will etch chromium at 1000 Å/min. Unfortunately, this etchant deteriorates with time, making control difficult.

HCl, in various concentrations in water, is a very effective etchant for chromium, but it requires a catalyst, like powdered zinc, to provide an electrical potential to breakdown the thin native oxide that occurs on chromium. HCl is supplied as a 35% solution, which results in an etch rate of 12,000 Å/min. Dilute solutions can be used to etch as slowly as 1000 Å/h. It is desirable to complete the etch once it has been started, without removing the chromium from the HCl, since it is difficult to restart the etch.

Chromium can be etched in a plasma reactor using Cl_2, CCl_4, or CCl_4/Ar gas mixtures. A barrel reactor is commonly used.

Etching of Tantalum Nitride, Nichrome®, Copper, and Gold. The thin-film technology developed by Bell Laboratories (67) uses tantalum nitride as the resistor material, and three layers of metal, nichrome for adhesion, copper for high conductivity, and gold to protect the copper from oxidation and to provide a suitable material for wire bonding. The films are deposited over the entire surface of the substrate, and the patterns are defined photolithographically.

The conductor pattern is defined in a photoresist layer as described previously. The gold and copper are etched in a solution containing KI and I. Using the same photoresist pattern, the nichrome layer is removed in a three-step process. The first etchant is dilute HCl, followed by $CuCl_2$ in solution, and NH_4OH. The function of the NH_4OH is to stop the action of the $CuCl_2$.

The resistor pattern is defined by a photoresist process that uses a mask containing both the conductor and resistor patterns, since both of these patterns must be protected during this step. The tantalum nitride is etched using a solution of HF and HNO_3.

Photoresist Stripping (70, 71)

Photoresists can be removed by several techniques. These include solvent strippers, liquid phase oxidizing strippers, and plasma oxidizing strippers.

The solvent strippers are proprietary mixtures of chlorinated hydrocarbons, phenolic-type compounds and wetting agents. A popular example is J-100 supplied by Indust-Ri-Chem. They are usually used at temperatures between 90 and 150°C. The stripping action is a result of absorption of the solvent, causing a swelling of the photoresist and a breakdown of its adhesion to the substrate. It is sometimes difficult to remove all traces of organic residue after using a solvent stripper. A typical stripping and cleaning sequence includes a 10-min soak in the hot stripper, organic solvent rinses using methanol or trichloroethylene followed by acetone, a flowing deionized water wash, an oxidizing acid soak (hot HNO_3 or CrO_3 dissolved in H_2SO_4), and a deionized water rinse. Solvent strippers can be used for almost all types of photoresist and are particularly useful for stripping photoresists from aluminum, which is readily oxidized by acid strippers.

Positive photoresists that have been hard-baked at temperatures below 120°C can be stripped in acetone. Because of this, an alternative to etching for defining metal patterns is possible. This technique is called "lift-off." Before metalization, the substrate is coated with positive photoresist and a pattern is defined such that the areas in which metal is desired are not covered by the developed photoresist. After the metal has been deposited, the photoresist is stripped in acetone with ultrasonic agitation. The undesired metal comes off with the photoresist, leaving the metal in the desired areas.

Liquid-phase oxidizing strippers include hot H_2SO_4 (180°C), hot chromic acid (CrO_3 dissolved in H_2SO_4), and mixtures of H_2SO_4 and 30% H_2O_2 (typically 1 : 3). These solutions are used for the oxidation of negative photoresists, resulting in CO_2 and water. After stripping, the substrates are washed in flowing deionized water.

Oxygen plasma strippers are effective techniques for removing photoresists. A barrel reactor at 10 torr is used for this operation. A typical process takes 30 min. The end of the process is signified by a change in the color of the plasma, which can be detected automatically (70). There is some evidence that plasma stripping results in mobile ion contamination of the gate oxide in MOS structures (58), indicating that other stripping procedures should be employed for this operation.

SUMMARY

Photolithography masks for microelectronics are made from high-resolution plates and thin films of hard surface materials like chromium and iron oxide on glass. Patterns are generated on these masks by the layout of oversized artwork which is photographically reduced and step-and-repeat reproduced at final size. Computers are used to assist in the generation of these patterns, and some patterns are exposed directly on substrates by computer controlled *e*-beam systems. Submicrometer patterns are produced by *e*-beam systems and x-ray lithography. Both positive and negative photoresists have been formulated for the protection of substrate materials during the process of transferring the patterns to the substrates. Even the relatively large patterns associated with thick-film hybrid microcircuits are transferred to screens using photographic processes. Photolithography is the most important process for the mass production of microelectronic circuits.

REFERENCES

1. L. Lewis, *Introduction to Photographic Principles* (New York: Dover, 1965).
2. R. M. Rose et al., *The Structure and Properties of Materials, Vol. IV* (New York: Wiley, 1966).
3. D. Abbott, *Inorganic Chemistry* (London: Mills and Boon, 1965).
4. K. G. Clark, *Solid State Technology* 15, 6 (1972).
5. *Kodak High Resolution Plate*, Kodak Pamphlet No. P-47, Rochester, N.Y., 1974.
6. *Kodak High Resolution Plate, Type 2*, Kodak Pamphlet No. P-226, Rochester, N.Y., 1970.
7. *Techniques of Microphotography*, Kodak Data Book P-52, Rochester, N.Y., 1970.
8. *Kodak Plates and Films for Science and Industry*, Kodak Data Book P-9, Rochester, N.Y., 1967.
9. J. B. Tong, *Solid State Technology* 11, 7 (1968).
10. T. G. Maple, *Solid State Technology* 9, 8 (1966).
11. W. S. DeForest, *Photoresist* (New York: McGraw-Hill, 1975).
12. K. G. Clark, *Solid State Technology* 16, 12 (1973).
13. T. C. Lekas, *Solid State Technology* 16, 12 (1973).
14. D. J. Sykes, *Solid State Technology* 16, 8 (1973).
15. R. L. Bersin, *Solid State Technology* 13, 6 (1970).
16. D. L. Spears and H. I. Smith, *Solid State Technology* 15, 7 (1972).
17. L. F. Thompson, *Solid State Technology* 17, 7 and 8 (1974).
18. E. I. Gordon and D. R. Herriott, *IEEE Transactions on Electron Devices* ED-22, 7 (1975).
19. H. S. Cole et al., *IEEE Transactions on Electron Devices* ED-22, 7 (1975).
20. J. S. Greeneich, *IEEE Transactions on Electron Devices* ED-22, 7 (1975).
21. F. H. Dill, *IEEE Transactions on Electron Devices* ED-22, 7 (1975).
22. J. M. Shaw, *IEEE Transactions on Electron Devices* ED-25, 4 (1978).
23. J. H. McCoy, *Developments in Semiconductor Microlithography II*, Society of Photo-Optical Instrumentation Engineers, Vol. 100, 1977.
24. G. P. Hughes, *Solid State Technology* 20, 5 (1977).
25. F. H. Dill et al., *IEEE Transactions on Electron Devices* ED-22, 7 (1975).
26. D. G. Kelemen, *Solid State Technology* 19, 8 (1976).
27. G. M. Henriksen *Developments in Semiconductor Microlithography II*,

Society of Photo-Optical Instrumentation Engineers, Vol. 100, 1977.

28. T. G. Maple, *Semiconductor Products and Solid State Technology* 9, 8 (1966).
29. Kodak Publication, *Techniques of Microphotography*, P-52, 1970.
30. R. Cast, *Solid State Technology* 14, 2 (1971).
31. B. P. Piwczk, *Electronic Packaging and Production* 18, 5 (1978).
32. R. E. Szupillo, *Solid State Technology* 12, 8 (1969).
33. A. R. Janus, *Solid State Technology* 16, 6 (1973).
34. D. L. Spears and H. I. Smith, *Solid State Technology* 15, 7 (1972).
35. G. P. Hughes, *Solid State Technology* 20, 5 (1977).
36. J. S. Greeneich, *IEEE Transactions on Electron Devices* ED-22, 7 (1975).
37. D. Maydan et al., *IEEE Transactions on Electron Devices* ED-22, 7 (1975).
38. E. Bassous et al., *Solid State Technology* 19, 9 (1976).
39. D. L. Atherton and A. A. Peck, *Semiconductor Products and Solid State Technology* 8, 8 (1965).
40. D. W. Hamer and J. V. Biggers, *Thick Film Hybrid Microcircuit Technology* (New York: Wiley-Interscience, 1972).
41. C. A. Harper, ed., *Handbook of Thick Film Hybrid Microelectronics* (New York: McGraw-Hill, 1974).
42. *Shipley Photo Resist Technical Manual*, Shipley Company Inc., Newton, MA., January 1978.
43. G. F. Damon, *Collected Papers from Kodak Seminars on Microminiaturization 1965–66*, Kodak Publication p-195, 1969.
44. K. Kerman and M. Stelter, *Semiconductor Products and Solid State Technology* 10, 4 (1967).
45. K. R. Dunham, *Solid State Technology* 14, 6 (1971).
46. K. G. Clark and R. G. Turner, *Solid State Technology* 12, 7 (1969).
47. K. G. Clark, *Solid State Technology* 14, 2 (1971).
48. J. Wilson and S. C. Bottomley, *Solid State Technology* 15, 6 (1972).
49. K. G. Clark and E. M. Juleff, *Solid State Technology* 16, 6 (1973).
50. J. D. E. Beynon et al., *Solid State Technology* 15, 11 (1972).
51. K. G. Clark and K. Okutsu, *Solid State Technology* 19, 4 (1976).
52. P. A. Denning, *Solid State Technology* 19, 5 (1976).
53. J. D. Cuthbert, *Solid State Technology* 20, 8 (1977).
54. R. Ruddell et al., *Solid State Technology* 21, 5 (1978).

55. J. P. Scott, *Solid State Technology* 20, 5 (1977).
56. *Circuits Manufacturing* 18, 4 (1978).
57. A. T. Bell, *Solid State Technology* 21, 4 (1978).
58. R. L. Maddox and H. L. Parker, *Solid State Technology* 21, 4 (1978).
59. R. L. Bersin, *Solid State Technology* 21, 4 (1978).
60. R. A. H. Heinecke, *Solid State Technology* 21, 4 (1978).
61. A. Jacob, *Solid State Technology* 19, 9 (1976).
62. R. L. Bersin and R. F. Reichelderfer, *Solid State Technology* 20, 4 (1977).
63. R. Kumar et al., *Solid State Technology* 19, 10 (1976).
64. H. Wolf, *Semiconductors* (New York: Wiley-Interscience, 1971).
65. K. E. Bean and P. S. Gleim, *Proceedings of the IEEE* 57, 9 (1969).
66. D. B. Lee, *Journal of Applied Physics* 40, 10 (1969).
67. R. W. Berry et al., *Thin Film Technology* (New York: Van Nostrand Reinhold, 1968).
68. P. G. Gloersen, *Solid State Technology* 19, 4 (1976).
69. L. D. Bollinger, *Solid State Technology* 20, 11 (1977).
70. W. B. Stafford and G. J. Gorin, *Solid State Technology* 20, 9 (1977).
71. S. M. Irving, *Solid State Technology* 14, 6 (1971).

PROBLEMS

2-1. The resolution limit of High Resolution Plate is 2000 lines/mm, and the peak sensitivity is 525 nm. What f-number lens is required to produce images with the maximum resolution on High Resolution Plate?

2-2. The reduction ratio of a camera, R, is given by $R = L_2/L_1$, where L_2 is the lens-to-artwork distance and L_1 is the lens-to-image distance. If L_1 is fixed at 36.06 cm, and R is to be maintained at 50 : 1 within ±0.1%, what is the allowable variation in L_2?

Chapter 3
Crystal Growth and Epitaxy

The electronic functions of almost all of the devices used in monolithic integrated circuits rely on the transport of charge carriers, electrons and holes, through single-crystal semiconductor materials. Silicon is the dominant material, but germanium, silicon carbide, gallium arsenide, indium antimonide, and others have also been used. The growth of large single crystals of silicon has evolved into a very successful process. In some cases, it is desirable to deposit a thin layer of single-crystal silicon on a single-crystal substrate. This process is called epitaxy. The crystal orientation of the deposited layer is established by the crystal structure of the substrate. If the substrate is of the same material as the deposited layer, the process is called homoepitaxy. It is possible, if there is a sufficient match between the lattice parameters of the two materials, to deposit a single-crystal of one material on a single-crystal substrate of another material. This process is called heteroepitaxy. Using this process, it is possible to deposit single-crystal layers of silicon on crystalline alumina for the silicon-on-sapphire technology. In this chapter, the processes involved in the growth of single-crystal silicon are described.

THE CRYSTAL STRUCTURE OF SILICON (1, 2)

Silicon is a Group IV element in the periodic table. It has four valence electrons and crystallizes in the diamond lattice structure. The bonding is covalent with each silicon atom sharing its four valence electrons with four other silicon atoms such that each atom appears to be surrounded by a closed valence shell of eight electrons.

The basic building block of the silicon lattice is a tetrahedron with each atom surrounded by four atoms at equal distances, as shown in Figure 3-1. The distance between lattice sites is 2.35 Å.

Instead of considering the basic building block, it is customary to consider a somewhat larger structure for the diamond lattice which exhibits cubic symmetry. Note that four of the five atoms in Figure 3-1 are located such that they represent the corner and three face centers for a cube with dimensions of 1. The total structure can be represented as two face-centered cubes, with one of the cubes shifted relative to the other along the diagonal to the location ($\frac{1}{4}$, $\frac{1}{4}$, $\frac{1}{4}$). This representation of the diamond lattice is shown in Figure 3-2. In this case, the cube edge for silicon is 5.428 Å.

It is convenient to represent the crystal planes in a cubic structure by a notation called Miller indices. These are defined as the reciprocals of the intercepts of the planes with the principal axes, normalized so that all of the indices are whole numbers. Figure 3-3 illustrates three of the basic crystal planes and their Miller indices. It may be necessary to shift the cube in the

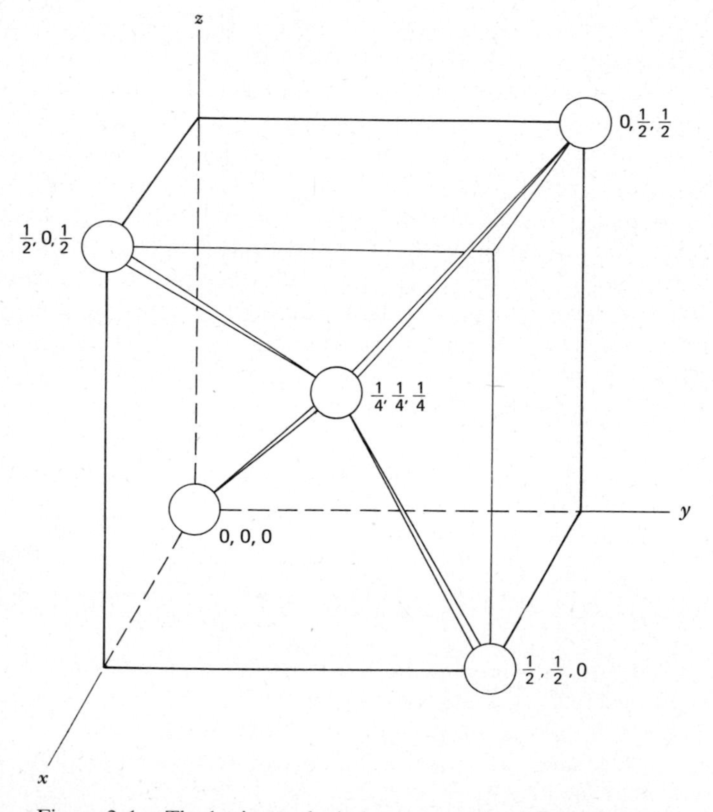

Figure 3-1. The basic tetrahedral structure of the diamond lattice.

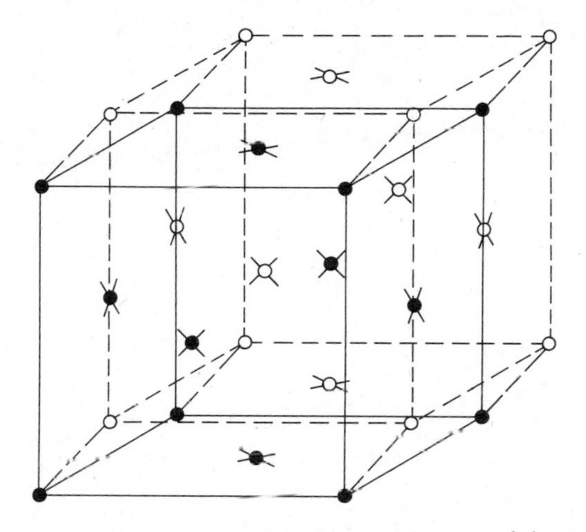

Figure 3-2. The diamond lattice as represented by two interpenetrating face centered cubic structures.

figure by one unit along a principal axis to obtain a meaningful set of indices if the plane of interest passes through the origin. Keep in mind that a crystal is assumed to be made up out of an infinite number of identical cubes. Consider the plane indicated in Figure 3-4*a*. This plane is equivalent to a (010) plane or a $(0\bar{1}0)$ plane (note that it is conventional to place a negative sign above rather than in front of a Miller index). Due to the symmetry of a cubic crystal, the planes represented by (100), (010), (001), $(\bar{1}00)$, $(0\bar{1}0)$, and $(00\bar{1})$ are equivalent. For convenience, this family of planes is designated {100}. It is also useful to define a set of directions that are perpendicular to the crystal planes. For a cubic crystal, the [111] direction is perpendicular to the (111) plane, and so forth. The set of directions for an equivalent set of planes is designated ⟨111⟩.

Careful study of the diamond lattice in Figure 3-1 indicates that the

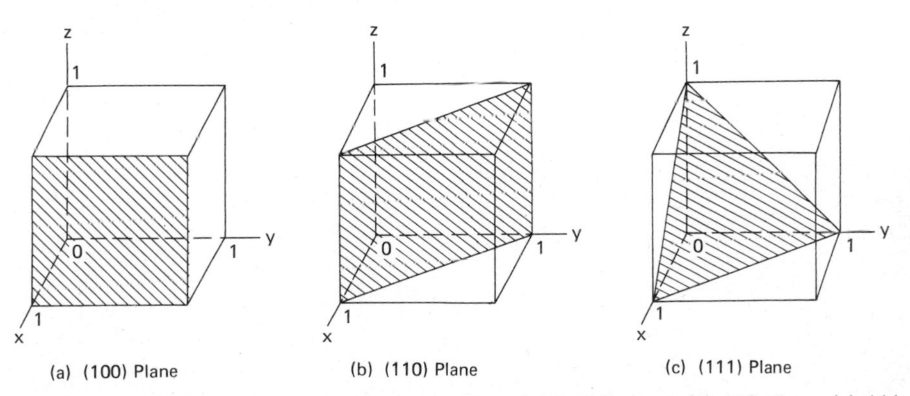

Figure 3-3. The Miller indices for low index planes. (*a*) 100 plane. (*b*) 110 plane. (*c*) 111 plane.

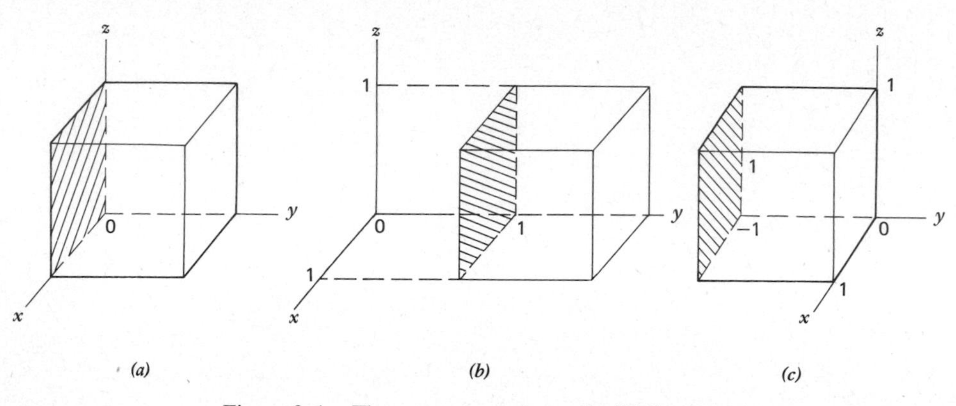

Figure 3-4. Three representations of a (010) plane.

central atom in the tetrahedron is offset from the corner of the lattice in the [111] direction. Figure 3-5 shows a diamond lattice as viewed parallel to the (111) planes. The lattice appears to consist of layers of atoms separated by single-bond distances. The assembly of silicon atoms into a crystal is most easily accomplished by rapid growth along (111) planes and slow growth in the [111] direction. Crystals are also grown in the [100] direction.

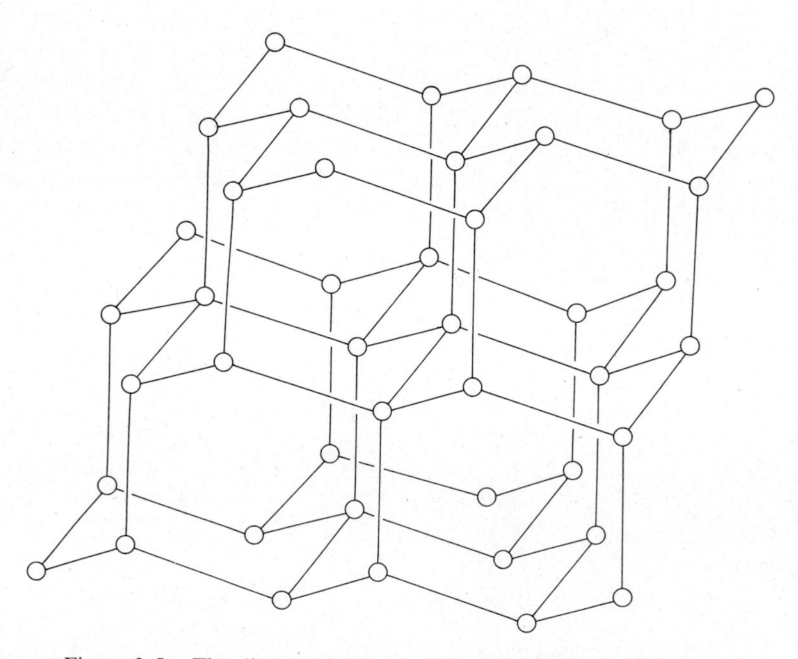

Figure 3-5. The diamond lattice as viewed parallel to a (111) plane.

IMPERFECTIONS IN CRYSTALS

It is essentially impossible to grow absolutely perfect crystals. The defects that occur are classified as point defects and line defects (dislocations). The point defects can be vacancies, interstitial atoms, impurity atoms substituted for the host atoms in the crystal lattice, or interstitial impurity atoms. Line defects include edge and screw dislocations. Gross defects like stacking faults, twin planes, and grain boundaries are also possible in crystals.

The most elementary point defect is the vacancy, which occurs when thermal fluctuations result in the movement of an atom from a lattice site to the surface. This is called a Schottky defect, and the energy of formation, E_S, is 2.3 eV in silicon. The equilibrium density of Schottky defects, n_S, is given by (1)

$$n_S = N \exp\left(-\frac{E_S}{kT}\right) \tag{3-1}$$

where N is the density of crystal atoms ($5.02 \times 10^{22}\ \mathrm{cm}^{-3}$ for silicon), k is Boltzmann's constant (8.62×10^{-5} eV/K), and T is the temperature in degrees kelvin. Since this is an equilibrium density, it may be necessary to anneal the crystal to obtain this level of defects.

An interstitial atom is one that is located in a normally vacant void within the crystal lattice. There are five voids of this type in the cube representation of the diamond crystal lattice. The energy of formation of an interstitial atom is comparable to that of a vacancy.

The vacancy-interstitial pair is called a Frenkel defect. This occurs when thermal agitation causes a lattice atom to move from a lattice site to an adjacent interstitial void. The energy of formation of this type of defect, E_F is between 0.5 and 1 eV, which is considerably lower than that of a Schottky defect. The equilibrium density of Frenkel defects, n_F, is given by

$$n_F = \sqrt{NN'} \exp\left(-\frac{E_F}{2kT}\right) \tag{3-2}$$

where N' is the density of available interstitial sites ($3.14 \times 10^{22}\ \mathrm{cm}^{-3}$ for silicon). This equation only applies under equilibrium conditions. The density of Frenkel defects in a real crystal will be considerably larger than the equilibrium value.

The formation of a vacancy results in the breaking of four covalent bonds. The formation of a vacancy adjacent to an existing vacancy, called a divacancy, only requires the breaking of two additional bonds, and thus requires less energy. The divacancy is frequently encountered in silicon.

Impurity atoms can enter the silicon lattice either in the interstitial voids or as substitutes for the silicon atoms. Some types of impurities are usually found in interstitial sites. These include nickel, zinc, copper, iron, cobalt, and maganese. Others, like gold, are found both in interstitial sites (approximately

10%) and in substitutional sites. There is an important class of impurities which, in dilute concentrations, are usually found in substitutional sites. These are the elements from Groups III and V of the periodic table, particularly boron, aluminum, gallium, indium, phosphorus, arsenic, and antimony. These elements are responsible for the wide range of electrical conductivity available in silicon, and the possibility of forming regions in which the principal conduction method is due either to holes or to electrons.

The substitution of impurity atoms for silicon atoms results in physical stress within the crystal lattice. This is due to the effective physical "size" of the atoms within the crystal. The distance between nearest neighbor lattice sites is 2.35 A. If the crystal is assumed to be constructed of hard spheres, the radius of these spheres would be 1.18 Å. Substituting an impurity atom of a different effective size will distort the crystal lattice. The effective physical size of an atom within the crystal is due to the bonding forces exerted on adjacent atoms. A perfect crystal lattice occurs when the atomic spacing produces a minimum energy condition. The effective radii of the substitutional dopants in crystalline silicon and the resultant lattice misfits are listed in Table 3-1. Note that arsenic is the only substitutional impurity that has a perfect fit with the silicon lattice.

From a metallurgical viewpoint, there is a limit to the concentration of impurities which can enter a crystal lattice without seriously disrupting the structure. This is called the solid solubility limit for the impurity. Figure 3-6 shows the solid solubilities of the substitutional impurities in silicon.

Dislocations or line defects are formed by the effects of stresses on the crystal as it solidifies from the melt. The following treatment considers a simple cubic lattice, rather than the complex diamond lattice, but the general results are similar.

An edge dislocation is shown in Figure 3–7. The plane indicated by $ABCD$ is an extra half-plane of atoms in the upper portion of the crystal lattice. The energy associated with the formation of an edge dislocation is approximately 30 eV/atom length.

A screw dislocation is shown in Figure 3-8. The distance B–B' represents one lattice spacing. The energy associated with the formation of a screw dislocation is between 10 and 20 eV/atom length. The energy associated with dislocations is much greater than that associated with point defects.

Stacking faults occur when there is a deviation from the normal order of

Table 3-1. **Effective Radii and Lattice Misfit for Substitutional Dopants in Silicon (1)**

	Dopant						
	B	Al	Ga	In	P	As	Sb
Effective radius, Å	0.88	1.26	1.26	1.44	1.10	1.18	1.36
Misfit factor	0.254	0.068	0.068	0.22	0.068	0	0.153

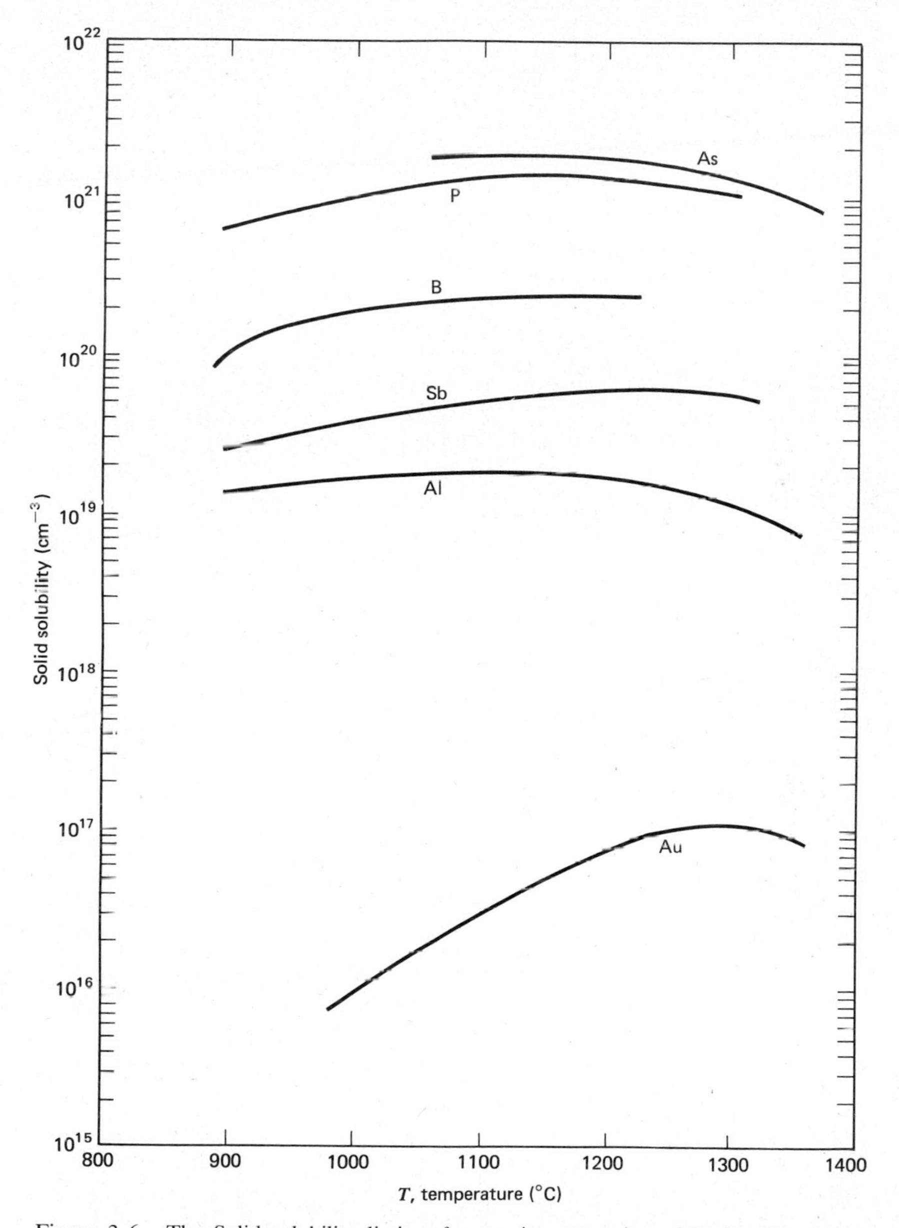

Figure 3-6. The Solid solubility limits of some important impurities in silicon as a function of temperature (9, 13). Adapted from material copyrighted in 1960 by the American Telephone and Telegraph Company and in 1969 by The Electrochemical Society. Reprinted by permission from the *Bell System Technical Journal* and of the publisher, The Electrochemical Society, Inc.

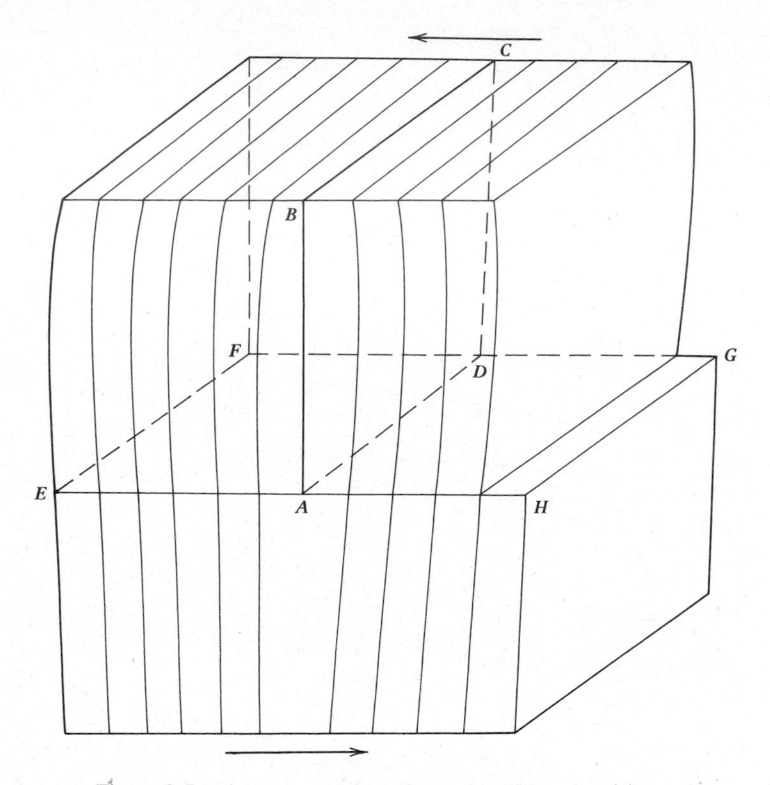

Figure 3-7. A representation of an edge dislocation (1).

the assembly of planes in a crystal. Figure 3–9 shows a view of a (111) plane in a silicon lattice with projections of the next two successive planes in the normal sequence of crystal growth. The base plane is designated "*A*", with the next plane designated "*B*" and the third plane designated "*C*". The preferred sequence is *ABCABC*. Stacking faults occur when the sequence is *ACAB-CABC* or *ABCBABC*. This situation can arise in epitaxial growth or when defect clusters occur in crystals during high-temperature processing after crystal growth.

Twinning occurs when two portions of the crystal with different orientations share a common plane of atoms. This can occur if the silicon is restricted in its growth from the melt.

Grain boundaries occur between single-crystal regions of polycrystalline material. Grain boundaries have significant effects on processing and electrical characteristics of semiconductor materials. Polycrystalline silicon depositions are used in the fabrication of certain MOS devices.

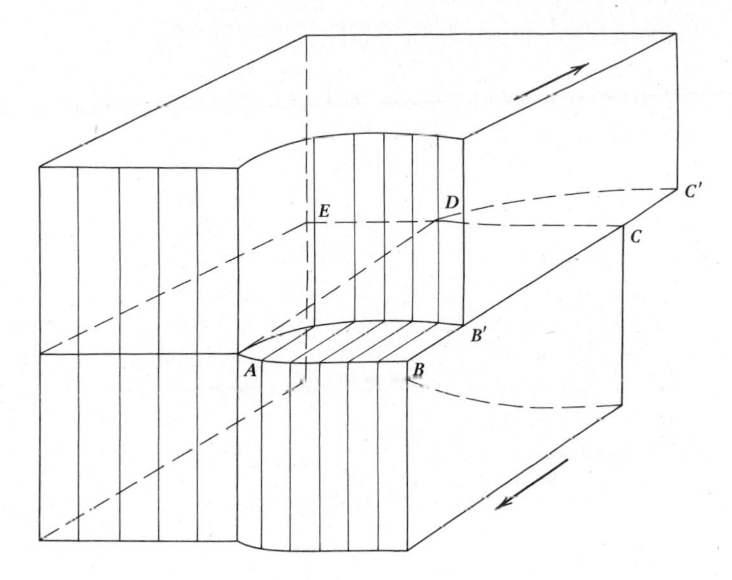

Figure 3-8. A representation of a screw dislocation (1).

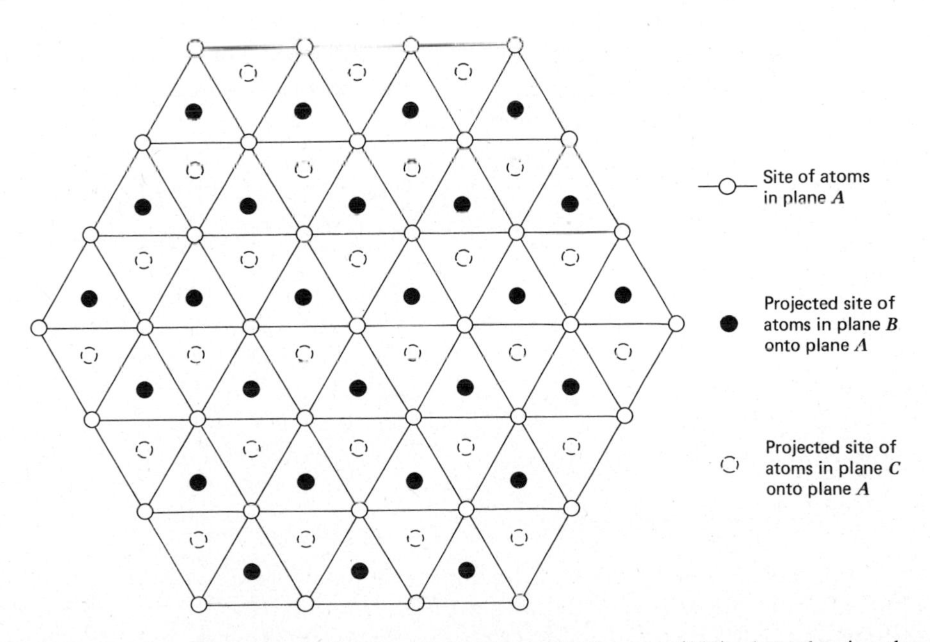

Figure 3-9. A silicon lattice structure viewed perpendicular to a (111) plane showing the normal stacking order (12).

THE PURIFICATION OF SILICON (3)

Silicon is the second most abundant element in the earth's crust. It occurs primarily in combination with oxygen in SiO_2 and various silicates. In order to prepare silicon for the growth of semiconductor-grade single crystals, it is necessary to remove certain electrically active impurities to the point where their concentrations are less than one part per billion on an atomic basis. Boron is the most difficult of these impurities to remove.

The first step in purifying silicon is the reduction of SiO_2 to silicon with carbon in a submerged electrode arc furnace. The result is metallurgical grade silicon with a purity of 98%. This type of silicon is used to improve the mechanical properties of aluminum, produce silicone chemicals, improve the characteristics of transformer steel, and serve as the source for semiconductor silicon. The semiconductor industry uses only a small fraction of the metallurgical grade silicon produced.

In order to purify the metallurgical grade silicon, it is converted to SiH_4, $SiCl_4$, $HSiCl_3$, or H_2SiCl_2. This is accomplished in a fluidized bed reaction with HCl. $HSiCl_3$ is the preferred silicon compound because it reacts at a lower temperature and a faster rate than the others. The $HSiCl_3$ is purified by conventional distillation techniques.

Semiconductor-grade polycrystalline silicon is obtained by reacting the $HSiCl_3$ with H_2 at temperatures between 1000 and 1200°C. The polycrystalline silicon deposits on a thin, high purity rod of polycrystalline silicon, which is used as a resistance heater. For large diameter depositions, typically 200 mm, the deposition time is several hundred hours.

THE GROWTH OF SINGLE-CRYSTAL SILICON (1–3)

Single-crystal silicon, the starting material for monolithic integrated circuits, is grown by two principal techniques. Most of the material is grown by a method based on that developed by Czochralski for the growth of single crystals of metals in 1917. The second method, the float-zone technique, results in higher purity crystals, which are used primarily in the fabrication of power devices, but there is a limited use of float-zone material in microcircuits.

The Czochralski Process

The Czochralski process consists of dipping a small single-crystal seed into molten silicon and slowly withdrawing the seed, while rotating it simultaneously. Figure 3-10 is a drawing of a Czochralski crystal growing apparatus. The crucible is made from graphite with a fused silica lining. Power is supplied from an RF source, or, in larger systems, by resistance heating. The atmosphere is

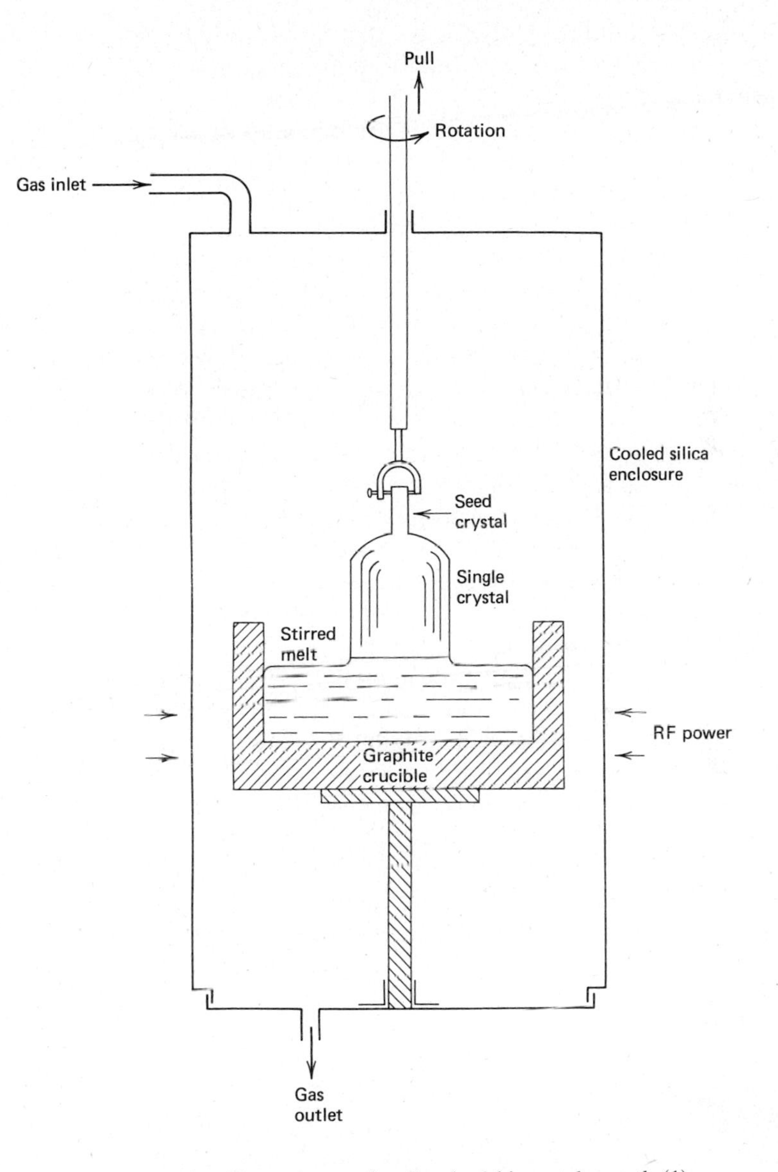

Figure 3-10. The apparatus for Czochralski crystal growth (1).

usually argon at either slightly higher than atmospheric pressure or a reduced pressure of 1 to 50 torr. Czochralski crystals are grown in either (100) or (111) crystal orientations.

Industry standard crystals range in diameter from 75 to 125 mm after grinding to produce uniform crystals with diameters within $\pm 50\ \mu m$. Typical crystals are 1 m long. Automatic diameter control makes use of an infrared sensor focused on the bright meniscus in the silicon melt, just adjacent to the solidified crystal. The pull rate or furnace power is automatically adjusted to keep the diameter constant. Silicon is introduced in the system as one-piece polycrystalline charges between 10 and 24 kg, or in the form of polycrystalline nuggets. The fused silica liner cracks if the melt is allowed to solidify. To avoid this, systems have been developed to continuously feed nuggets into the crucible while it is still hot.

The crystal perfection of modern Czochralski-grown silicon is excellent. A technique developed by Dash in 1958 has been employed to produce crystals which are free of dislocations. The seed crystal is 5 to 8 mm in diameter. After the seed is dipped into the melt, it is pulled at a rate that reduces the diameter of the crystal for a short distance, and then the crystal is allowed to rapidly increase in diameter until it reaches final size. This "necking" process allows any dislocations to grow out and start the crystal free of them.

Impurities, both intentional and unintentional, are introduced into silicon grown by the Czochralski technique. The intentional impurities are the dopants used to determine the electrical properties of the crystal, and are discussed below. The most important unintentional impurity is oxygen, which comes from the fused silica crucible liner. The oxygen is present in concentrations of 10 to 50 ppm. This is close to the solubility limit of oxygen in silicon. This oxygen will be electrically active in the form of donor levels unless the crystal is annealed between 600 and 700°C and rapidly cooled in the range between 500 and 300°C. The oxygen can also precipitate at defects like vacancies and interstitials after temperature cycling. These defect clusters can be important in determining the yield of large-scale integrated circuits.

Dopants are usually added to the melt in the form of powders of heavily doped silicon. The dopant is kept reasonably uniformly distributed in the melt by the stirring action of the rotation of the seed and the crucible. As a crystal solidifies the concentration of impurity incorporated in the crystal is different from the concentration in the melt. The ratio of the impurity concentration in the solid, N_S, to that in the liquid, N_L, is called the segregation or distribution coefficient, k. The value of k depends on the impurity element. The segregation coefficients for the substitutional impurities in silicon are listed in Table 3.2. The impurity distribution, as a function of axial position, x, along the crystal, is given by

$$N_S\left(\frac{x}{l}\right) = kN_{L0}\left(1 - \frac{x}{l}\right)^{k-1} \tag{3-3}$$

***Table 3-2.* Segregation Coefficients for Substitutional Impurities in Silicon**

Element	k
B	0.80
Al	0.0020
Ga	0.0080
In	0.0004
P	0.35
As	0.30
Sb	0.023

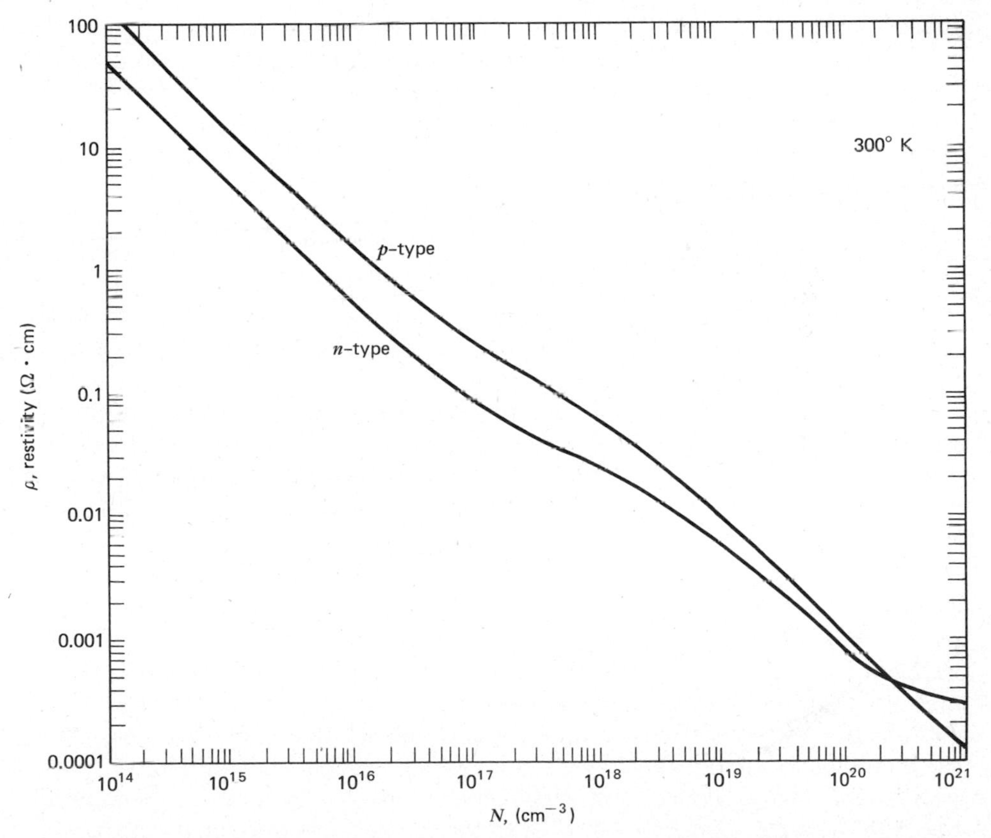

Figure 3-11. Resistivity as a function of doping concentration for silicon at 300°K (11). Copyright © 1962 American Telephone and Telegraph. Reprinted by Permission from *The Bell System Technical Journal.*

where N_{L0} is the initial impurity concentration in the melt and l is the total length of the crystal. The crystal doping is not uniform along the axis. Since boron has the highest segregation coefficient, the axial dopant variation is less in boron doped crystals than in those doped with the other substitutional dopants. Axial doping variations are not particularly important for microcircuit applications, since the crystal is sliced into thin wafers perpendicular to the axis, and sorted according to resistivity. The dependence of resistivity on doping concentration for silicon is shown in Figure 3-11. Axial variation of doping concentration is important from the viewpoint of the yield of useful wafers from a crystal. One technique for reducing this variation is to program the pressure under which the crystal is grown. As the pressure is reduced, volatile dopants are depleted from the melt, permitting a more uniform doping concentration in the crystal. Of more importance is the variation of doping concentration as a function of radial distance from the axis. Striations of the dopant are observed in Czochralski-grown silicon. These are due to the nonuniformity of the thermal environment during crystal growth. Since the microscopic growth rate is dependent on the local temperature, the amount of dopant incorporated varies slightly with the location within the crystal. The resistivity across a typical wafer cut from a Czochralski-grown crystal is within 20% of the nominal value. These resistivity variations can have an adverse effect on the yield of large-area integrated circuits.

The Float-Zone Process

The float-zone process has some advantages over the Czochralski process for the growth of certain types of silicon crystals. The molten silicon in the float-zone apparatus is not contained in a crucible, and, therefore, is not subject to the oxygen contamination present in Czochralski-grown crystals. If the resultant crystal must have a resistivity greater than $25\ \Omega \cdot \text{cm}$, the float-zone process is necessary. It is difficult to obtain crystals larger than 75 mm in diameter with the float-zone process. Float-zone crystals can be grown with either (100) or (111) orientations.

The apparatus for float-zone crystal growth is shown in Figure 3–12. The atmosphere is usually argon at reduced pressure. A polycrystalline silicon rod up to 100 cm in length is clamped in a chuck at the top. A seed crystal is clamped in a chuck at the bottom. An RF coil is positioned to produce a melt at the bottom of the polycrystalline rod. The molten zone is held in place by a combination of surface tension and levitation effects due to the RF field. The seed is brought into contact with the molten silicon and withdrawn, with rotation, to form a reduced diameter section, followed by a tapered section to the final desired diameter. This has the same effect as the neck in a Czochralski-grown crystal to produce a dislocation-free crystal. The molten zone is passed through the length of the polycrystalline rod until the crystal

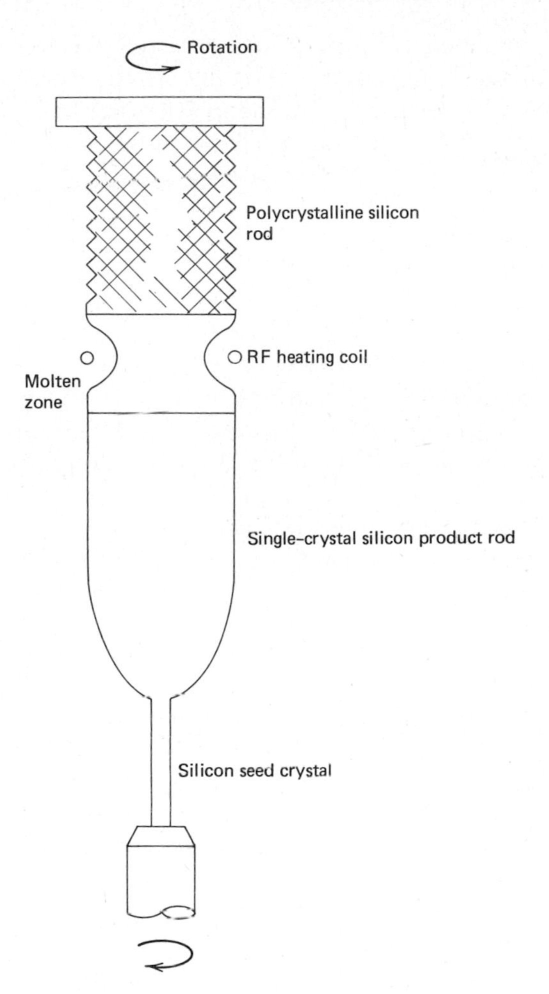

Figure 3-12. The apparatus for float-zone crystal growth (1).

is grown. Multiple passes of the molten zone under vacuum can be used to reduce the impurity content of the crystal by volatizing the dopant in the molten zone. Crystals with resistivities up to 30,000 $\Omega \cdot$ cm have been obtained by this technique.

Dopants can be added to float-zone crystals by several methods. The dopants can be included uniformly in the polycrystalline starting material, or by inserting heavily doped silicon into slots machined periodically along the polycrystalline rod. More uniform doping can be obtained by including a programmed concentration of phosphine (PH_3) or diborane (B_2H_6) in the atmosphere surrounding the growing crystal. Because of the unstable thermal environment in the molten zone, radial resistivity variations of 40% are common in float-zone silicon. It is possible to produce a highly uniform phosphorus doping by subjecting an undoped crystal to a stream of thermal neutrons which transmute some of the silicon atoms into phosphorus.

EPITAXIAL GROWTH

For mechanical strength, silicon wafers are typically 160 to 500 μm thick, depending on the diameter of the wafer. The active portion of a bipolar integrated circuit is located in an *n*-type layer within 3 to 10 μm of the surface. For a typical junction-isolated circuit, the substrate is 10 Ω · cm *p*-type. A convenient way to fabricate this structure is to grow the *n*-type layer on the *p*-type substrate by homoepitaxy. This results in a high quality single-crystal extension of the substrate with an approximately uniform doping distribution. For integrated injection logic circuits, the *n*-type layer is grown on a heavily doped *n*-type (designated n^+) substrate. For complementary MOS circuits on sapphire substrates, it is customary to grow a 1-μm layer of lightly doped *p*-type silicon heteroepitaxily. The crystal orientation of the silicon on sapphire is (100).

The basic processes of homo- and heteroepitaxy are similar. The process consists of passing a gaseous compound of silicon over hot substrates. The high temperature causes the silicon compound to decompose, either by chemical reaction or by pyrolysis, and the silicon atoms are permitted to arrange themselves on the crystal lattice of the substrate, if the proper conditions exist. The silicon compounds most often used for this process are: $SiCl_4$, $HSiCl_3$, H_2SiCl_2, and SiH_4. For the chlorine compounds, the decomposition is a chemical reduction of the compound with hydrogen, and occurs on the substrate. For silane (SiH_4), the decomposition is pyrolytic (temperature aided) and occurs in the gas stream above the substrate.

The apparatus used for epitaxial growth takes one of three forms. The substrates are placed on silicon carbide coated carbon plates called susceptors. The three basic configurations are shown in Figure 3-13. Resistance heating is sometimes used for pancake systems, but RF heating is the primary power source for epitaxy.

As an example of a typical epitaxial system, a horizontal reactor using silane as a silicon source will be considered. The fused silica reactor chamber is water cooled to prevent the decomposition of silicon on the walls. The RF heated susceptor is mounted at a slight angle to the gas flow to provide for more uniform deposition along the length of the susceptor. The carrier gas is ultrahigh-purity hydrogen, which is used in large quantities, typically 100 l/min to produce a reasonable gas velocity. As the gas passes over the hot susceptor (typically 1050°C) a stagnant layer is formed (6), the thickness of which is dependent on the velocity. If it is assumed that all of the temperature change

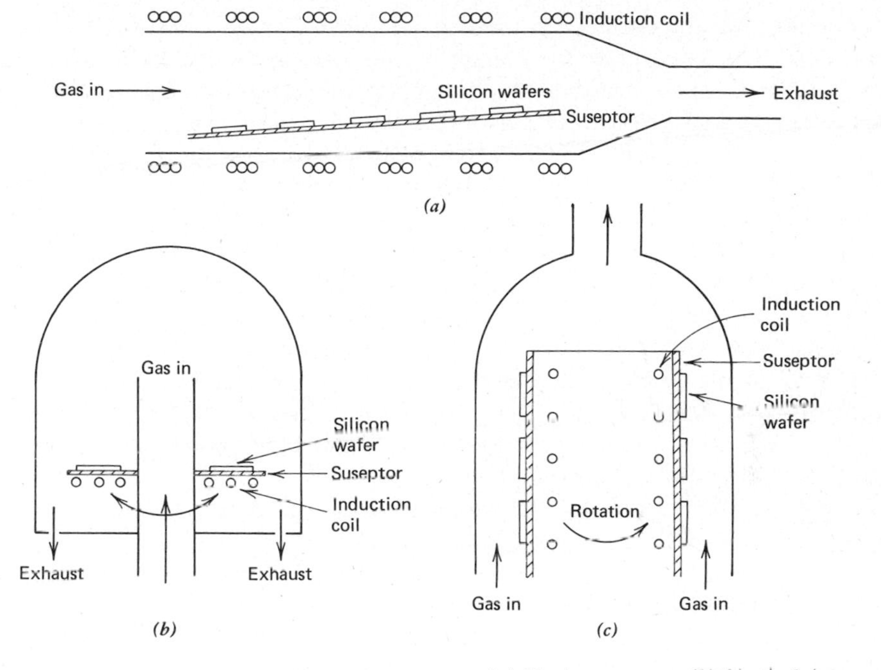

Figure 3-13. Three types of epitaxial reactors (*a*) Horizon reactor. (*b*) Vertical (pancake) reactor. (*c*) Cylinder displacement flow reactor.

occurs across the stagnant layer, the mechanism for deposition is the diffusion of silane molecules through the stagnant layer. As these molecules reach the decomposition temperature (approximately 500°C), the silicon-hydrogen bonds break. The silicon atoms descend to the substrate, and the hydrogen molecules return to the gas stream. If the concentration of the decomposing molecules becomes too large, the silicon atoms coalesce while still in the stagnant layer, with some of them rising by convection and being swept along in the moving gas stream above the stagnant layer, while the remainder fall on the substrates, causing major defects in the growing layers. This gas phase nucleation limits the growth rate available in a silane system. The efficiency of silane depositions approaches 35% for low concentrations.

Since the concentration of silane decreases as the gas is passed along the length of the susceptor, it is necessary to decrease the thickness of the stagnant layer as a function of position along the susceptor to obtain a uniform deposition rate for the entire length of the susceptor. This can be accomplished by tilting the susceptor at a small angle to the gas flow. This has the effect of reducing the cross-sectional area perpendicular to the gas flow, thus increasing the gas velocity and reducing the thickness of the stagnant layer. For a given susceptor tilt angle, there is an optimum gas flow for uniform growth. A rectangular reactor chamber is used to minimize growth variations across the susceptor. The growth rate for a silane system can be controlled linearly over a range from 0.1 to 1 μm/min by controlling the ratio of silane to hydrogen. Silane depositions are usually performed between 1050 and 1100°C.

Epitaxy using $SiCl_4$, $HSiCl_3$, or H_2SiCl_2 in hydrogen results in the formation of gaseous HCl, which is an etchant for silicon at the deposition temperatures of 1100 to 1150°C. This does not cause a degradation of the quality of the layer if sufficient hydrogen is present. The highest decomposition rates of quality epitaxial layers are obtained from systems using H_2SiCl_2. This type of system is also the least sensitive to deposition temperature.

The doping of epitaxial layers is accomplished by including controlled quantities of the hydrides of the dopants in the gas stream. The most commonly used dopant gases are phosphine (PH_3), arsine (AsH_3), and diborane (B_2H_6). Doping is a linear function of the ratio of dopant gas molecules to silicon source molecules in the gas flow over a wide range. There is also the possibility of the out-diffusion of impurities from the substrate during epitaxial growth.

A typical epitaxial growth process includes steps for the in situ preparation of the substrates. If the substrates are silicon, they will have a thin native oxide layer. This can be removed by reduction with hydrogen at 1250°C for 10 to 20 min. This is followed by a dilute gaseous HCl or SF_6 etch for 5 min at 1250°C to remove the top layers of silicon before the substrates are lowered to the deposition temperature.

If epitaxial layers are to be grown on (111) oriented silicon, it is customary to use slightly misoriented substrates (6, 7). The misorientation is typically 3 to 4° toward the nearest $\langle 110 \rangle$ direction from a (111) plane.

The major effects of the misorientation are to increase the growth rate by providing nucleation sites, and make it possible for steps in the substrate to be faithfully reproduced in the epitaxial layer. The steps in the substrate occur during the selective diffusion of heavily doped n^+ layers, which become "buried" under the epitaxial layer during monolithic bipolar processing. It is necessary that the masks for succeeding processes be aligned to the pattern for the buried layer. Rectangular patterns of steps on the substrate are distorted into trapezoids on an epitaxial layer grown on a perfectly oriented (111) substrate due to the difference in growth rates along different crystal facets.

Epitaxial layers will generally contain the same defects as the substrate. The presence of foreign particles on the substrate will usually lead to stacking faults.

SUMMARY

Large single crystals of silicon can be grown by either Czochralski or float-zone techniques. These crystals are essentially free of dislocations. Czochralski-grown crystals have a large oxygen content, which can produce undesirable defect clusters after additional high-temperature processes. Intentionally introduced impurities permit the growth of silicon crystals with resistivities that span the range from 30,000 Ω · cm to 0.001 Ω · cm. Thin layers of single-crystal silicon can be deposited on single-crystal substrates of silicon and sapphire by the epitaxial process.

REFERENCES

1. S. K. Ghandhi, *The Theory and Practice of Microelectronics* (New York: Wiley, 1968). Figures reprinted by permission of the publisher.
2. W. R. Runyan, *Silicon Semiconductor Technology* (New York: McGraw-Hill, 1965).
3. L. D. Crossman and J. A. Baker, *Semiconductor Silicon 1977*. The Electrochemical Society, Princeton, 1977.
4. R. M. Burger and R. P. Donovan, eds., *Fundamentals of Silicon Integrated Device Technology, Vol. I* (Englewood Cliffs: Prentice-Hall, 1967).
5. M. L. Hammond and W. P. Cox, *Silicon Device Processing*, C. P. Marsden, ed., National Bureau of Standards Special Publication 337, November 1970.
6. F. C. Eversteyn et al., *Journal of the Electrochemical Society* 117, 7 (1970).
7. D. C. Gupta and P. Wang, *Solid State Technology* 11, 10 (1968).
8. J. Nishizawa, T. Terasaki, and M. Shimbo, *Journal of Crystal Growth* 17, (1972).
9. F. A. Trumbore, *Bell System Technical Journal* 39, 1 (1960).
10. R. K. Jain and R. J. Van Overstraeten, *Solid State Electronics* 16, 2 (1973).
11. J. C. Irvin, *Bell System Technical Journal* 41, 2 (1962).
12. H. Henderson and W. Whitten, *Epitaxial Film Growth*, Motorola Semiconductor Products Division, Interim Report No. 2, AD 42126, NTIS, 1963.
13. G. L. Vick and K. M. Whittle, *Journal of the Electrochemical Society*, 116, 8 (1969).

PROBLEMS

3-1. If it is assumed that the atoms in a crystal lattice are hard spheres with a radius equal to one-half the distance between the centers of nearest neighbors, find the ratio of the volume occupied by atoms to the total volume available in the diamond crystal lattice. Compare this to the same ratio for a simple cubic lattice.

3-2. For a substitutional impurity atom to move through a crystal lattice it is necessary for a vacancy to occur adjacent to the lattice site occupied by the impurity atom. To determine which type of defect is most likely to participate in the movement of impurity atoms and the temperature dependence of the vacancy density, calculate the equilibrium density of Schottky and Frenkel defects for 27, 400, 800, and 1200°C. (Assume that $E_F = 0.85$ eV.)

3-3. Calculate and plot the impurity distribution of boron as a function of axial position in a 30-cm-long silicon crystal that has been pulled from a melt with an initial impurity concentration of 10^{17} cm^{-3}. Repeat this calculation and plot if the impurity is arsenic.

Chapter 4
Substrate Preparation

The substrate for a microelectronic circuit is the base upon which the circuit is fabricated. Substrates must have sufficient mechanical strength to support the circuit during the fabrication process. The electrical characteristics desirable for a substrate depend on the type of microcircuit being fabricated. In general, hybrid microcircuits are deposited *on* substrates, and monolithic integrated circuits are formed *within* substrates.

The substrates used in the manufacture of hybrids are usually dielectric materials, like ceramics, glasses, or single-crystal insulators, but, in some cases, conductors or semiconductors coated with a dielectric layer are used. In most microwave hybrids, the substrate participates as a dielectric in the operation of the circuit.

The substrates used for monolithic integrated circuits are silicon wafers sliced from large single crystals, except in the case of special fabrication processes like silicon on sapphire. In most monolithic microcircuits, the substrate provides both electrical and mechanical functions.

In this chapter, the methods used for substrate preparation and typical cleaning processes associated with the basic types of microelectronics are described.

SILICON WAFER PRODUCTION (1, 2)

Silicon single crystals are grown by the processes described in Chapter 3. The transformation from a large crystal to a silicon wafer ready for integrated circuit processing requires a number of steps. These include

1. Diameter sizing.

2. Orientation.
3. Slicing.
4. Etching.
5. Polishing.
6. Cleaning.

Diameter Sizing

The industry has adopted certain wafer diameters as standards so that automated processing equipment can also be standardized. Standard wafer sizes include 50, 75, and 100 mm. Due to fluctuations in the growth process, the as-grown crystals will have variations in diameter, and it is customary to grow the crystals several millimeters larger than the desired size. The crystal is then ground to a diameter approximately 0.4 mm larger than the standard size, either by centerless or center-type grinding systems. The grinding can be accomplished using a belt or a wheel with silicon carbide, alumina, or diamond abrasive. This grinding leaves a damaged surface on the crystal which must be removed by subsequent processing.

Orientation

In order to provide an alignment reference and to identify the crystal type, each crystal has an accurately positioned flat or flats ground on specific crystal planes. The orientation of the crystal is determined by an x-ray diffraction technique. The reference flat is larger than the identification flat. The standard flats for silicon wafers are shown in Figure 4-1. Production systems maintain the orientation of flats within $\pm 0.5°$. The initial pattern defined on a substrate is aligned relative to the orientation flat. This assures that the wafer will fracture along natural cleavage planes during the die separation process.

At this point, the grinding damage is removed by an isotropic silicon polishing etch (HF, HNO_3, and CH_3COOH). The result is a polished silicon crystal with the diameter within ± 1 mm of the desired size.

Slicing

The most common method for slicing silicon wafers makes use of an inside-diameter saw blade supported around the outside edge. A thinner blade can be used in this type of equipment than that using a hub-mounted blade with cutting on the outside diameter. The blade consists of a stainless steel core, 100 to 150 μm thick, with a diamond/nickel matrix plated on the inside diameter. Each slicing operation results in a loss of at least a 250-μm-slice of silicon, equivalent to losing more than 30% of the silicon in a crystal. Techniques to reduce this slicing loss (kerf) are constantly being sought.

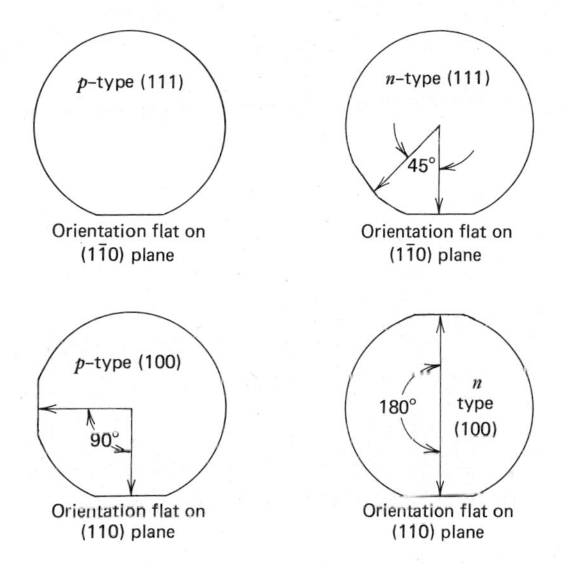

Figure 4-1. The standard orientation and identification flats for silicon wafers (2). Reprinted by permission of *Solid State Technology* from "Silicon Wafer Technology-State of the Art" by R. B. Herring, May 1976, Vol. 19, No. 5. Copyright © Cowan Publishing Corp.

In some cases, the edges of the wafer are rounded at this point, typically with a 50- to 100-μm radius. This prevents possible chipping of the sharp edges during subsequent processing, and reduces edge buildup in photoresist spinning operations.

Etching

The sawing process results in a rough surface characterized by scratches littered with crystal particles. This damage is removed by an isotropic silicon etch (HF, HNO_3, and CH_3COOH). This process removes 50 to 80 μm of silicon. A certain amount of edge rounding will also occur during this process. Since polishing is usually performed only on the front side of the wafer, the etching process determines the quality of the finish on the back side of the wafer.

The wafers are sorted for thickness and resistivity at this point in the process. Because of the axial inpurity gradient of the crystals and the time consumed in resistivity measurements, it is common practice to crop the crystals into sections by resistivity range before slicing, rather than measure the resistivity of each wafer. Automatic sorting by wafer thickness, using noncontact measuring techniques, is also common in the industry.

In some cases, special treatment of the back side of the wafers is performed. The production of a uniformly distributed defect density on the back of the wafer by sand blasting or chemical means, has been demonstrated to reduce the effects of oxidation-induced stacking faults in wafers by providing sites for these defect clusters at the backs of the wafers.

Polishing

The final surface finish on the front side of the wafer is performed by a process called "chem-mechanical" polishing. The wafers are mounted on a carrier using wax or resin, or by friction against a silicone rubber compound. The polishing compound is a colloidal suspension of fine SiO_2 particles in an aqueous, alkaline solution with a pH between 10 and 12. The polishing process consists of two steps, with 50 μm being removed in the first step, and negligible silicon removal in the second step. Different polishing pads are used for the two steps.

The advent of projection printing in photolithography has introduced the requirement for better control of the flatness of the wafers. The overall flatness requirement for projection printing is ± 6 μm.

Postpolishing Cleaning

At the conclusion of the polishing process, the polishing compound and mounting medium must be removed from the wafers. The cleaning process usually includes solvents, mechanical scrubbing using detergents, rinses in deionized water, and drying in nitrogen. The wafers are then inspected under oblique illumination with a collimated light source to reveal microscopic scratches, pits, or haze.

SAPPHIRE WAFER PRODUCTION (3)

Single-crystal wafers of alumina, Al_2O_3, are used in MOS silicon-on-sapphire circuits, and as substrates for some microwave thin-film hybrids. The substrates are usually referred to as sapphire, which, in its natural form is single-crystal alumina containing titanium as an impurity. The method of sapphire wafer production for electronic use is similar to that for silicon.

Most of the alumina crystals are grown by the Czochralski technique. The process for growing low dislocation density alumina crystals was developed for the laser industry, since alumina doped with chromium is the basis of the ruby laser. The orientation of the sapphire crystals for heteroepitaxy is designated $(1\bar{1}02)$, which are Miller indices for hexagonal crystal structures. The first three of these indices refer to three axes in the hexagonal base plane, which are oriented 120° apart. The fourth index refers to directions along the crystal axis. These are indicated in Figure 4-2. The mismatch between this crystal plane and the (100) plane in silicon is relatively small. The crystals are sliced with a

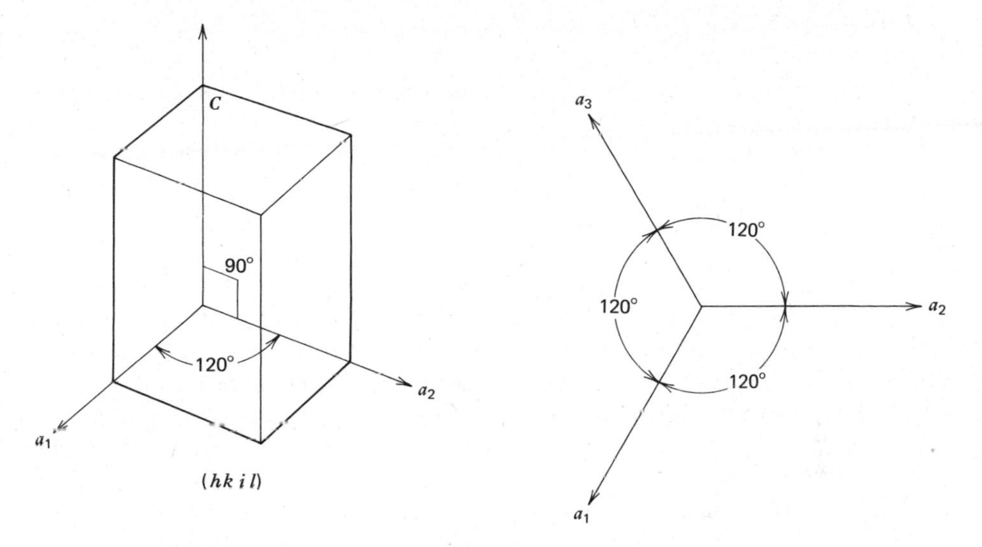

Figure 4-2. Miller indices for an hexagonal crystal structure. Note: $h+k=-i$.

$$h=\frac{1}{\text{intercept on } a_1}$$

$$k=\frac{1}{\text{intercept on } a_2}$$

$$i=\frac{1}{\text{intercept on } a_3}$$

$$l=\frac{1}{\text{intercept on } C}$$

diamond saw and mechanically polished using fine diamond particles. In some cases, alumina particles are used for obtaining the final finish. Sapphire can be etched in hot (180°C) phosphoric acid, or in hydrogen at 1150°C.

Some of the sapphire used for microelectronics applications is grown in ribbon form by the edge-defined film growth process. This consists of pulling the crystal through a silicon carbide die, which forces the crystal to grow as a single-crystal ribbon of the desired width and thickness for use as rectangular substrates.

CERAMIC SUBSTRATE PRODUCTION (4–7)

Most of the substrates used for hybrid microcircuits consists of ceramic materials. Thick-film substrates are usually 96% alumina or 99.5% beryllia. Typical thin-film substrates are glass and glazed 99.5% alumina. The need for inexpensive large substrates has led to the development of steel substrates coated with a ceramic glaze. Ceramic substrates are available in a variety of standard sizes and shapes, or may be custom fabricated to special complex patterns.

The main function for the substrate for a hybrid microcircuit is to provide mechanical support. The ideal substrate would be strong, light, inexpensive, dimensionally stable, electrically insulating, thermally conducting, and able to withstand high temperature processing. The surface should have the proper finish either for screen printing or thin-film deposition, and the dielectric characteristics should be suited to the particular application.

Alumina ceramics rate high in most of the desirable substrate characteristics. The major disadvantage of alumina ceramics is low thermal conductivity, between 4 and 7% of that of copper. Beryllia is an outstanding substrate material, with a thermal conductivity higher than many metals and approximately 60% that of copper, but it is much more expensive than alumina, and the powder form of this material is toxic and dangerous to handle. The dielectric constants of these ceramics are low, typically 9.2 for alumina and 6.4 for beryllia. If a high dielectric constant substrate is desired for a particular application, barium titanate ($BaTiO_3$), with a relative permittivity between 1200 and 6500, is a popular substrate material. The surface finish of 96% alumina is excellent for screen printing, with peak to valley distances of approximately 15,000 Å. This provides a surface that is smooth enough for screen printing, but rough enough and containing a large microscopic surface area to permit good adhesion between the film and the substrate. For high-frequency applications or thin-film depositions, a smoother surface is desirable, like that of 99.5% alumina, glazed alumina or beryllia, single-crystal alumina, or glass.

The alumina ceramic substrates are formed from Al_2O_3 particles that average 3 to 5 μm in diameter. The remaining materials are silicon dioxide, magnesium oxide, and various silicate glasses. The glasses act as a glue to hold the alumina particles together. There are two popular methods for the formation of ceramic substrates. The first technique makes use of hydraulic pressure to compact the particles into the desired shape. The second consists of mixing the ceramic powder with an organic vehicle and casting the mixture in the form of a tape. The casting process is followed by a drying step to remove the organic vehicle. Shapes (including holes) can easily be cut in the ceramic while it is in its unfired or "green" state. The final step in either process is air firing at 1700°C. In some cases, creases are formed in the "green" substrates so that they may be snapped apart after hybrid elements are formed on the substrate.

Beryllia ceramic substrates are usually made with a purity level of 99.5%. The average grain size is between 15 and 25 μm. Green forming is accomplished by hydraulic pressing or tape casting. The substrates are fired in a controlled atmosphere at 1500°C. The inhalation of beryllia dust is hazardous. For this reason, machining of beryllia ceramics, including laser scribing or resistor trimming must be performed in fume hoods.

Porcelained steel substrates are a form of ceramic coated material. The process is essentially the same as that used for making kitchen appliances. The steel core provides mechanical strength far in excess of that of any of the other substrates. The steel can be formed in a variety of shapes with holes for mounting or to facilitate the addition of components. The porcelain coating is applied by spraying and is hardened by baking. The firing temperature for screen printed components on porcelainized steel is limited to 650°C due to the glassy nature of the coating. This is approximately 200°C less than the typical firing temperature of thick-film inks formulated for use on alumina or beryllia substrates, and, therefore, requires special inks.

SMOOTH SURFACE SUBSTRATE PRODUCTION (8)

For thin-film depositions, it is desirable to use substrates with atomically smooth surfaces. The substrates that most closely approximate these conditions are glass, glazed alumina ceramics, polished alumina ceramics, and single-crystal alumina (sapphire). The most popular glass substrates are those made from alumino-borosilicate glass produced by the sheet glass process. Sheet glass is made by melting the ingredients in a tank-type furnace. The raw materials are added at one end, and a ribbon of glass is drawn out at the other end. The surface smoothness obtained by this process is represented by a peak-to-valley distance of approximately 10,000 Å. Since thin-films range in thickness between 50 and 50,000 Å, fine-grain alumina must either be polished or coated with a lead borosilicate glaze, typically 75 μm thick. Glazed beryllia substrates are used when high thermal conductivity is required.

The mechanical strength and thermal conductivities of the substrate materials must be balanced against the cost. Glass is the least expensive of the thin-film substrates, but also is the most subject to breakage. Single-crystal alumina has the most impressive characteristics for a thin-film substrate, but its cost eliminates its consideration in many cases. Thermally oxidized silicon also provides an excellent surface for thin-film depositions but is more expensive than its ceramic alternatives.

SUBSTRATE CLEANING (8–15)

Substrate preparation is one of the most important steps in the fabrication of microcircuits. The cleaning steps that are performed at many times during the manufacturing process have a significant effect on the number of acceptable circuits which are produced. Nowhere is the old adage, "cleanliness is next to godliness," more appropriate than in the production of microcircuits. The emphasis on cleanliness applies to the environment in which the circuits are fabricated, the chemicals (including water) used in the processing, and the substrates.

Clean Rooms

Airborne particles produce defects in microcircuits. The classification of clean areas is based on a mixed unit system. The basic unit is the number of particles which are greater than 0.5 μm in diameter per cubic foot of air space. Thus, a *Class 100* clean area has less than 100 such particles per cubic foot. This unit does not transform in a useful way to the metric system since *Class 100* implies approximately 900 particles per cubic meter. The smallest particles visible to the human eye are approximately 10 μm in diameter, which is about one-tenth the diameter of a human hair. Cigarette smoke particles are approximately 0.5 μm in diameter. A typical factory environment represents a *Class 100,000* or higher situation, with particles in the form of dirt, fibers, and smoke. Monolithic integrated circuits are customarily produced in *Class 100* environments. In order to obtain this type of environment, it is necessary to circulate the air through a series of filters, with the final filtration guaranteeing the desired particle count. The air in clean rooms is usually arranged to flow from the ceiling to the floor in a laminar-flow condition. The ultraclean area is frequently surrounded by a region of vertical laminar flow with a particle count near 1000 per cubic foot. The clean room concept was originated in order to provide germ-free atmospheres for people with a low tolerance for disease.

In addition to cleaning the incoming air, it is necessary to eliminate particle sources within the clean area. Among the items banned from clean rooms are paper, smoke, makeup, and many types of external clothing. Special smocks or coveralls using nonshedding synthetic cloths have been developed for use in clean rooms. In this case, the term "coverall" includes shoes, hair, hands, and, in some cases, faces, as well as the arms, legs, and torso. The areas in which the cleanliness standards are most strictly enforced are those in which photolithography takes place. Air showers and shoe cleaners are commonly encountered in the anterooms to clean rooms.

In some situations, laminar-flow work stations are used in place of, or inside of, clear rooms. They are self-contained units that are available in either vertical or horizontal air flow.

Chemicals

The chemicals used in microelectronics are highly refined products that exceed the purity requirements for reagent grade. "Electronic grade" chemicals have very low levels of particular impurities which might influence the electronic properties of devices processed with their use. The stringent requirements of MOS processing have led to the development of special "low mobile ion grade" electronic chemicals. The distribution of these chemicals is limited to companies having a *bona fide* use for them.

Perhaps the most important chemical used in microelectronics production is ultrahigh purity water. This water is usually referred to as "deionized water," but it has been subjected to a number of purification steps in addition to the removal of ionized impurities. The criteria used to describe the quality of water are resistivity, particle size, and bacteria count. A typical system for converting city water into ultrahigh purity water includes a number of filter steps. The first filter is an activated charcoal filter, which removes chlorine, organic materials, and large particles. This is followed by a particle filter, which removes all particles larger than 10 μm. Dissolved minerals are removed by a process called "reverse osmosis," followed by deionizing beds. Osmosis is the spontaneous passage of a liquid from a dilute solution to a concentrated solution across a semipermeable membrane that allows the passage of water but not the solute. The process continues until the pressure is large enough to prevent any further transfer of the water to the more concentrated solution. Reverse osmosis makes use of pressure on the contaminated side of the membrane to force water through the membrane to the relatively pure side. Under optimum conditions, reverse osmosis removes 90% of the dissolved minerals. Further deionization is accomplished by using mixed bed ion-exchange cartridges. The process involves resins that have the property of exchanging ions from the resins with those in solution. Anions in solution are exchanged for hydroxyl ions from one type of resin. Cations in solution are exchanged for hydrogen ions from the other type of resin. The result is water and a reduction in the ability of the resins to remove ions. The ion-exchange cartridges can be replaced or regenerated when they are no longer capable of producing the quality of water desired. The combination of reverse osmosis and ion-exchange is a cost-effective method for producing deionized water. Bacteria is controlled in a water system by exposing the water to UV radiation. The final filtration is accomplished by passing the water through membranes that remove all particles larger than 0.22 μm. The theoretical resistivity of pure water is 18.3 M$\Omega \cdot$cm at 25°C. The water obtained from a system like the one described above has a resistivity of 18 M$\Omega \cdot$cm with less than 1.2 colonies of bacteria per milliliter and particles less than 0.22 μm in diameter.

Cleaning Processes

The processes used for "cleaning" substrates are not intended to produce atomically clean surfaces. Instead, it is desirable to be able to produce a surface that is reproducible and stable. It is essential that the results of a particular operation will be predictable each time that operation is performed. For these reasons, certain cleaning operations, using electronic grade chemicals in clean room environments, have been adopted by the microcircuit manufacturers. Each company has its own favorite cleaning process for each particular operation. In many cases these processes are proprietary.

Silicon Wafer Cleaning. The processes used for cleaning silicon wafers are similar to those used for removing photoresists. The initial step is immersion in a strong oxidizing agent. Typical solutions for this step include: 70% HNO_3 at 80°C, a mixture of 9 parts concentrated H_2SO_4 and 1 part HNO_3 at 180°C, Cr_2O_3 dissolved in H_2SO_4 at 80°C, H_2O_2 and H_2SO_4, H_2O_2 and HCl, and the combination of dilute solutions of NH_4OH–H_2O_2 followed by HCl–H_2O_2. The function of the oxidizer is to remove organic contaminants. It will also result in the formation of an oxide on the silicon. This oxide is usually removed by a dip in an HF-deionized water (1:10) solution. This is followed by a rinse in flowing deionized water until the resistivity of the water leaving the bath is greater than $12\,M\Omega \cdot cm$. The wafer is then dryed by spinning under a nitrogen blow-off. Some companies have employed plasma oxidation in place of the chemical oxidant. Scrubbing, using a detergent and water with a nylon or mohair brush, is also popular. This is followed by an NH_4OH rinse and a water rinse. The water can be removed by isopropyl alcohol, either by immersion and a spin dry or by alcohol vapor drying.

Thick-Film Substrate Cleaning. Thick-film ceramic substrates require less cleaning than those for the other microelectronics technologies, primarily because the patterns are larger. A common technique is vapor degreasing or ultrasonic agitation in isopropyl alcohol. A very effective cleaning treatment for these substrates is air firing at 1000°C. Halogen-based solvents are avoided in most thick-film facilities because of the detrimental effects of the vapor on thick-film resistors. The vapor forms an acid at the firing temperature and etches away the glass.

Thin-Film Substrate Cleaning. Contamination on the surface of thin-film substrates can seriously affect the adhesion of the deposited films. A typical substrate cleaning process includes ultrasonic agitation in a detergent solution, a deionized water rinse, oxidation of organic impurities in boiling H_2O_2, followed by a deionized water rinse. The substrates are dried in hot nitrogen just before they are placed in the vacuum system. In situ sputter etching is often used to assure that the surface is well prepared to receive the deposited film.

SUMMARY

Substrates provide the mechanical support for microcircuits. For monolithic integrated circuits, the substrates are silicon wafers cut from large single crystals. The crystallographic orientation of the crystals is determined by x-ray diffraction techniques, and reference flats are ground on the crystal before slicing. The wafers are polished to a mirror finish that is free of work damage.

Sapphire substrates for heteroepitaxy or thin-film depositions are prepared in the same manner. Ceramic substrates for hybrid microelectronics are made from alumina or beryllia. To make these substrates, powders are formed into the desired shape by hydraulic pressure or tape casting. This "green" ceramic is then fired at a high temperature to form the hard material used as a substrate. Thin-film substrates are made from glass or glazed ceramics.

The environment for microcircuit processing is usually a "clean room" with extremely well filtered air and strict operating rules to reduce the possibilities of contamination. The chemicals used for processing meet very strict requirements for purity. Water undergoes treatment that includes the removal of organic contaminants, dissolved minerals, bacteria, and particulate matter. Substrates are subjected to chemical treatments that assure reproducible surface conditions for each process in the fabrication of microcircuits.

REFERENCES

1. A. C. Bonora, *Semiconductor Silicon 1977*, The Electrochemical Society, Princeton, 1977.
2. R. B. Herring, *Solid State Technology* 19, 5 (1976).
3. G. A. Keig, *Solid State Technology* 15, 9 (1972).
4. D. W. Hamer and J. V. Biggers, *Thick Film Hybrid Microcircuit Technology* (New York: Wiley-Interscience, 1972).
5. C. A. Harper, ed., *Handbook of Thick Film Hybrid Microelectronics* (New York: McGraw-Hill, 1974).
6. P. L. Fleischner, *Solid State Technology* 20, 1 (1977).
7. G. Lane, *Circuits Manufacturing* 18, 2 (1978).
8. R. W. Berry et al., *Thin Film Technology* (Princeton, N.J.: Van Nostrand, 1968).
9. A. Mayer and D. A. Puotinen, *Silicon Device Processing*, C. P. Marsden, ed., National Bureau of Standards Special Publication 337, 1970.
10. M. K. Brumbaugh, *Solid State Technology* 15, 2 (1972).
11. C. K. Monzeglio, *Solid State Technology* 15, 2 (1972).
12. D. Lafeville, *Solid State Technology* 18, 1 (1975).
13. R. Kohout, *Solid State Technology* 17, 6 (1974).
14. W. G. Savola and J. T. Wallace, *Solid State Technology* 16, 11 (1973).
15. D. Tolliver, *Solid State Technology* 18, 11 (1975).

Chapter 5
The Oxidation of Silicon

The conversion of silicon into silicon dioxide (SiO_2) by thermal oxidation is an important process for both bipolar and MOS monolithic integrated circuits. Thermal oxides provide a uniform, stable protective layer on silicon. In bipolar technology, the oxide is used to assure repeatable junction characteristics by its presence where the junction intersects the surface of the silicon. Another important application for thermal oxides is based on the ability of the oxide to act as blocking barrier to the diffusion of most of the electrically active substitutional impurities used for the doping of silicon. This permits the doping of regions of a silicon wafer, specified by removing the oxide in selected regions by photolithography, while protecting the remaining parts of the wafer with oxide. In MOS circuits, the oxide acts as the dielectric material between the gate electrode and the silicon. In this chapter, the theory of oxide growth, the processes used for growing oxides, the redistribution of impurities during oxidation, and the methods for determining oxide thickness are considered.

THE THEORY OF OXIDE GROWTH (1–5)

The thermal oxidation of silicon results in a random three-dimensional network of silicon dioxide constructed from tetrahedral cells consisting of four oxygen ions surrounding a silicon ion. The bonding distance between silicon and oxygen ions is 1.62 Å and the distance between oxygen ions is 2.65 Å (2). The tetrahedra are joined to one another by bridging oxygen ions, each of which is common to two tetrahedra. In single-crystal SiO_2 (quartz), all of the oxygen ions perform the bridging function. In thermal SiO_2, some of the oxygen ions belong to only one tetrahedron, and are designated nonbridging ions. The degree of cohesion and etch rate, for a particular thermal oxide, depend on the ratio of nonbridging to bridging oxygen ions. A possible conception of a thermal oxide, including typical defects, is shown in Figure 5-1.

The most common oxidizing species used in the growth of thermal oxides are dry oxygen and water vapor. The chemical reactions are

$$Si + O_2 \rightarrow SiO_2$$

and

$$Si + 2H_2O \rightarrow SiO_2 + 2H_2$$

The reaction takes place at the interface between the silicon and the oxide, requiring the oxidizing species to be transported through the growing oxide.

During the oxidation process, a portion of the silicon wafer is consumed in the growing oxide. The thickness of silicon, x_s, consumed in the growth of an oxide layer of thickness x_{ox}, is given by

$$x_s = \frac{x_{ox} N_{ox}}{N_s} \tag{5-1}$$

where N_{ox} is the density of oxide molecules, $2.3 \times 10^{22}\ cm^{-3}$, and N_s is the

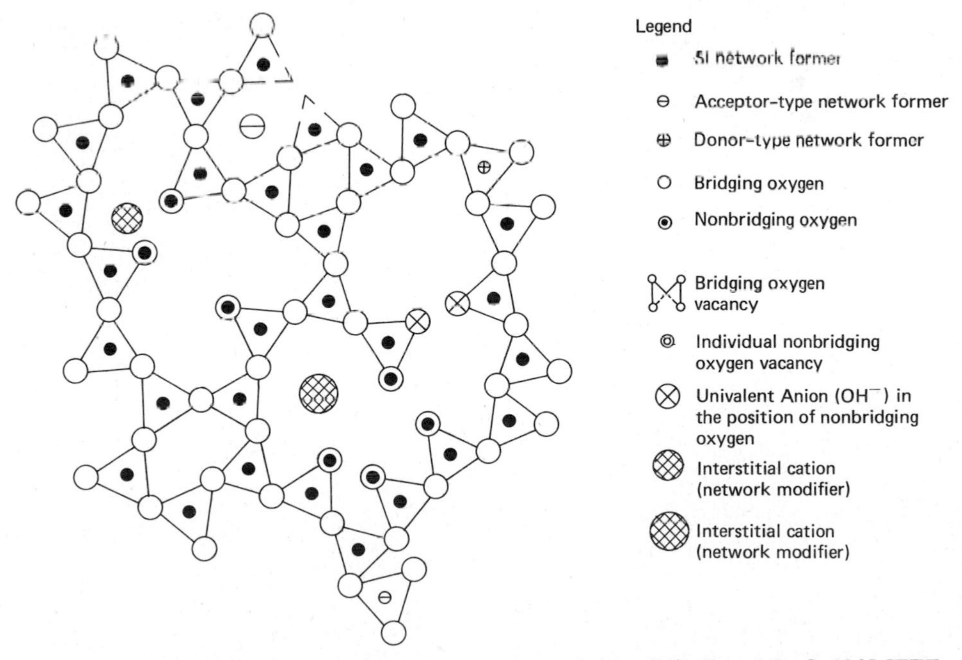

Figure 5-1. Typical bonding structures in a thermal oxide film (10). Copyright © 1965 IEEE. Reprinted from *IEEE Trans. on Electron Devices*, March 1965, p. 98.

density of silicon atoms, 5.0×10^{22} cm^{-3}. The result is

$$x_s = 0.46x_{ox} \tag{5-2}$$

A model for the kinetics of oxide growth has been developed by Grove (2). In order to be incorporated in an oxide layer, the oxidizing species must be transported from the bulk of the gas to the oxide surface, diffuse through the existing oxide, and react at the silicon surface. This model is indicated in Figure 5-2. Under steady-state conditions, the three fluxes must be equal.

The individual fluxes can be approximated as follows. The gas phase flux, F_1, is given by

$$F_1 = h_G(C_G - C_S) \tag{5-3}$$

where C_G is the concentration of the oxidizing species in the bulk of the gas, C_S is the concentration of the oxidizing species at the gas-oxide interface, and h_G is the mass-transfer coefficient, based on the concept of diffusion through a stagnant layer near the interface.

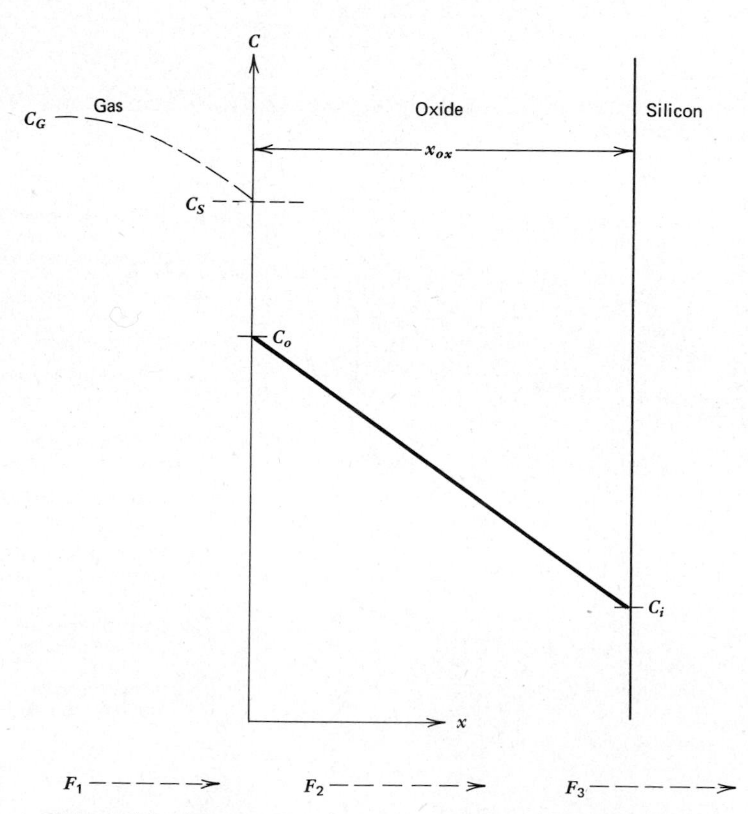

Figure 5-2. The geometry for the theoretical treatment of the thermal oxidation of silicon (2).

It is assumed that the concentration of a species within a solid is proportional to the partial pressure of that species in the surrounding gas. This is an equilibrium condition called Henry's law. The concentration of the oxidant at the outer surface of the oxide, C_o, is assumed to be proportional to the partial pressure of the oxidant right next to the surface, p_S, and is given by

$$C_o = Hp_S \tag{5-4}$$

where H is Henry's constant. The equilibrium concentration of the oxidant in the oxide, C_A, is given by

$$C_A = Hp_G \tag{5-5}$$

where p_G is the partial pressure of the oxidant in the bulk of the gas stream. From the ideal gas law,

$$C_G = \frac{p_G}{kT} \quad \text{and} \quad C_S = \frac{p_S}{kT} \tag{5-6}$$

which leads to

$$F_1 = h(C_A - C_o) \tag{5-7}$$

where h is the gas-phase mass-transfer coefficient related to the concentrations in the solid, and is given by

$$h = \frac{h_G}{HkT} \tag{5-8}$$

The flux across the oxide, F_2, is assumed to be due to diffusion, and is given by

$$F_2 = D\,\frac{C_o - C_i}{x_{ox}} \tag{5-9}$$

where D is the diffusion coefficient for the oxidant, and C_i is the concentration of the oxidant at the oxide-silicon interface.

The flux of oxidant at the oxide-silicon interface, F_3, is determined by assuming that the oxidation rate is proportional to the concentration of the oxidant at the interface. This is given by

$$F_3 = k_S C_i \tag{5-10}$$

where k_S is the chemical surface-reaction rate constant for oxidation.

Since all three fluxes are equal in steady state, C_i and C_o can be determined from the above relationships.

$$C_i = \frac{C_A}{1 + k_S/h + k_S x_{ox}/D} \tag{5-11}$$

$$C_o = \frac{(1 + k_S x_{ox}/D)C_A}{1 + k_S/h + k_S x_{ox}/D} \tag{5-12}$$

In order to calculate the growth rate, it is necessary to determine the quantity of oxidant incorporated in a unit volume of oxide, N_1. Since there are 2.3×10^{22} SiO_2 molecules per cubic centimeter in the oxide, for growth in dry oxygen, $N_1 = 2.3 \times 10^{22}\ cm^{-3}$, with one oxygen molecule per SiO_2 molecule. For growth in water vapor, $N_1 = 4.6 \times 10^{22}\ cm^{-3}$, with two water molecules per SiO_2 molecule.

The differential equation for oxide growth is given by

$$N_1 \frac{dx_{ox}}{dt} = k_s C_i = \frac{k_S C_A}{1 + k_S/h + k_S x_{ox}/D} \tag{5-13}$$

With an initial condition of $x_{ox}(0) = x_i$, the solution of this equation may be written

$$x_{ox}^2 + A x_{ox} = B(t + \tau) \tag{5-14}$$

where

$$A \equiv 2D\left(\frac{1}{k_S} + \frac{1}{h}\right)$$

$$B \equiv \frac{2DC_A}{N_1}$$

and

$$\tau \equiv \frac{x_i^2 + A x_i}{B}$$

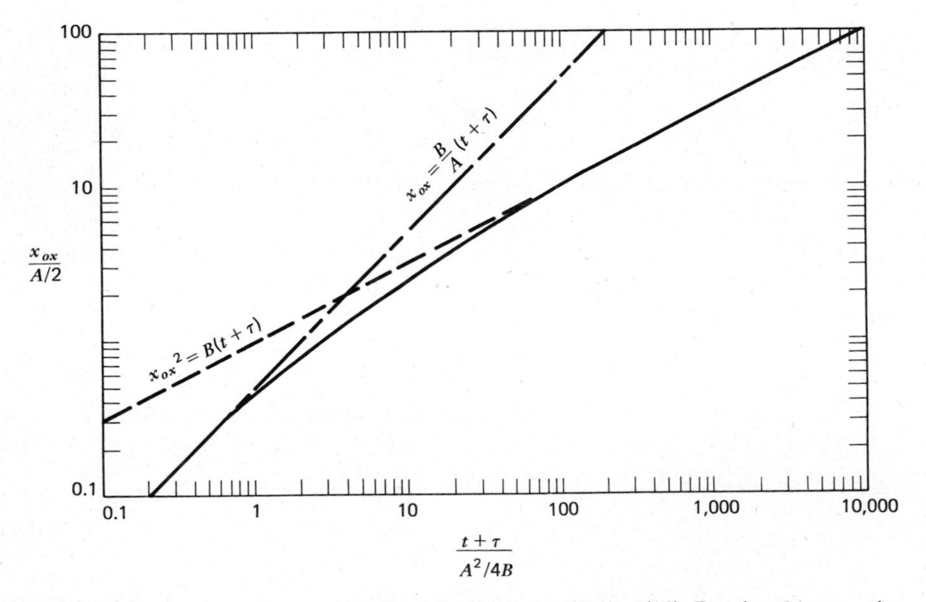

Figure 5-3. Asymptotic approximations for thermal oxidation (11). Reprinted by permission of the author.

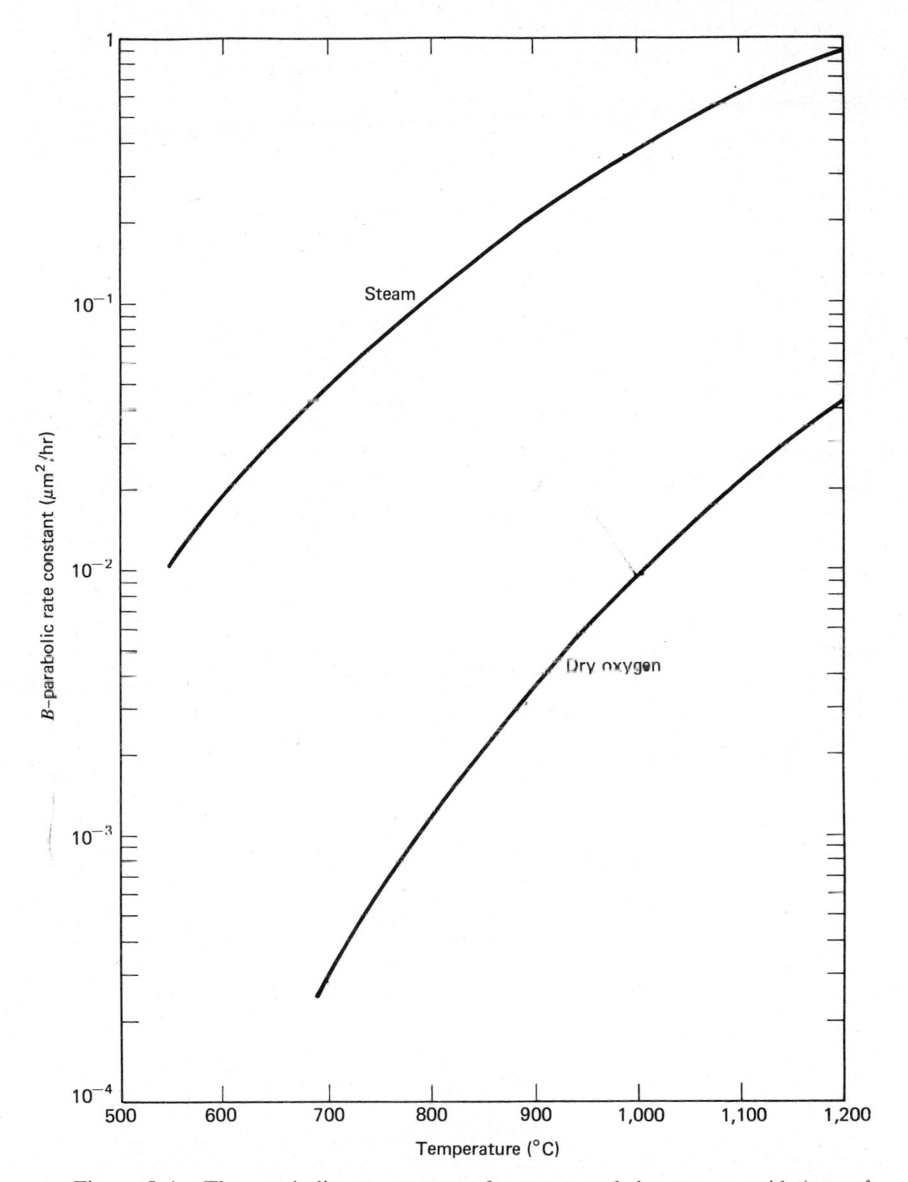

Figure 5-4. The parabolic rate constant for steam and dry oxygen oxidations of silicon (11). Reprinted by permission of the author.

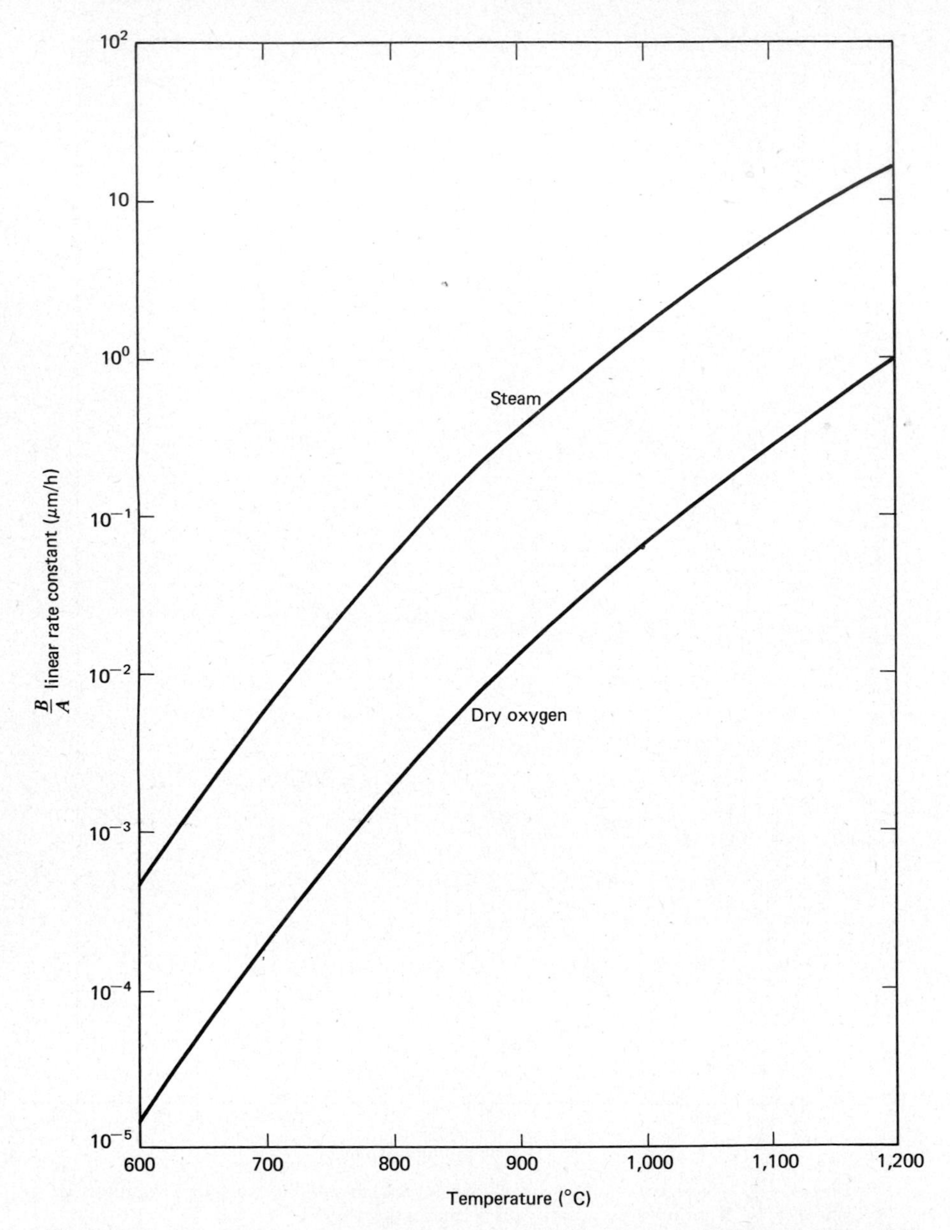

Figure 5-5. The linear rate constant for steam and dry oxygen oxidations of silicon (11). Reprinted by permission of the author.

The inclusion of the initial condition permits the estimation of oxide thickness when multiple oxidations are performed.

The general solution leads to

$$x_{ox} = \frac{A}{2}\left\{\left[1 + \frac{(t+\tau)4B}{A^2}\right]^{1/2} - 1\right\} \tag{5-15}$$

For long times, this may be approximated by

$$x_{ox}^2 = Bt \tag{5-16}$$

and B is called the parabolic rate constant. For short times, when

$$(t+\tau) \ll \frac{A^2}{4B}$$

$$x_{ox} \approx \frac{B}{A}(t+\tau) \tag{5-17}$$

and B/A is referred to as the linear rate constant. There is general experimental agreement with this expression, as shown in Figure 5-3. This implies that the Henry's law assumption was reasonably correct, and that there is essentially no dissociation of the molecules at the gas-oxide interface. Thus, the oxidant, whether oxygen or water vapor, is transported through the oxide in molecular form. The rate constants for water vapor and dry oxygen atmospheric pressure oxide growths, as obtained experimentally, are shown in Figures 5-4 and 5-5. Oxide thickness as a function of time for oxygen and water vapor is shown in Figure 5-6.

The oxidation rate can be altered by increasing the pressure of the oxidizing species (5). For a pressure of 60 atm and a temperature of 700°C, the oxidation rate is limited only by the reaction rate at the oxide-silicon interface. Under these conditions, the diffusion coefficient of water vapor is sufficiently large that the water vapor is almost uniformly distributed throughout the oxide, and is in equilibrium with water vapor in the gas phase. Hence, the oxidation rate is constant with time and varies directly with pressure. Systems for the oxidation of silicon under 10 atm of steam have been developed in which the oxidation rate at 920°C is comparable to that at 1200°C in 1 atm and that at 795°C is comparable to 920°C, 1-atm oxidation.

The oxidation rate is also a function of crystal orientation, due to the number of silicon bonds available for the reaction. The oxidation rates increase approximately in a ratio of 1:2:3 for (100), (110), and (111) oriented silicon.

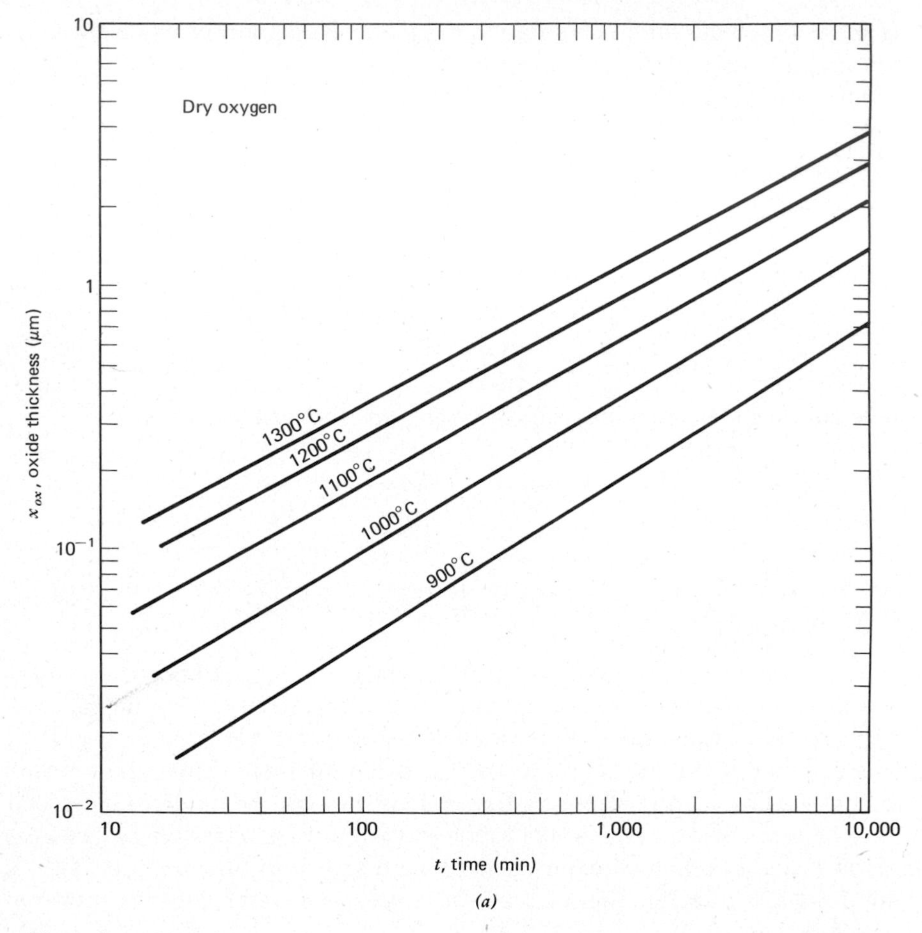

Figure 5-6. Experimental results for thermal oxidations of silicon as a function of time for (*a*) dry oxygen, and (*b*) atmospheric steam (3).

OXIDATION PROCESSES (6-8)

The thermal oxidation of silicon is usually performed in an open-tube furnace system with the oxidant passing over the wafers. The gases are in ultrahigh purity form, frequently from liquid sources passed through molecular sieves. The cleanliness of the system is particularly important for MOS processing.

The oxidation furnace is the same type of resistance heated system as is used in diffusion processes. It is a three-zone system with precise control of the temperature of the center zone, and the end zones controlled with reference to the center zone. Systems are available which will maintain a flat-temperature profile over a 50-cm-length to within ±1/10°C. Low-mass furnace elements are

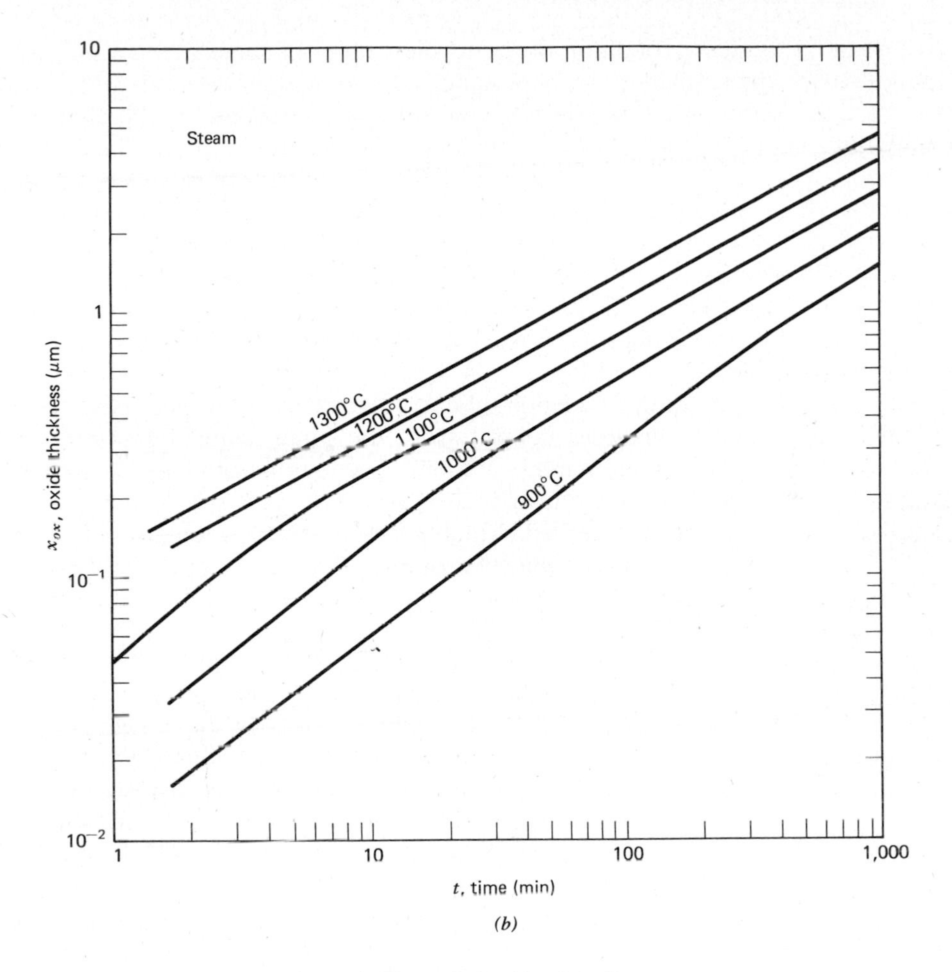

Figure 5-6. (*Continued*)

used, in conjunction with fast control systems to minimize the response time to the transient caused by inserting a boat load of more than 100 silicon wafers.

The hardware for the oxidizing system is most commonly fabricated from fused silica (SiO_2 glass), but some systems use polycrystalline silicon or silicon carbide. The gas control system, sometimes referred to as the "jungle," may be fabricated from fused silica or carefully cleaned stainless steel. The oxidation tube is placed within the furnace element, and the openings between the tube and the insulating blocks at the ends of the furnace are packed with an insulating material consisting of fine fibers of alumina, to prevent thermal leakage. Fused silica glass devitrifies when cycled above and below 1000°C, resulting in a sagging condition of the furnace tube. In some cases, a nonsagging furnace liner, made from a ceramic material containing both silica and alumina, is used between the furnace element and the oxidation tube. The use of polycrystalline silicon or silicon carbide for the oxidation tube eliminates this problem. The wafers are placed in a slotted fused silica carrier, called a boat, which is usually placed on a fused silica wheeled cart. An automatic boat puller is often used to insert the cart into a predetermined position in the oxidation tube at a programmed rate. After the oxidation process has been completed, the boat puller withdraws the cart from the furnace at a programmed rate. The insertion end of the furnace tube is either in a laminar-flow clean room or a clean work station.

The oxidants enter the oxidation tube in gaseous form. Oxygen is available in ultrahigh purity form or as boil-off from the liquid state. Purging gases include nitrogen, helium, and argon. Water vapor may be obtained in several ways. Boiling deionized water in a fused silica flask is a simple and useful source of water vapor. Controlled dripping of deionized water on a heated fused silica surface is another effective source. For high-purity oxidations, steam is generated inside the oxidation tube by the burning of hydrogen in oxygen.

The electrical properties of the oxide can be substantially improved by including a small amount (2 to 6%) of anhydrous HCl in the oxidizing gas stream (6). This has the effect of cleaning ("gettering") both the growing oxide and the oxidation tube of sodium ion contamination. Sodium ions are mobile in thermal oxides and readily drift under bias to within a short distance of the silicon interface, where they remain, effectively increasing the interface charge density. This presents difficulties in MOS devices. It is desirable to reduce sodium ion contamination to the lowest possible level. A typical oxidation process consists of a dry/wet/dry cycle, with most of the oxide growth occurring during the wet part of the process. Dry oxidations result in denser oxides. If HCl is added to the wet and final dry cycles, the sodium problem is significantly reduced. It has been demonstrated that HCl cleaned oxidation tubes produce higher quality oxides, even if no HCl is included in the oxidation step. Gaseous chlorine or nitrogen bubbled through trichloroethylene (C_2HCl_3) added to the oxidant will also result in clean oxides. It should be noted that oxygen must be present in trichloroethylene processes to avoid the deposition of carbon. Some of the sodium contamination occurs during the deposition of metal for the conductor and gate patterns. HCl gettering can not be used at this point in processing. It has been demonstrated, however, that an oxide grown with dry oxygen and HCl incorporates chlorine within the oxide, which helps to produce a more stable oxide.

A problem frequently encountered in the oxidation of dislocation-free silicon wafers is the creation of oxidation-induced stacking faults. This is due to the precipitation of interstitial oxygen already present in the wafers as a result of the crystal growth process. These stacking faults can be essentially eliminated by oxidations at temperatures above 1230°C or by dry oxidations with chlorine compounds present in the oxidizing atmosphere (7, 8). The stacking fault sizes are typically 10 to 20 μm in length for dry oxidations between 1100 and 1200°C. These stacking faults shrink in size if a subsequent oxidation is performed with nitrogen bubbled through trichloroethylene added to the gas stream. The use of a chlorine additive during the entire oxidation process virtually eliminates the creation of oxidation-induced stacking faults.

IMPURITY REDISTRIBUTION DURING OXIDATION (1–4, 9)

As a thermal oxide is grown on a doped silicon wafer, the impurity concentration in the silicon in the vicinity of the silicon-oxide interface is altered. This change depends on the ability of the impurity to be incorporated in the oxide as the growth takes place.

A distribution or segregation coefficient, m, is defined by

$$m = \frac{\text{equilibrium concentration of impurity in silicon}}{\text{equilibrium concentration of impurity in } SiO_2} \qquad (5\text{-}18)$$

The value of m for boron is 0.3, while it is 10 for phosphorus, antimony, and arsenic. All of these elements have low diffusion coefficients in SiO_2. Gallium has a segregation coefficient of 20, but has a relatively high diffusion coefficient in SiO_2.

A value of m greater than 1, with a low diffusion coefficient in SiO_2, results in a rejection of the impurity in the growing oxide, and a resulting buildup of the impurity in the silicon near the interface. This is the case for phosphorus, arsenic, and antimony. A value of m less than 1, with a low diffusion coefficient in SiO_2, results in a buildup of the impurity in the oxide, and a depletion of the impurity in the silicon near the interface. This is the case for boron. The four possible conditions are illustrated in Figure 5-7. The importance of impurity redistribution during oxidation is discussed in Chapter 7.

OXIDE THICKNESS MEASUREMENTS (3)

Optical methods are used for the measurement of thermal oxide thickness. These include color, interference techniques, ellipsometry, and mechanical techniques.

The easiest method for estimating oxide thickness is to observe the film under illumination by a white light. Uniform color is evidence of a uniform oxide thickness. The sensation of a single color is due to the absence of

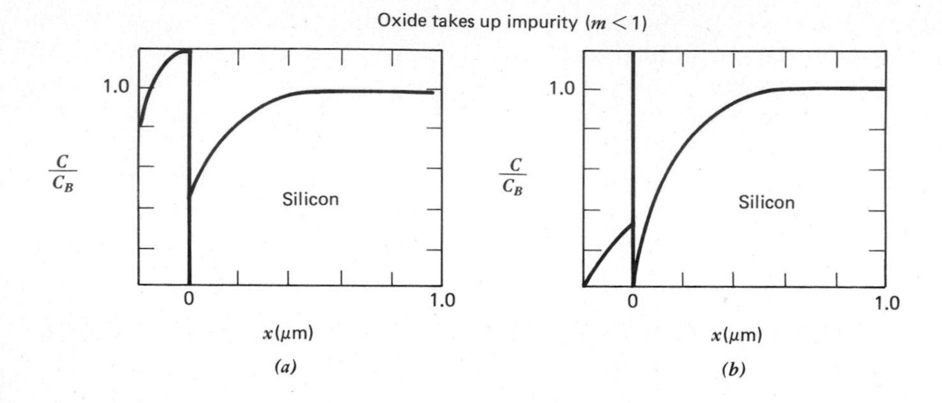

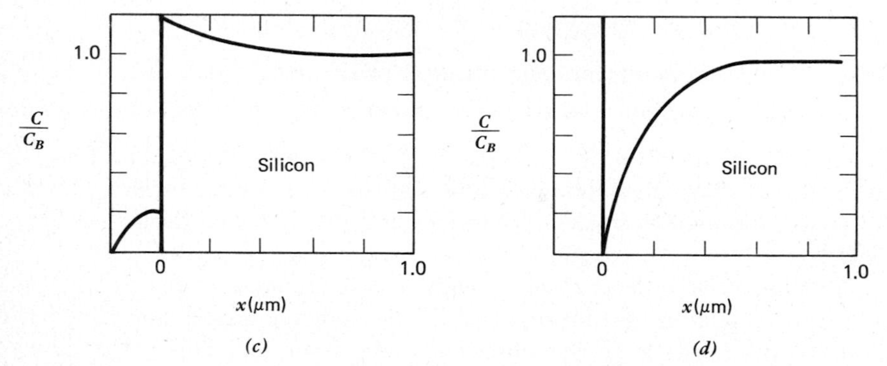

Figure 5-7. The effects of oxidation on impurity profiles (2). (*a*) Diffusion in oxide slow (boron). (*b*) Diffusion in oxide fast (boron with H_2 ambient). (*c*) Diffusion in oxide slow (phosphorus). (*d*) Diffusion in oxide fast (gallium).

particular frequencies in white light. The missing frequencies are due to destructive interference as the light is reflected from the oxide-air interface. Destructive interference occurs when

$$2d = \frac{(2k-1)}{2}\frac{\lambda}{n} \tag{5-19}$$

where d is the thickness of the film (for perpendicular illumination), $k = 1, 2, 3, \ldots, \lambda$ is the wavelength of the light undergoing destructive interference, and n is the index of refraction of the film. The same color can occur for different thicknesses of oxide due to the factor k in the expression. Oblique illumination changes the path length, and, thus, the observed color. The colors of various

Table 5-1. **Observed Colors for Different Oxide Thicknesses for Perpendicular Illumination with White Light (3)**

	Thickness (Å)			
Color	1	2	3	4
Grey	100			
Tan	300			
Brown	500			
Blue	800			
Violet	1000	2800	4600	6500
Blue	1500	3000	4900	6800
Green	1800	3300	5200	7200
Yellow	2100	3700	5600	7500
Orange	2200	4000	6000	
Red	2500	4400	6200	

oxide thicknesses are indicated in Table 5-1. Color determination is very subjective, and the effect of the same color appearing more than once in the table makes this an inaccurate technique. It is useful to make up a display containing oxidized wafers whose thickness have been accurately determined, and comparing wafers to this display for a quick check on process reproducibility.

Interference techniques are based on the fringes formed when a small wedge between a partially transmitting mirror and an oxidized wafer are illuminated with monochromatic light. These interference fringes are called Fizeau fringes. In practice, the oxide is etched from a portion of the wafer to form a step or a wedge (a wedge makes it easier to see the fringe shift), and the wafer is coated with aluminum to provide the same reflecting surface on both levels. A sodium vapor lamp provides the monochromatic illumination. The optical system is shown in Figure 5-8, along with a typical pattern as seen by the observer. The thickness of the oxide, x_{ox}, is determined by

$$x_{ox} = N\frac{\lambda}{2} \tag{5-20}$$

where N is number of fringes shifted between the top surface and the bottom surface (1.5 in the figure), and λ is the wavelength of the monochromatic light (5890 Å for sodium vapor). The wedge width is important in the measurement of thicker oxide films. A gradual slope can be obtained by masking with Apiezon W® dissolved in trichloroethylene and etching the oxide in 48% HF.

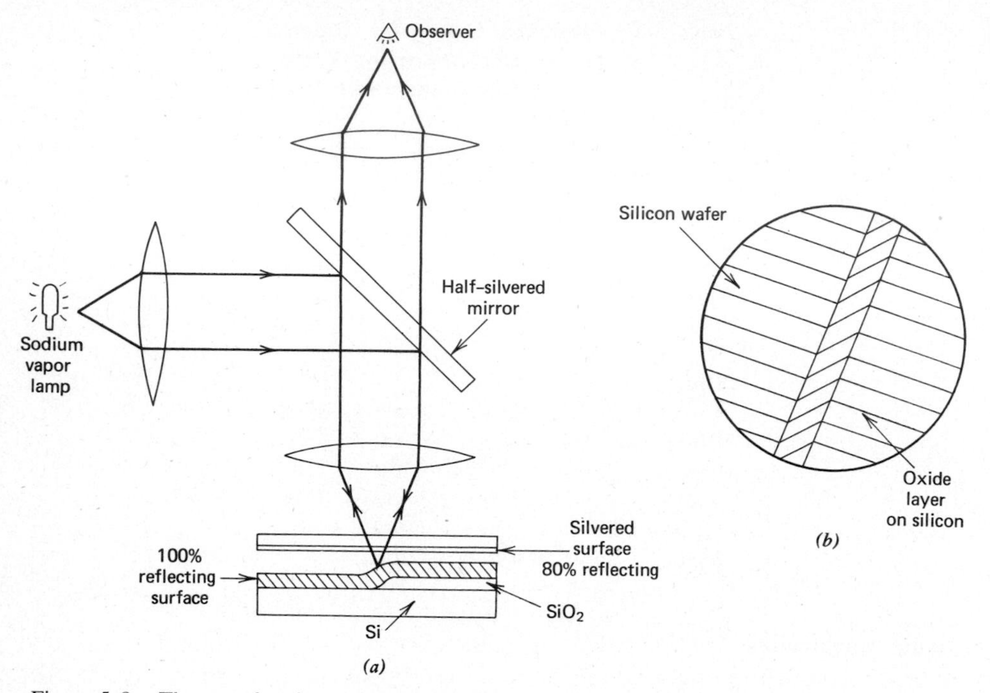

Figure 5-8. The use of an interference microscope for measuring oxide thickness (3). (*a*) Optical system. (*b*) Fizeau fringes.

Ellipsometry is the most sophisticated and the most accurate method for measuring oxide thickness. It is based on the phase shifts that occur when non-normal polarized monochromatic illumination reflects from both the oxide surface and the underlying silicon surface. That portion of the incident wave which is normal to the plane of incidence is reflected differently than that portion which is contained in the plane of incidence, as expressed by the Fresnel formulas. The plane of incidence is the plane containing both the propagation vector of the incident wave and that normal to the illuminated surface. In ellipsometry, a linearly polarized wave, polarized at 45° to the plane of incidence is reflected from the surfaces. The reflected waves are elliptically polarized. The ellipticities of the reflected waves are measured and related to the properties of the film. Graphs of the solutions of the ellipsometry equations are used to determine the thickness of the film based on the index of refraction. Since the thickness scale is periodic, the thickness is known within an approximate range of 2400 Å. Therefore, if the thickness is known approximately, ellipsometry can be used to accurately determine the thickness. Conventional optical devices like Nichol prisms, quarter-wave plates, polarizers, and analyzers are used in ellipsometers.

It is possible to use a mechanical technique to measure oxide thickness.

The instruments for this purpose make use of a diamond stylus, which is mechanically scanned across the surface of the sample at a low tracking force (typically 15 mg). The oxide must be selectively etched to provide a step for measurement. These instruments are capable of detecting step heights between 25 Å and 100 μm, with the most reliable readings associated with the larger step heights.

SUMMARY

Thermal oxides are grown on silicon wafers in open tube furnaces using either dry oxygen or water vapor. Oxides grown in dry oxygen are denser than those grown in water vapor, but the oxidation rate is much lower in dry oxygen. The oxidation rates can be determined by formulas or from graphs determined experimentally. Oxidation under high pressure results in a higher oxidation rate. The inclusion of chlorine compounds in the oxidizing atmosphere significantly improves the quality of the oxide by immobilizing sodium contamination, and reduces the creation of oxidation-induced stacking faults. The impurity concentration in the silicon at the interface is altered by oxidation. Oxidation thickness can be estimated by the color of the oxide, or by more accurate techniques like interference microscopes or ellipsometry.

REFERENCES

1. S. K. Ghandhi, *The Theory and Practice of Microelectronics* (New York: Wiley, 1968).
2. A. S. Grove, *Physics and Technology of Semiconductor Devices* (New York: Wiley, 1967). Figures reprinted by permission of the publisher.
3. R. M. Burger and R. P. Donovan, *Fundamentals of Silicon Integrated Device Technology, Vol. 1* (Englewood Cliffs, N.J.: Prentice-Hall, 1967). Figure 5-6 reprinted from pp. 41 and 49 by permission of the publisher.
4. D. J. Hamilton and W. G. Howard, *Basic Integrated Circuit Engineering* (New York: McGraw-Hill, 1975).
5. R. Champagne and M. Toole, *Solid State Technology* 20, 12 (1977).
6. R. J. Kriegler, *Semiconductor Silicon 1973,* The Electrochemical Society, Princeton, 1973.
7. C. W. Pearce and G. A. Rozgonyi, *Semiconductor Silicon 1977*, The Electrochemical Society, Princeton, 1977.
8. T. Hattori, *Journal of the Electrochemical Society* 123, 6 (1976).
9. H. F. Wolf, *Silicon Semiconductor Data* (Oxford: Pergamon Press, 1969).
10. A. G. Revesz, *IEEE Transactions on Electron Devices ED*-12, 3 (1965).

11. B. E. Deal and A. S. Grove, *Journal of Applied Physics* 36, 12 (1965).
12. B. E. Deal et al., *Journal of the Electrochemical Society* 112, 3 (1965).

PROBLEMS

5-1. An oxide is grown at 1000°C for 15 min in dry oxygen. Estimate the thickness of the oxide using Equation 5-15 and compare the result with the approximations of Equations 5-16 and 5-17 and the empirical data in Figure 5-6. Assume that the initial oxide thickness was 50 Å.

5-2. The wafer of Problem 5-1 is subjected to a further oxidation in atmospheric steam at 1200°C for 15 min. Estimate the oxide thickness and the thickness of silicon consumed in the oxidation. Under perpendicular illumination with white light, what color does the oxide appear to have?

Chapter 6
Film Depositions

Films of various materials are used in all types of microelectronics circuits. These films may be deposited by a number of different techniques, including evaporation, sputtering, plating, anodization, chemical vapor deposition, and screen printing. The selection of the deposition technique is based on the material to be deposited, the thickness desired, the substrate characteristics, and the electrical and mechanical properties required. Films range in thickness from a few hundred angstroms to tens of micrometers. They serve as conductors, insulators, resistors, and, under certain conditions, active devices. In this chapter, the deposition equipment and processes typical in the microelectonics industry are described.

VACUUM PROCESSING EQUIPMENT

Many film depositions take place under reduced pressure conditions. The most common unit used as an indication of pressure in vacuum systems is the torr, which is 1 mm of mercury in a barometer, (760 mm of mercury corresponds to atmospheric pressure at sea level at 0°C). The unit is named after Torricelli, a pioneer in vacuum technology. The SI unit for pressure is the pascal which is a newton per square meter (1 pascal $= 7.5 \times 10^{-3}$ torr). The torr will be used throughout this book.

Mechanical pumps with oil seals are commonly used for rapid pumping to pressures of 10^{-2} torr. These pumps are frequently called roughing or fore-line pumps. Most of these pumps are of the vane or rotary piston type. A disadvantage of mechanical pumps is the possible contamination of the evacuated chamber by oil.

An oil-free technique for rough-pumping makes use of the adsorption and absorption properties of zeolites (5). Zeolites are alkali metal aluminosilicates which can be dehydrated without changing their crystal lattice. As a result, molecules of certain gases can occupy the spaces left by the removal of water. The zeolites are activated by immersing their container in liquid nitrogen. These pumps are used for volumes less than 50 l. The same materials are used as molecular sieves for the purification of gases.

If lower pressures are required, a different type of pump is used. Since high-vacuum pumps are only effective if the starting pressure is already low, a roughing pump is used to obtain a pressure of approximately 10^{-2} torr. A system of valves is then employed to close off the roughing line and open the high-vacuum line. An effective mechanism for obtaining pressures near 10^{-7} torr is the diffusion pump. A diagram of a diffusion pump is shown in Figure 6-1. A low-boiling point hydrocarbon fluid is heated in the bottom of the pump. The molecules of the vapor are ejected downward by the annular jets. This downward motion of the vapor molecules is also imparted to gas molecules, which collide with the vapor molecules. These gas molecules are

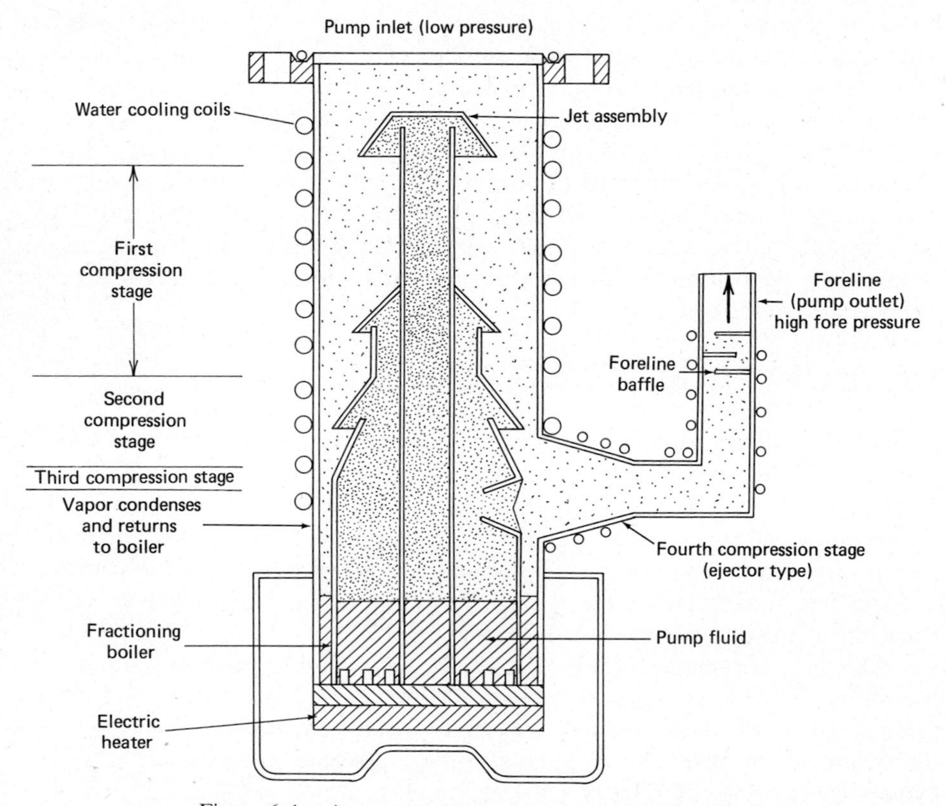

Figure 6-1. A cross-sectional view of a diffusion pump.

swept into the fore-line through the ejector jet. This reduces the gas concentration in the pump volume, resulting in diffusion of gas molecules from the chamber to the pump. This process continues until the gas pressure in the chamber reaches the desired level. Some of the pump vapor molecules will be deflected upward after colliding with gas molecules. To prevent contamination of the chamber, one or more baffles (traps) are inserted between the pump and the chamber. The baffles consist of metal chevrons, which do not permit line-of-sight paths between the pump and the chamber. The chevrons are cooled by water, freon, or liquid nitrogen. The cold baffle surfaces condense the hot pump vapor molecules on contact. A typical bell jar-diffusion pump vacuum system is shown in Figure 6-2.

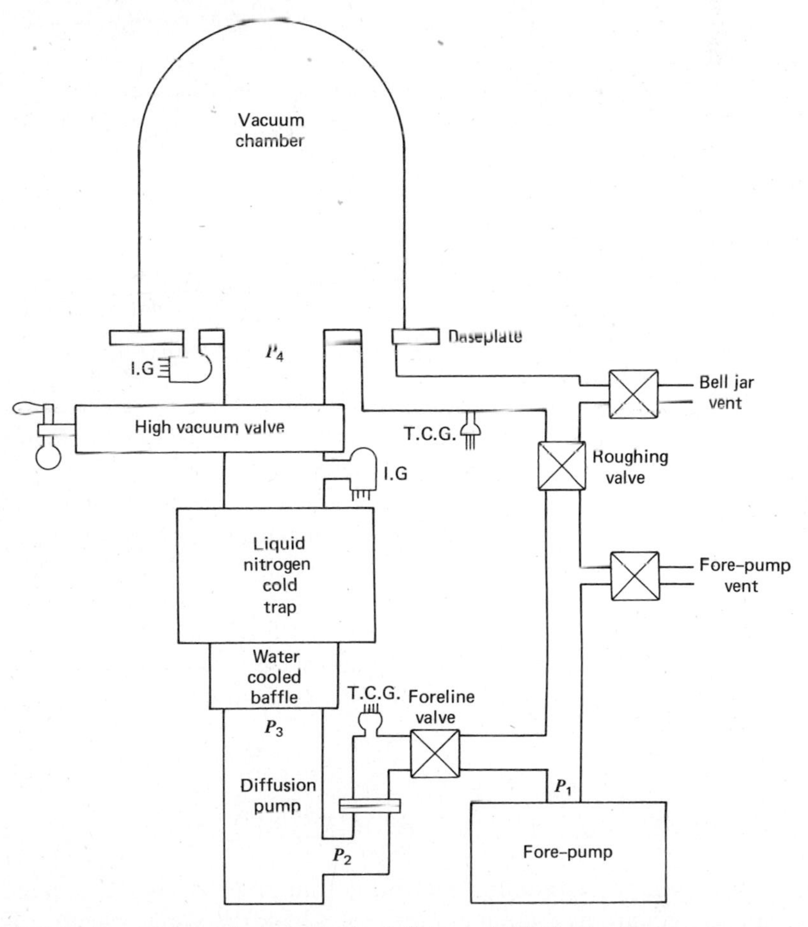

Figure 6-2. A schematic drawing of a bell-jar and diffusion pump vacuum system. I.G. indicates positions of ionization gauges. T.C.G. indicates positions of thermocouple or pirani gauges.

To obtain lower pressures than a diffusion pump system, an ion pump system can be used. The ion pump makes use of a high-voltage dc discharge between titanium plates. Gas molecules are ionized as they pass between the plates. A permanent magnet establishes a magnetic field that acts, in combination with the electric field, to confine the ions in paths between the plates until they spiral into the cathode. Ion pumps are only efficient at low pressures. A typical system includes a roughing pump (preferably of the zeolite adsorption type for a cleaner system), a getter pump, and an ion pump. A getter pump is a system that evaporates fresh layers of titanium onto surfaces within the system. The newly deposited titanium is an effective absorber of gas molecules.

The gauges used for indications of low pressure take several forms. For the roughing cycle, a thermocouple gauge is frequently used. The output of this type of gauge depends on the ability of the atmosphere surrounding a heated filament to transfer heat from the filament to a cylinder, which surrounds the filament at a prescribed distance. As the number of molecules in the atmosphere decreases, the rate of heat transfer also decreases. The voltage generated in a thermocouple in contact with the cylinder is used as a measure of the pressure. The most commonly encountered gauge in a diffusion pump system is the Bayard-Alpert ionization gauge. It consists of a heated filament, which emits electrons, a helical grid, which is positive with respect to the filament, and a thin wire plate, which is negative with respect to the filament. This assembly is exposed to the atmosphere within the chamber. Electrons from the filament are accelerated by the grid, and most of them pass through the opening in the grid and are repelled by the plate. These electrons then go into oscillation around the grid until they strike a grid wire or encounter a gas molecule. If a collision occurs between an electron and a gas molecule with sufficient energy to ionize the molecule, the positive gas ion will be attracted either to the filament or the plate, depending on the location of the collision. Only those ions collected by the plate register as current in the gauge. Adsorbed gas molecules on the plate and the walls of the gauge contribute to erroneous indications. The gauge can be degassed by heating the grid for several minutes. The combined effects of the hot filament and the hot grid will produce the desired baking effect. The plate current is calibrated to indicate the pressure. In an ion pump system, the current in the pump power supply is used as a pressure indicator.

VACUUM DEPOSITION PROCESSES

The most widely used method for depositing thin films consists of heating solid materials in a vacuum to the temperature at which the vapor pressure is 10^{-2} torr, and condensing the vapor on a cooler substrate. For quality and uniformity of the deposited films, it is desirable that the mean free path between collisions of the evaporated atoms or molecules with gas molecules be long compared to the distance between the source of the material being vaporized and the substrate upon which it is being condensed. This mean free path depends on the pressure of the atmosphere surrounding the source. At 10^{-4} torr, the mean free path is 45 cm, and at 10^{-6} torr, the mean free path is 45 m. It is recommended that the maximum pressure for an evaporation system be 10^{-5} torr. In general, the lower the pressure, the better the quality of the film. Since most pumping systems become very inefficient as they approach their ultimate low pressure, there is a compromise between pressure and

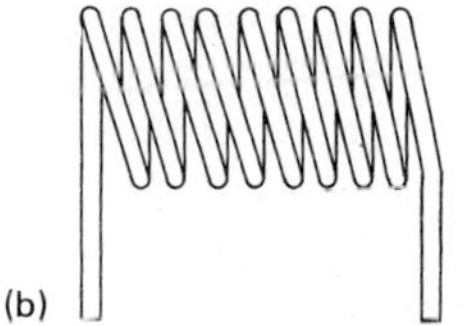

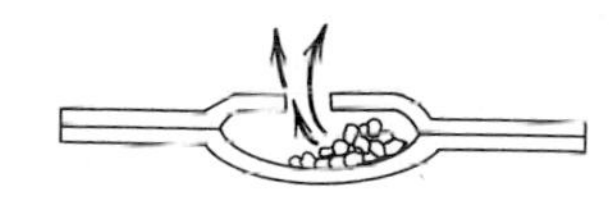

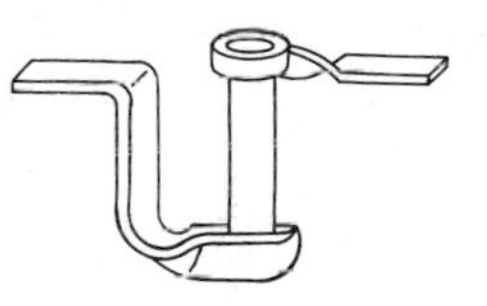

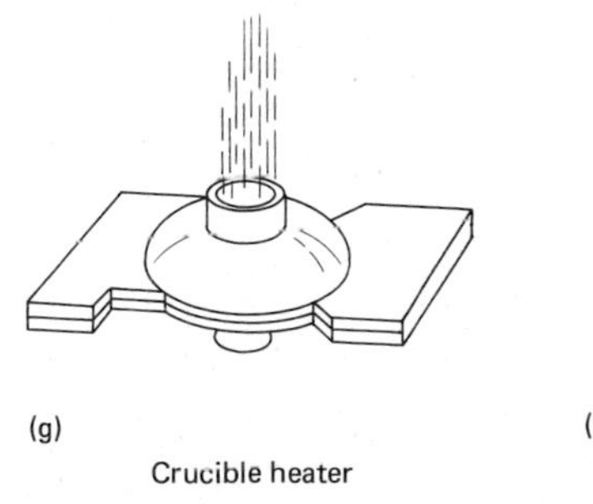

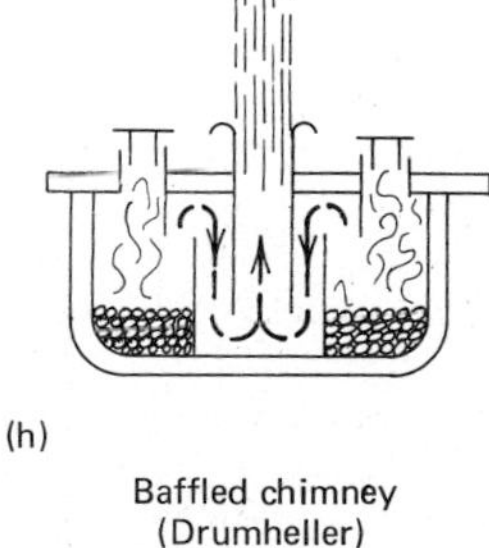

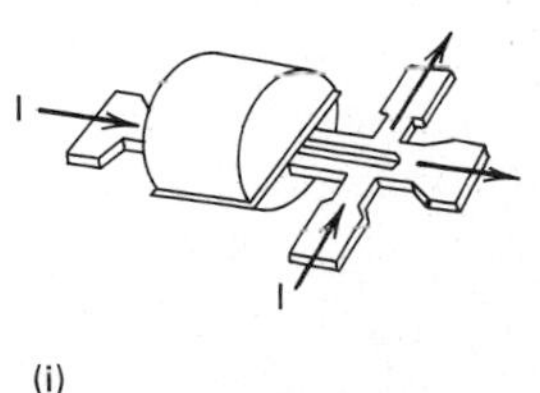

Figure 6.3. Various sources for the thermal evaporation of materials by resistance heating (2). From *Thin Film Phenomena* by K. L. Chopra. Copyright © 1969. Used with permission of McGraw-Hill Book Company.

pumping time for specific processes so that acceptable films are produced in reasonable times.

The weight of material evaporated per unit time is proportional to the square root of the molecular weight of the material. The rate of deposition of a film depends on many factors, including: separation distance from source to substrate, relative angular position of the source to the substrate, the geometry of the source, and the condensation coefficient for the evaporated species. Both deposition rate and film thickness can be monitored during the deposition process by an electronic system featuring a crystal controlled oscillator. The crystal is mounted, inside the evaporation chamber, in the same area as the substrates. The deposited film changes the mass of the crystal, and, hence, the frequency of the oscillator. This type of system is used to automatically control the deposition.

The most common method for heating the source is resistance heating. The materials used to fabricate the heaters are refractory metals like tungsten, molybdenum, tantalum, and niobium. Some typical forms for thermal evaporation sources are shown in Figure 6-3. To provide uniform coverage for a large number of substrates, it may be necessary to use an array of sources and a mechanical substrate holder with rotation capabilities in two directions. Many metals, like aluminum, gold, copper, chromium, and nickel, are readily deposited by thermal evaporation. Alloys may be difficult to evaporate because of

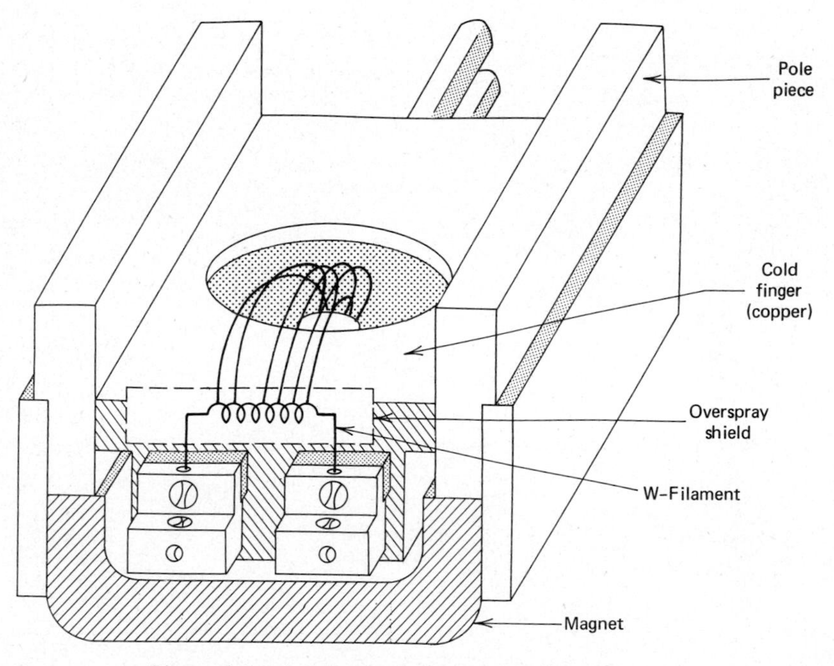

Figure 6-4. An electron beam evaporation system.

the different evaporation rates of the constituents. The usual practice is to simultaneously deposit the various elements from separate sources. It is also possible to deposit alloys by flash evaporation, in which a powder of the alloy is dropped onto a heated surface and evaporates almost immediately.

Another approach to thermal evaporation is to use a focused electron beam as the heat source. An *e*-beam evaporation source is shown in Figure 6-4. The copper cold finger is water cooled. The energy is concentrated in a spot approximately 3 mm in diameter. *e*-beam evaporation is a clean technique, and can be used for the deposition of a wide variety of materials, including compounds and alloys.

SPUTTERING

Sputtering is the process of removing surface atoms or molecules from the surface of a material by bombardment with energetic ions. The types of sputtering systems used for microelectronics depositions include dc, RF, and magnetron. Sputtering is performed in vacuum type equipment, but with pressures in the range between 25 and 75×10^{-3} torr.

Sputtering

Conducting materials can be sputtered in a system in which the ionizing energy is supplied from a dc power source. A typical system for dc sputtering is shown in Figure 6-5. The material to be deposited is called the target and forms the cathode of the system. The anode is spaced between 1 and 12 cm from the target and connected to the positive terminal of a high-voltage dc power supply, with the negative terminal connected to the target.

A dc sputtering system operates in the following manner. The substrates are placed on the anode, or on a substrate heater surrounded by the anode. The chamber is pumped to a high-vacuum condition, and then back-filled with ultrahigh-purity argon to the desired sputtering pressure. A gaseous glow discharge is established between the anode and the target. The atoms sputtered from the target fall on the substrates and form a film. Each material has a characteristic "sputtering yield." This is the number of atoms removed from the target per impacting ion, which is a function of the energy of the ion. The deposition rate on substrates at the anode depends on the sputtering yield and the ion current.

Insulators can be deposited by a process called reactive dc sputtering. A reactive gas, like oxygen, is mixed with the sputtering gas so that a compound is formed during the process. It is difficult to form a stoichiometric compound in this manner. It is frequently necessary to anneal an oxide formed by reactive sputtering in an oxidizing atmosphere at an elevated temperature for a period

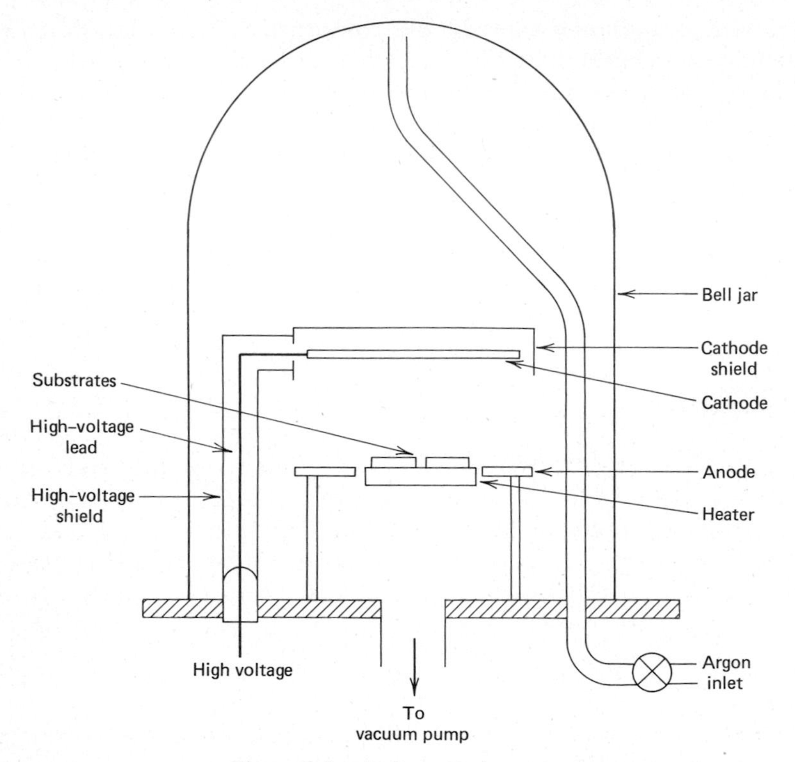

Figure 6-5. A dc sputtering system.

of time to obtain the desired result. This is sometimes performed within the sputtering system after the deposition has been performed.

It is usually desirable to perform an in situ cleaning process within the sputtering system before carrying out the deposition. This process is called back-sputtering. The polarity of the power supply is reversed so that the sputtering occurs on the substrates rather than the target. This produces a very clean substrate for the deposition.

RF Sputtering

Insulators and conductors can also be deposited by a process called RF sputtering. The geometry of the system is essentially the same as that for a dc sputtering system. An RF field is applied between the anode and cathode. To confine the discharge to the target area, coils are placed around the bell jar to create an axial magnetic field. If the target is an insulator, the RF sputtering system operates as follows. Since the electrons and ions created in the discharge have much different masses, more electrons strike the target during the half-cycle when the target is positive than positive ions when the target is negative. Under steady-state conditions, a net negative charge builds up on the insulator, creating a negative bias between the target and the anode. Sputtering occurs in a manner similar to the dc case. If a conductor is the target, it must be capacitively coupled to the cathode, so that a dc bias can be established. The overall efficiency of an RF sputtering system for depositing a conductor is lower than that of a dc sputtering system.

Magnetron Sputtering (6–8)

The deposition rate of a dc sputtering system can be significantly improved by using magnetic fields to intensify the gas discharge. The magnetic fields are produced by permanent magnets and are oriented such that they are approximately parallel to the exposed surface of the target. Planar and conical targets are used in this type of system, and the targets are water cooled. The substrates are mounted on a planetary fixture, like that in an evaporation system, which is independent of the anode. High deposition rates for aluminum, aluminum/2% silicon alloys, and aluminum/4% copper/2% silicon alloys are possible with dc magnetron sputtering. These alloys are used to improve the characteristics of certain types of integrated circuits, as discussed in Chapters 9 and 10. Because of the major differences between the evaporation rates of aluminum and silicon at the same temperature, co-evaporation is necessary, even in *e*-beam evaporation systems. Magnetron sputtering represents, therefore, a significant improvement for the deposition of these aluminum alloys.

RF magnetron sputtering can be used for the deposition of SiO_2. In conventional RF sputtering, electron and positive ion bombardment of the substrates causes degradation of the deposited film and heating of the substrates. The addition of a strong magnetic field parallel to the substrates protects them from charged particle bombardment, and permits a higher deposition rate.

PLATING AND ANODIZATION (3)

In certain applications, it is desirable to build up the thickness of a conductor film to reduce its resistance. It is usually more economical to deposit this material by electroplating. A similar process, anodization, can be used to form a dielectric film.

Electroplating makes use of aqueous solutions of metal salts as the electrolyte. A dc source is connected so that the substrate is the cathode and the anode is the metal to be plated. The deposition rate is dependent on the current density in the bath.

Anodization is used to form quality thin dielectric films on certain metals.

In this case, the substrate is connected to the positive terminal of the dc power supply. Anodization is limited to the metals or semiconductors that most readily form coherent films of metal oxides, like aluminum, tantalum, and silicon. Tantalum is a popular thin-film resistor material. It can be anodized in aqueous solutions containing anions like acetates, citrates, borates, and sulfates. The cation is usually hydrogen or ammonium. This process can be used for resistor trimming or to form a capacitor dielectric. Aluminum can only be anodized in electrolytes with a pH range between 6 and 9, since aluminum oxide is soluble in acids outside that pH range. The anodization of silicon is part of an evaluation process to determine impurity profiles.

The anodization process is frequently performed at a constant current density. This requires an increasing voltage as the oxide grows, resulting in a constant electric field. If a capacitor dielectric is being formed, the last part of the anodization process is performed at a constant voltage. This provides for a slow growth, which yields a layer of high-quality dielectric.

CHEMICAL VAPOR DEPOSITION (10–13)

Chemical vapor deposition (CVD) is the term applied to the process in which a film is deposited by a chemical reaction or pyrolytic decomposition in the gas phase in the neighborhood of the substrate. Epitaxial growth (Chapter 3) is a typical example of CVD. Various systems have been developed for CVD at low temperatures, low pressures, and plasma enhancement. Typical applications for CVD in microelectronics include depositions of silicon dioxide (SiO_2), silicon nitride (Si_3N_4), and polycrystalline silicon.

The thermal oxidation of silicon produces high-quality films of SiO_2, but it is only applicable on silicon substrates. It is usually necessary to deposit SiO_2 on top of existing oxide layers, metals, or nitride layers. If a thick oxide layer is required, it can be deposited by the oxidation of silane (SiH_4) or one of the chlorosilanes ($SiCl_4$, $SiHCl_3$, SiH_2Cl_2) with either N_2O or CO_2 at temperatures between 800 and 1000°C. These are high-quality oxides, but inferior to thermal oxides. They are frequently used for field oxides in MOS structures.

If an oxide layer is to be used as an insulator between metal layers in a multilevel intraconnection pattern, the temperature of the deposition must be kept below 500°C. Silane or a chlorosilane can be reacted with oxygen between 350 and 500°C to form an oxide of reasonable quality.

Silicon nitride is used as a passivation layer in some bipolar integrated circuits, and also serves as a constituent in multilayer gate insulators in certain types of MOS circuits. A typical silicon nitride CVD system uses the reaction of silane and ammonia at temperatures between 600 and 800°C in a nitrogen carrier gas.

Low-temperature CVD can be performed in a system that provides for continuous deposition. The substrates are passed through a nitrogen curtain as they are heated to the deposition temperature. They are then transported into the reaction chamber, and then through another nitrogen curtain before leaving the system.

Atmospheric pressure CVD usually takes place in a cold-wall reactor. This reduces wall deposits to a minimum, and, thus, prevents the degradation of the film due to particles falling on the substrates from the walls. The substrates are placed flat on an RF heated susceptor.

Systems have been developed for CVD at pressures in the range between 0.5 and 1 torr. These systems use hot-wall reaction chambers with vertically mounted wafers, not unlike the furnaces used for thermal oxidation described in Chapter 5. The substrate throughput is much higher in this type of system, as compared to a typical atmospheric-pressure CVD system. In addition, the need for carrier gases is significantly reduced. The uniformity of deposition is also better in the low-pressure systems. These factors contribute to a substantial economic advantage of low-pressure CVD as compared to atmospheric-pressure CVD.

Plasma enhanced CVD has been demonstrated for silicon nitride and silicon dioxide depositions. An important characteristic of glow discharge plasmas is that the "electron temperature" in the plasma is typically 10 to 100 times that of the average gas molecule temperature. This means that the average gas molecules can be maintained at a relatively low temperature while the electron energy is sufficient to break molecular bonds, leading to the creation of chemically active species. The results of a plasma enhanced deposition at 240°C at a pressure of 0.2 torr in silane, ammonia, and nitrogen are similar to those of an 800°C deposition in a conventional reactor system.

THICK-FILM SCREEN PRINTING (14–20)

The films deposited by screen printing are significantly different from those deposited by the other techniques described in this chapter. The inks used for this process usually consist of three components, an organic vehicle, glass frit, and active materials. The active materials depend on the purpose intended for the film. They may be elemental metals or alloys for conductor films, semiconducting compounds or alloys for resistor films, or dielectric materials for insulating films. The glass frit serves as a bonding agent between active particles, and to provide bonding of the film to the substrate. Some conductor inks do not contain glass frit, and are referred to as "molecular bonding" or "fritless" inks. The organic vehicle provides the flow characteristics of the ink.

The rheology, or flow characteristic, of a thick-film ink is a complex situation. The shear rate is a nonlinear function of stress. Most thick-film inks exhibit a pseudoplastic behavior, in which the shear rate is very low for low stress, and increases sharply once a particular stress is reached. This is indicated in Figure 6-6, where a pseudoplastic material is compared with an ideal fluid. The pseudoplastic properties of thick-film inks prevent the inks from flowing through the screen until the squeegee applies sufficient stress to change their flow characteristics.

The screens are typically constructed of stainless steel woven mesh with a mesh count of 40 to 156 per centimeter, depending on the precision of the printing required. The percentage of open area is typically between 36 and 46. They are usually coated with a photosensitive emulsion with a buildup on the printing side of 8 to 15 μm. The pattern is defined photographically and the emulsion is removed from the regions in which a pattern is to be printed. This is shown in Figure 6-7. For reproducible printing, the screen tension must be uniform.

The screen printer is a precision apparatus that provides for mounting the screen, positioning the screen relative to the substrate in x, y, z, and angle, positioning the squeegee relative to the screen, and moving the squeegee with a controlled speed. Most screen printers are powered by air pressure. The printing process takes place as follows. The ink is spread on the screen by a spreading bar, while the squeegee is in a retracted position. In off contact printing, the screen is positioned at a precise distance above the substrate, typically 500 to 1000 μm. This distance is called the snap-off distance. The squeegee deforms the screen so that the screen is brought in contact with the substrate in the area under the squeegee. Typical squeegee materials are neoprene, urethane, and polyurethane. The squeegee blade must form a seal with the screen during the printing stroke, and maintain a constant angle of

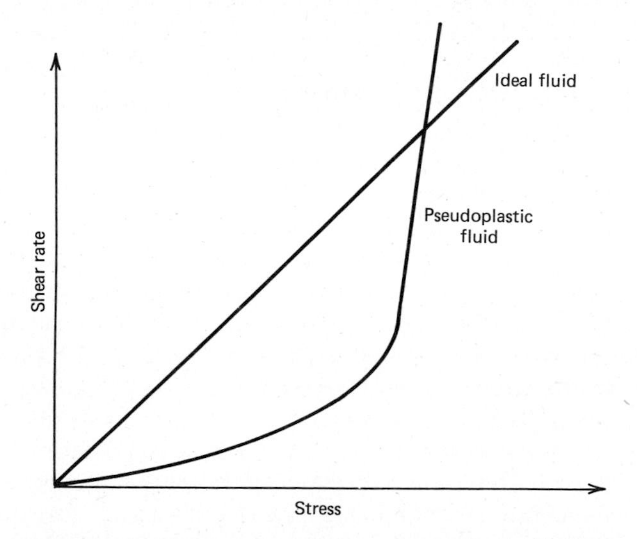

Figure 6-6. The flow characteristics of a pseudoplastic fluid.

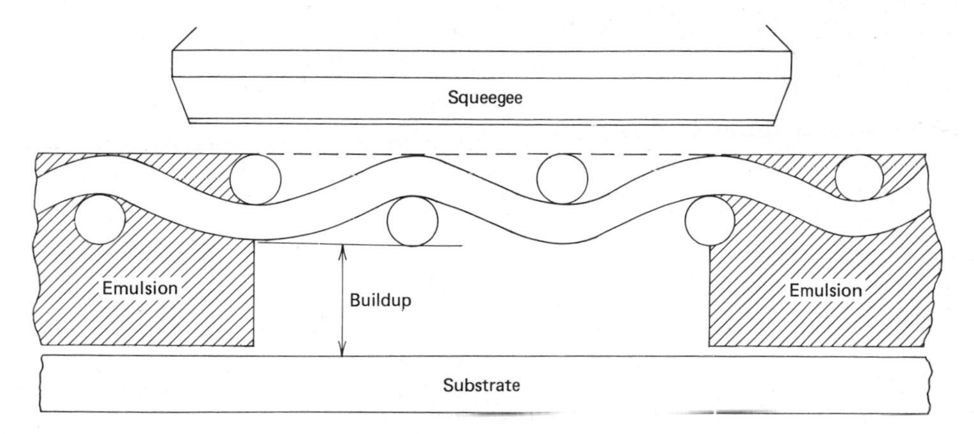

Figure 6-7. The emulsion buildup on a patterned screen.

attack so that the force exerted on the ink is constant. When the ink comes in contact with the substrate, it wets the ceramic. As the squeegee passes, the screen, due to its tension and the preset distance, snaps off the substrate, leaving the ink that was in the holes in the screen deposited on the substrate in the desired pattern. This process is indicated in Figure 6-8.

In some cases, contact printing is desirable. In this type of printing, the screen is placed in direct contact with the substrate and the ink is forced through the screen by the squeegee. The screen is then lifted vertically from the substrate, leaving the ink in place.

After printing, the ink must settle to fill in the gaps left by the wire mesh. If the formulation of the ink permits too much flow at this point, the pattern loses its desired shape. After settling, the ink is dried by infrared in a conveyer oven. The drying temperature is limited to 150°C.

The printed pattern must be fired to convert it into a usable film for microelectronics. The firing process takes place in a conveyer furnace with four or more resistance heated zones. The desired time-temperature profile is shown in Figure 6-9. The plateau at 350°C in the early part of the firing cycle is used to burn off the organic vehicle. Depending on the type of ink being fired, various events occur during the peak firing period, including glass flow and chemical reactions. Following a controlled quench to room temperature, the ink has become part of the ceramic. The atmosphere in the furnace is usually filtered dry air, which enters the furnace on the end where the substrates leave, and is withdrawn at the end where the substrates enter. This is to prevent the byproducts of the burning of the organic vehicle from being deposited downstream and polluting the circuits during peak firing. Some furnaces have provisions for controlled atmospheres in various zones so that reducing or inert atmospheres can be provided for firing non-noble metals.

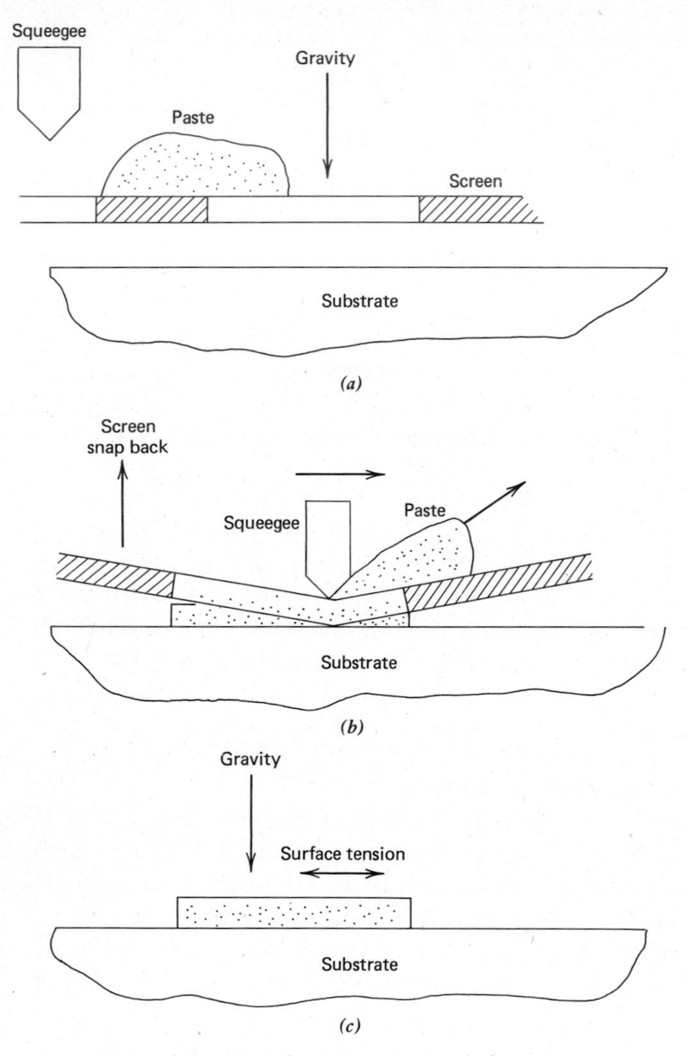

Figure 6-8. The off-contact screen printing process.

SUMMARY

Films can be deposited on substrates by a variety of techniques. Very clean thin-films are deposited by thermal evaporation under high-vacuum conditions, either by resistance heating or by the use of a focused electron beam. Sputtering is used to deposit materials, which are difficult to evaporate. Chemical vapor deposition is used to deposit films of silicon dioxide, silicon nitride, and polycrystalline silicon on monolithic integrated circuits. Electrochemical techniques are used to plate conductor films on some thin-film

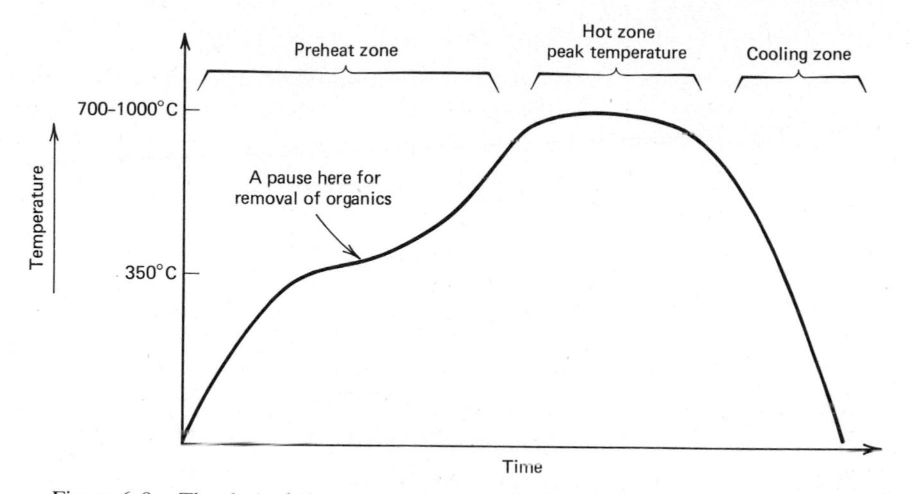

Figure 6-9. The desired time-temperature profile for the firing of thick-film inks (14).

circuits and to anodize conductors to form insulators. A much different technique, screen printing, is used to deposit thick-films of conductor, resistor, and insulating materials on ceramic substrates for thick-film hybrid microcircuits.

REFERENCES

1. C. M. van Atta, *Vacuum Science and Engineering* (New York: McGraw-Hill, 1965).
2. K. L. Chopra, *Thin Film Phenomena* (New York: McGraw-Hill, 1969).
3. R. W. Berry et al., *Thin Film Technology* (New York: Van Nostrand Reinhold, 1968).
4. R. R. LaPelle, *Practical Vacuum Systems* (New York: McGraw-Hill, 1972).
5. A. Roth, *Vacuum Technology* (Amsterdam: North-Holland, 1976).
6. K. Urbanek, *Solid State Technology* 20, 4 (1977).
7. V. Hoffman, *Solid State Technology* 19, 12 (1976).
8. T. van Vorous, *Solid State Technology* 19, 12 (1976).
9. P. J. Clarke, *Solid State Technology* 19, 12 (1976).
10. W. C. Benzing et al., *Solid State Technology* 16, 11 (1973).
11. R. S. Rosler and W. C. Benzing, *Solid State Technology* 20, 7 (1977).
12. R. S. Rosler, *Solid State Technology* 20, 4 (1977).

13. R. S. Rosler et al., *Solid State Technology* 19, 6 (1976).
14. D. W. Hamer and J. V. Biggers, *Thick Film Hybrid Microcircuit Technology* (New York: Wiley-Interscience, 1972). Figure reprinted by permission of the publisher.
15. C. A. Harper, ed., *Handbook of Thick Film Hybrid Microelectronics* (New York: McGraw-Hill, 1974).
16. M. L. Topfer, *Thick Film Microelectronics* (New York: Van Nostrand Reinhold, 1971).
17. R. A. Rilkoski, *Hybrid Microelectronic Circuits, The Thick Film* (New York: Wiley-Interscience, 1973).
18. R. E. Cote, *Solid State Technology* 18, (1975).
19. L. F. Miller, *Solid State Technology* 17, 10 (1974).
20. F. Franconville et al., *Solid State Technology* 17, 10 (1974).

Chapter 7
Selective Doping Techniques

The addition of impurities to a semiconductor changes most of its electrical properties, including majority carrier type, carrier concentration, carrier mobility, excess carrier lifetime, and internal electric fields. The energy levels of the electrically active impurities within the bandgap of silicon are shown in Figure 7-1. The impurities with energy levels near the valence and conduction band edges are those from Groups III and V of the periodic table, the substitutional impurities that are used to control the carrier concentrations and majority carrier type in the material. Although all impurities tend to reduce excess carrier lifetime, those impurities like gold, with energy levels near the middle of the bandgap, are effective recombination centers, and are introduced into silicon when it is desirable to reduce excess carrier lifetime.

Impurities, either substitional or interstitial, represent deviations from a perfect crystal lattice. Thermal agitations of the lattice atoms also represent deviations from the perfect lattice structure. These deviations cause scattering of the charge carriers, which determine the effective mobility of the carriers. Ionized impurities are particularly effective for scattering charge carriers. The carrier concentration is related to the net impurity concentration; for example, in an *n*-type semiconductor, the conduction electron concentration is related to the difference between the donor concentration and the acceptor concentration. This makes it possible to convert a *p*-type material to an *n*-type material by adding enough donor atoms to overcompensate the acceptor atoms. At room temperature, essentially all the substitional impurities are ionized, both donors and acceptors. In a compensated *n*-type material, the acceptors receive electrons from some of the donors, not from the valence band as in a noncompensated *p*-type material. In a compensated *p*-type semiconductor, the

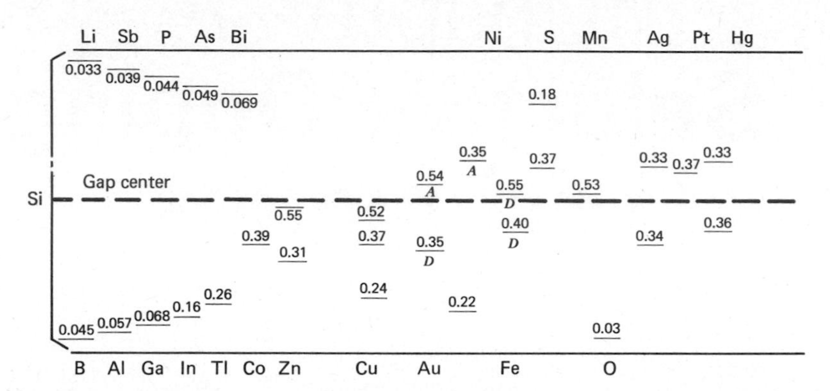

Figure 7-1. The energy levels within the forbidden gap for important impurities in silicon (31, 32). Reprinted with permission from *Solid-State Electronics*, Vol. 11, No. 6, S. M. Sze and J. C. Irvin, "Resistivity, Mobility and Impurity Levels in Ga, As, Ge, and Si at 300 K," Pergamon Press, Ltd.

situation is reversed. The important point is that the *net* impurity concentration determines the carrier concentration but the *total* impurity concentration determines the mobility and the excess carrier lifetime. Unfortunately, much of the data available on silicon, like the resistivity versus doping density (Figure 3-11), and excess carrier lifetime versus resistivity (Figure 7-2) were determined for uncompensated samples, and care must be exercised in using this information in the design of device structures. Figure 7-3 shows data relating

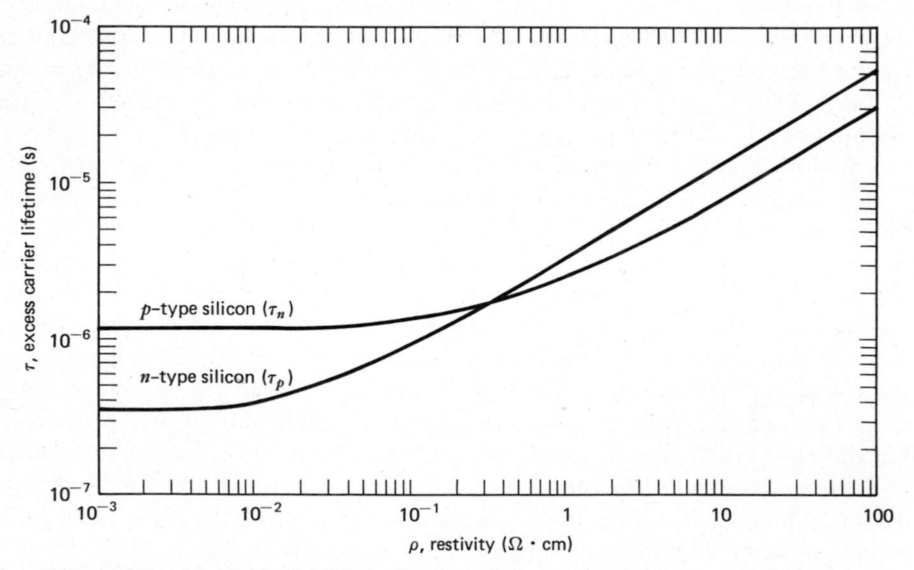

Figure 7-2. Excess carrier lifetime as a function of resistivity for silicon (9, 15, 33). Reprinted with permission of Pergamon Press Ltd.

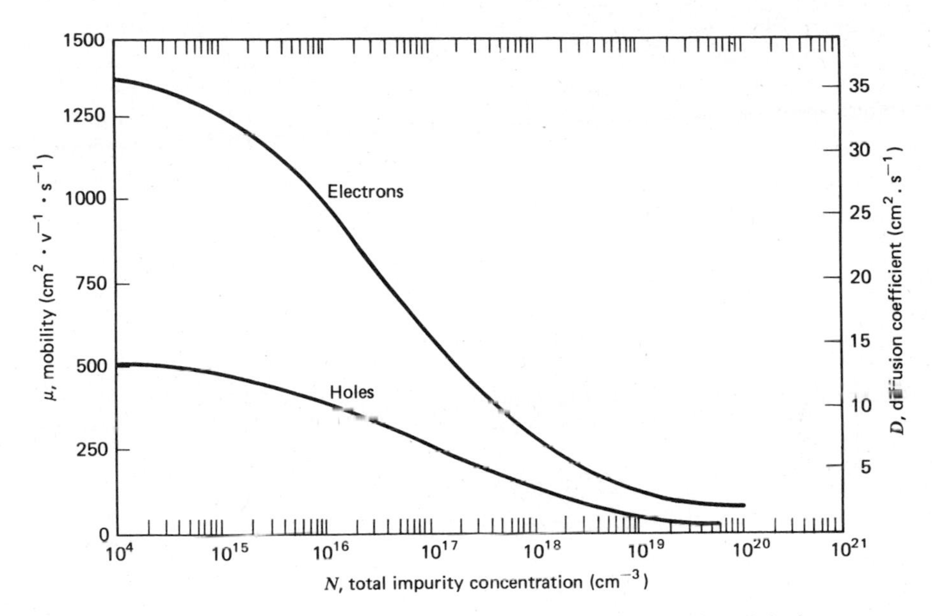

Figure 7-3. Carrier mobility and diffusion coefficient as a function of total doping concentration in silicon at 300°K (30).

carrier mobility and diffusion coefficient as a function of total impurity concentration, and must be used accordingly.

Impurities can be introduced into semiconductor materials in several ways. In Chapter 3, the doping of single crystals during crystal growth and epitaxy was discussed. Alloying is a form of doping by forming a molten region and allowing it to recrystallize. In this chapter, the selective introduction of impurity atoms by solid-state diffusion and ion implantation are discussed.

DIFFUSION

A very popular technique for the creation of localized junctions in semiconductors is called diffusion. Silicon dioxide acts as a mask for the diffusion of most of the substitutional impurities in silicon. By opening windows in the silicon dioxide layer covering thermally oxidized silicon using the photoengraving process, it is possible to introduce impurities into selected areas of the wafer, and produce *pn* junctions to form portions of semiconductor devices. Diffusion is also used to alter the profiles of dopants in semiconductors after the impurities have been introduced by diffusion, ion implantation, epitaxy, or crystal growth. In the following sections, the theory of diffusion, systems for performing diffusions, and the methods for evaluating diffused junctions are described.

Diffusion Theory

Diffusion is statistical in nature, and is related to the random motion of energetic particles. If there is a gradient in the concentration of energetic particles, there is a net flow of particles from regions of higher concentration to those of lower concentration. The mathematical relationships governing diffusion are called Fick's laws. The first of these laws relates the particle current density, J_N, to the concentration gradient, and, for a one-dimensional case, is written

$$J_N = -D\frac{\partial N}{\partial x} \tag{7-1}$$

where D is the diffusion coefficient, and N is the particle density. Fick's second law makes use of the continuity relationship for particle current,

$$\frac{\partial N}{\partial t} = -\frac{\partial J_N}{\partial x} \tag{7-2}$$

to obtain

$$\frac{\partial N}{\partial t} = +\frac{\partial}{\partial x}\left(D\frac{\partial N}{\partial x}\right) \tag{7-3}$$

which reduces to

$$\frac{\partial N}{\partial t} = D\frac{\partial^2 N}{\partial x^2} \tag{7-4}$$

if D is independent of x. Unfortunately, the diffusion coefficients for phosphorus, boron, and arsenic, the three most frequently used dopants in silicon, are dependent on the concentration of the diffusing species. There are also electrical interactions between diffusing impurity ions and the background impurity distribution, which further complicates the diffusion process. In addition, the diffusion coefficients for the important impurities in silicon are strong functions of temperature.

For ideal diffusants (D independent of x), Equation 7-4 has been solved for a number of boundary conditions. Two of these situations have been applied to the diffusion of impurities in semiconductors. The first of these is when the semiconductor wafer is surrounded by a continuously replenished concentration of impurity atoms. This is called a constant source diffusion. The second situation is when a thin layer of impurity atoms is deposited on the silicon wafer, and used as the diffusion source. This is called a limited source diffusion.

For a constant source diffusion, the solution is given by

$$N(x, t) = N_0 \operatorname{erfc}\left(\frac{x^2}{4Dt}\right)^{1/2} \tag{7-5}$$

where $N(x, t)$ is the impurity concentration at point x (the distance from the surface) at time t, N_0 is the impurity concentration at the surface ($x = 0$) and is maintained constant, and D is the diffusion coefficient. The complementary error function (erfc) is a tabulated function, and is plotted in Figure 7-4. For a limited source diffusion, the solution is

$$N(x, t) = \left[\frac{Q}{(\pi Dt)^{1/2}}\right] \exp\left(-\frac{x^2}{4Dt}\right) \tag{7-6}$$

where Q is the concentration of impurity atoms per unit surface area at $t = 0$. This solution is the well-known Gaussian, and is also plotted in Figure 7-4. Both solutions are based on classical diffusion theory for diffusion into a semi-infinite slab. This is a reasonable approximation for typical shallow diffusions in semiconductors. Even though the diffusants in silicon are not ideal, and, thus, do not strictly obey Fick's second law, Equations 7-5 and 7-6 can be used to estimate the doping profiles for practical diffusions.

The mechanism for diffusion of an impurity in a crystal lattice depends on whether the impurity is interstitial or substitutional. For interstitial impurities, the probability of a vacant site being adjacent to an occupied site, in the proper direction, is high, and the energy required for such a transition is low, approximately 1 eV, resulting in high diffusion coefficients. For substitutional impurities, it is necessary that a vacant lattice site appear on the proper side and adjacent to a site occupied by an impurity atom. The energy of formation of a vacancy (a Schottky defect) is relatively high, 2 to 3 eV, resulting in much lower diffusion coefficients for substitutional impurities as compared to interstitial impurities at a particular temperature. For example, gold, which diffuses by the interstitial mechanism, has a diffusion coefficient of $3.75 \times 10^{-5}\ \text{cm}^2 \cdot \text{s}^{-1}$ at 1000°C, compared to boron, a substitutional impurity, which has a diffusion coefficient of $1.7 \times 10^{-14}\ \text{cm}^2 \cdot \text{s}^{-1}$ at that temperature.

The general form for the diffusion coefficient for substitutional impurities in silicon is given by (1)

$$D = h\left[D_i^0 + D_i^-\left(\frac{n}{n_i}\right) + D_i^=\left(\frac{n}{n_i}\right)^2 + D_i^+\left(\frac{p}{n_i}\right)\right] \tag{7-7}$$

where h is a field enhancement factor given by

$$h = 1 + \left[1 + 4\left(\frac{n_i}{N}\right)^2\right]^{-1/2} \tag{7-8}$$

where n_i is the intrinsic carrier concentration at the diffusion temperature and N is the impurity concentration. The quantities, D_i^0, D_i^-, $D_i^=$, and D_i^+, represent the diffusion coefficients of a paired impurity ion and vacancy complex, with the charge state of the vacancy indicated by the superscript, under intrinsic conditions ($n = p = n_i$). The superscript 0 indicates a neutral

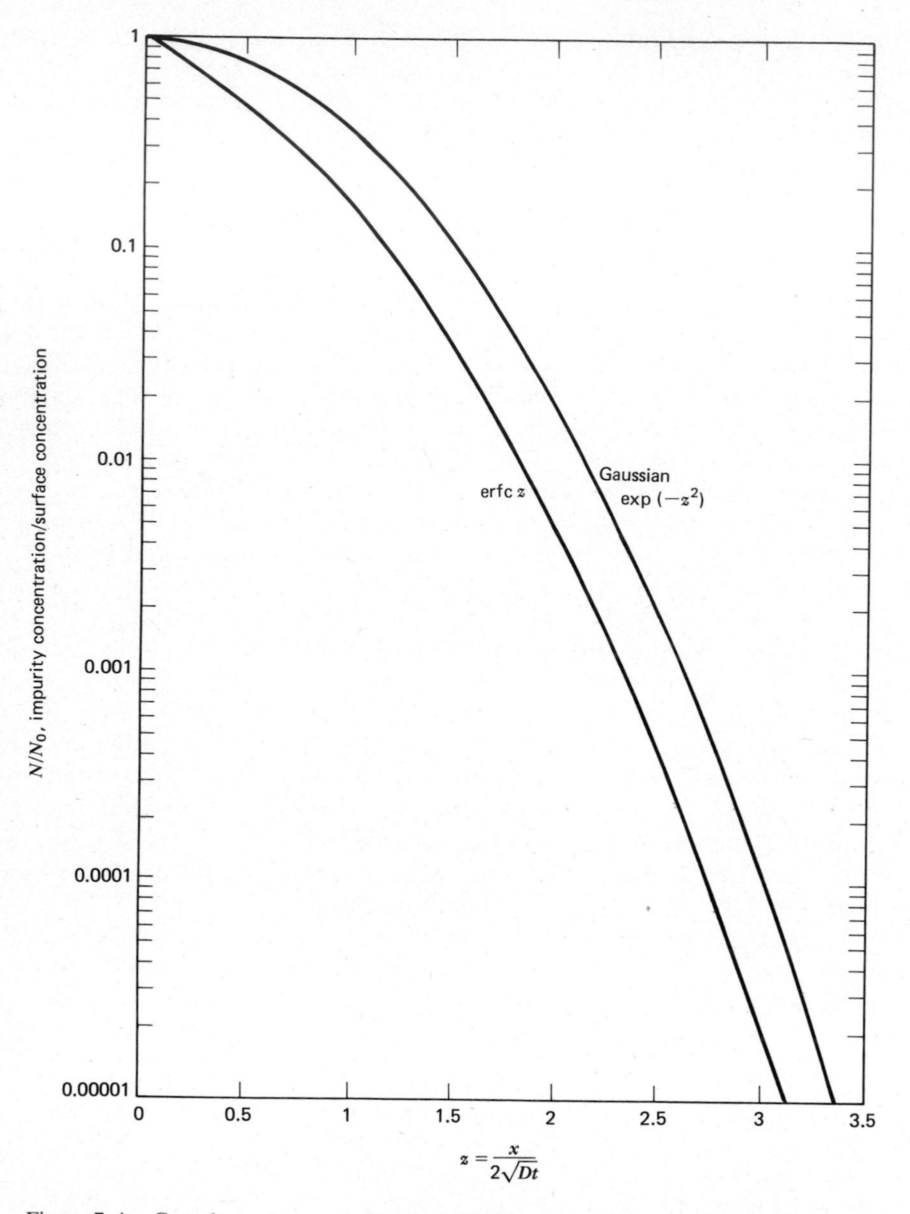

Figure 7-4. Complementary error function and Gaussian distributions normalized to the surface concentration.

vacancy, $-$ indicates a single negative charge, $=$ indicates a double negative charge, and $+$ indicates a positive charge. Each of the D_i quantities is a function of temperature and can be written

$$D_i = D_0 \exp\left(-\frac{E_a}{kT}\right) \tag{7-9}$$

where D_0 is the asymptotic value of the diffusion coefficient at infinite temperature and E_a is the activation energy for the process. The quantities, n and p, in Equation 7-7 are the electron and hole concentrations at the diffusion temperature.

The formation of the vacancy-impurity ion pairs in the different charge states depends on the particular impurity species. For boron, the neutral and positive charge states have been observed. Phosphorus forms neutral and both single and double negative-charge states. Arsenic forms the neutral and single negative-charge states. In Table 7-1, the values of D_0 and E_a for the observed charge states are listed. For convenience, the intrinsic diffusion coefficients for boron, phosphorus, and arsenic are plotted as a function of temperature in Figure 7-5. The intrinsic carrier concentration is plotted as a function of temperature in Figure 7-6.

Boron Diffusion Coefficients. Boron has a concentration dependent diffusion coefficient that can be determined from Equation 7-7. It can be written in the form (1)

$$D = h\left[D_i^0 + D_i^+\left(\frac{p}{n_i}\right)\right]$$

or

$$D = hD_i\left[1 + \left(\frac{D_i^+}{D_i^0}\right)\left(\frac{p}{n_i}\right)\right] \Big/ \left[1 + \left(\frac{D_i^+}{D_i^0}\right)\right] \tag{7-10}$$

Table 7-1. **D_0 and E_a for the Intrinsic Diffusion Coefficients of Substitutional Impurities in Silicon(1)**

Charge State		Boron	Phosphorus	Arsenic	Antimony
Neutral	D_0[a]	0.091	3.85	0.38	0.214
	E_a	3.36	3.66	3.58	3.65
Single negative	D_0	—	4.44	22.9	13
	E_a	—	4.0	4.1	4.0
Double negative	D_0	—	44.2	—	—
	E_a	—	4.37	—	—
Single positive	D_0	166.3	—	—	—
	E_a	4.08	—	—	—

[a] The units of D_0 are $cm^2 \cdot s^{-1}$ and E_a are eV.

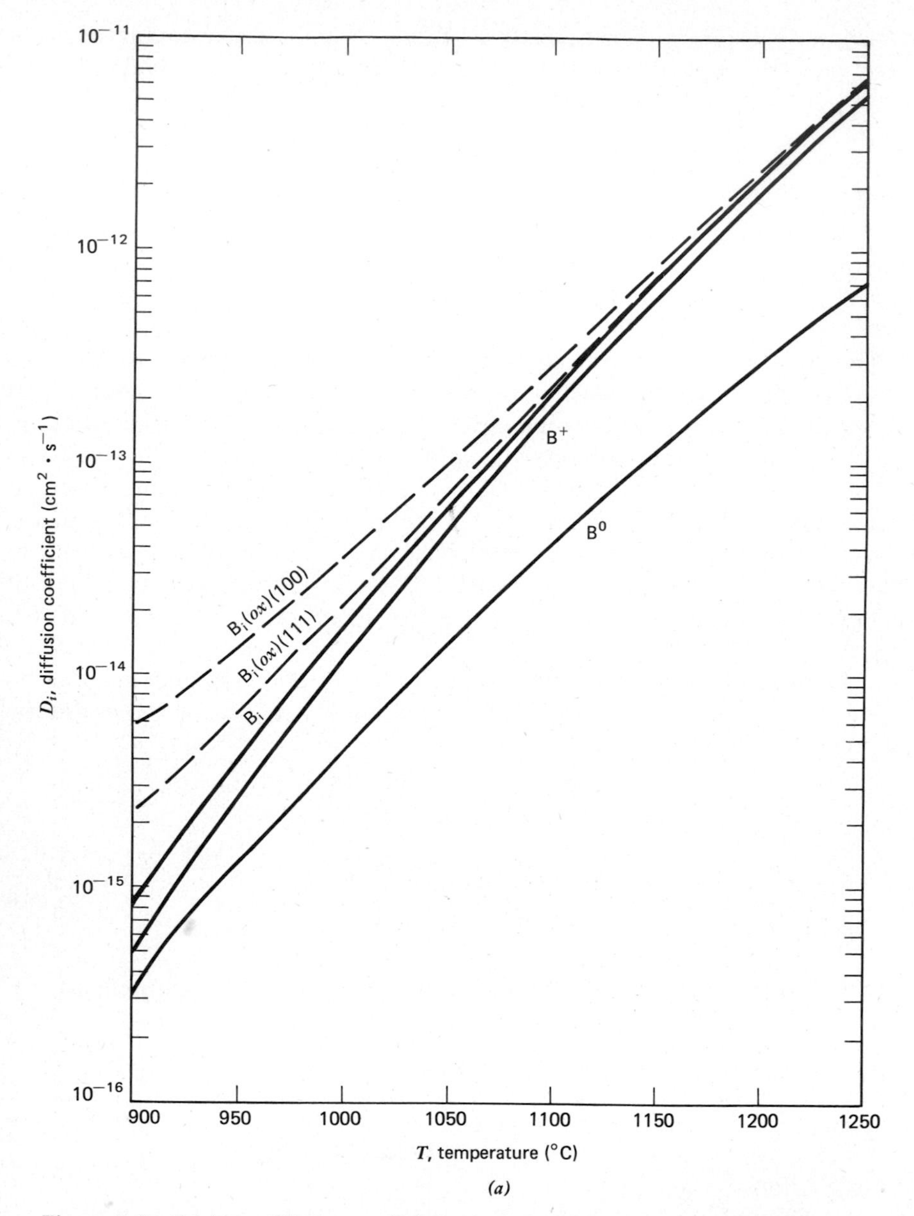

(*a*)

Figure 7-5. Intrinsic diffusion coefficients for substitutional impurities in silcon. (*a*) Boron (1, 5). (*b*) Phosphorus and arsenic (1).

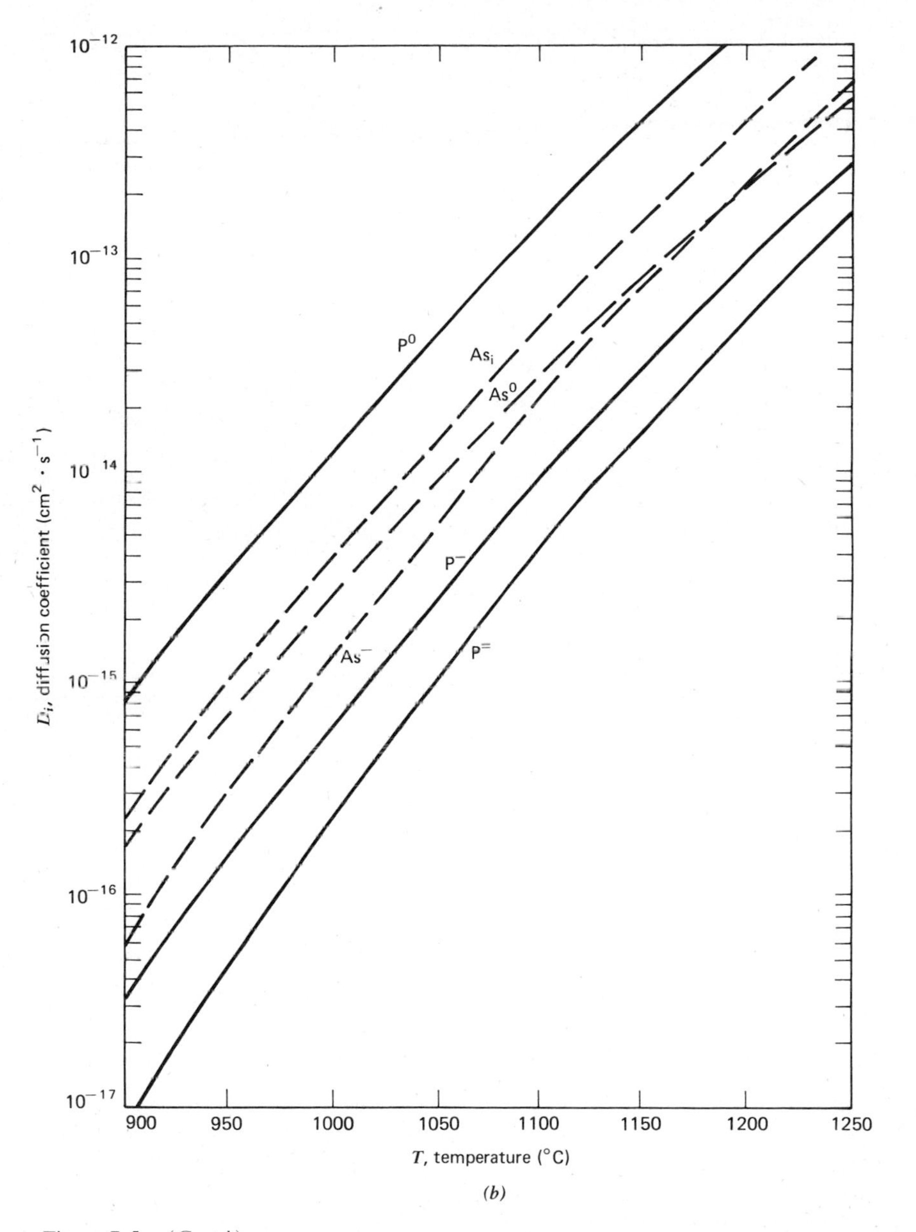

Figure 7-5. (*Contd.*)

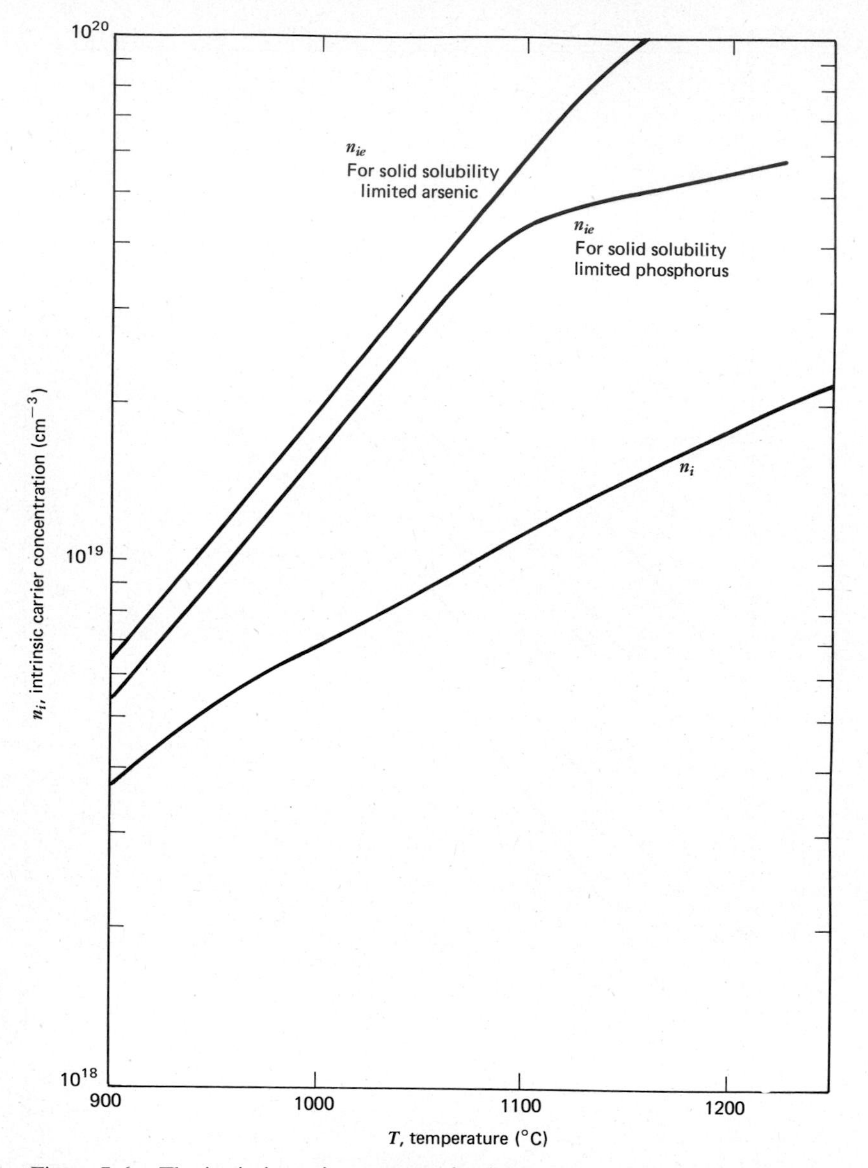

Figure 7-6. The intrinsic carrier concentration in silicon at temperatures in the range usually associated with the diffusion process, and the effective intrinsic carrier concentrations for solid solubility doped arsenic and phosphorus (3).

where $D_i = D_i^0 + D_i^+$ is called the intrinsic diffusion coefficient. This quantity is also plotted in Figure 7-5. In this equation, p is essentially equal to the boron concentration. The result of a concentration dependent diffusion coefficient is that the profile is not a Gaussian or complementary error function. From measured data (2), it has been demonstrated that a constant source boron diffusion profile can be approximated by

$$N(x, t) = N_0(1 - Y^{2/3}) \tag{7-11}$$

where N_0 is the surface concentration and Y has the form, when $N(x)$ is much less than N_0, of

$$Y = \left(\frac{x^2 n_i}{6N_0 D_i t}\right)^{1/m} \tag{7-12}$$

where m is a constant. If the boron has been diffused into a silicon wafer with an n-type uniform doping concentration, N_B (where the B represents background), a junction is formed at the point, x_j, where $N(x_j, t) = N_B$. If N_B is several orders of magnitude less than N_0, the junction location can be approximated by finding the point at which $N(x, t)$ goes to zero, or $Y - 1$. For this condition,

$$\frac{x_j}{2(D_i t)^{1/2}} = 1.225\left(\frac{N_0}{n_i}\right)^{1/2} \tag{7-13}$$

This result provides a means of estimating the effective constant diffusion coefficient that would result in a junction at the same location if a complementary error function profile were assumed. From Figure 7-4, a value of $z = x_j/2(Dt)^{1/2}$ can be determined for a particular ratio of N_0 to N_B. Then the effective diffusion coefficient for a complementary error function approximation, D_{effE}, is given by

$$D_{\text{effE}} = \left(\frac{1.225}{z}\right)^2 \left(\frac{N_0}{n_i}\right) D_i \tag{7-14}$$

For Gaussian diffusions, where the initial surface concentration is high, but the final surface concentration is on the order of or below the intrinsic carrier concentration, the effective diffusion coefficient, D_{effG}, may be approximated by

$$D_{\text{effG}} = h(N_{01}) D_i \tag{7-15}$$

where N_{01} is the initial surface concentration. For most cases, $h(N_{01}) \simeq 2$.

If a boron diffusion is performed in an oxidizing atmosphere, the diffusion coefficient is altered. There is also a dependence on crystal orientation. Empirical curves for these alterations are shown in Figure 7-5.

Arsenic Diffusion Coefficients. Arsenic also exhibits a concentration dependent diffusion coefficient, resulting in impurity profiles that do not conform to complementary error functions or Gaussians. The arsenic diffusion coefficient is given by

$$D = h\left[D_i^0 + D_i^-\left(\frac{n}{n_{ie}}\right)\right] \tag{7-16}$$

Note that, due to the possibility of bandgap narrowing for heavily doped materials, an effective intrinsic carrier concentration, n_{ie}, is used in this expression. A plot of n_{ie} as a function of temperature for solid solubility limited arsenic is included in Figure 7-6. The electron density, n, may be less than the doping density since not all of the arsenic atoms in heavily doped materials are electrically active. The solid solubility limits and electrically active limits for boron, arsenic, and phosphorus are shown in Figure 7-7.

Since most arsenic diffusions are performed at high temperatures, due to the relatively low-diffusion coefficient, both of the diffusion coefficients are essentially equal, and the equation for D can be written (1)

$$D \simeq hD_i^-\left(1+\frac{n}{n_{ie}}\right) \tag{7-17}$$

Because of the simplicity of this equation, it is possible to obtain a polynomial approximation to the diffusion profile given by

$$N(x, t) \simeq N_0(1-0.87Y-0.45Y^2) \tag{7.18}$$

where $Y=(x^2 n_{ie}/8N_0D_i t)^{1/2}$ for a constant source diffusion. Using the same technique that was used in the boron example, an effective diffusion coefficient can be obtained, resulting in

$$D_{\text{effE}} = \left(\frac{1.146}{z}\right)^2\left(\frac{N_0}{n_{ie}}\right)D_i \tag{7-19}$$

D_i for arsenic is also shown in Figure 7-5. For a Gaussian profile with a high initial surface concentration and a moderate final surface concentration,

$$D_{\text{effG}} = h(N_{01})D_i \tag{7-20}$$

The diffusion coefficient for arsenic is independent of crystal orientation and ambient.

Phosphorus Diffusion Coefficients. Phosphorus exhibits a more complicated anomalous diffusion profile than that of boron or arsenic. Some typical shallow diffusion profiles for phosphorus are shown in Figure 7-8. Note that the electrically active phosphorus is limited to approximately $5\times10^{20}\,\text{cm}^{-3}$ at 950°C. This represents the electron concentration at the surface, n_s, and is different from the total concentration at the surface, N_0. There is another

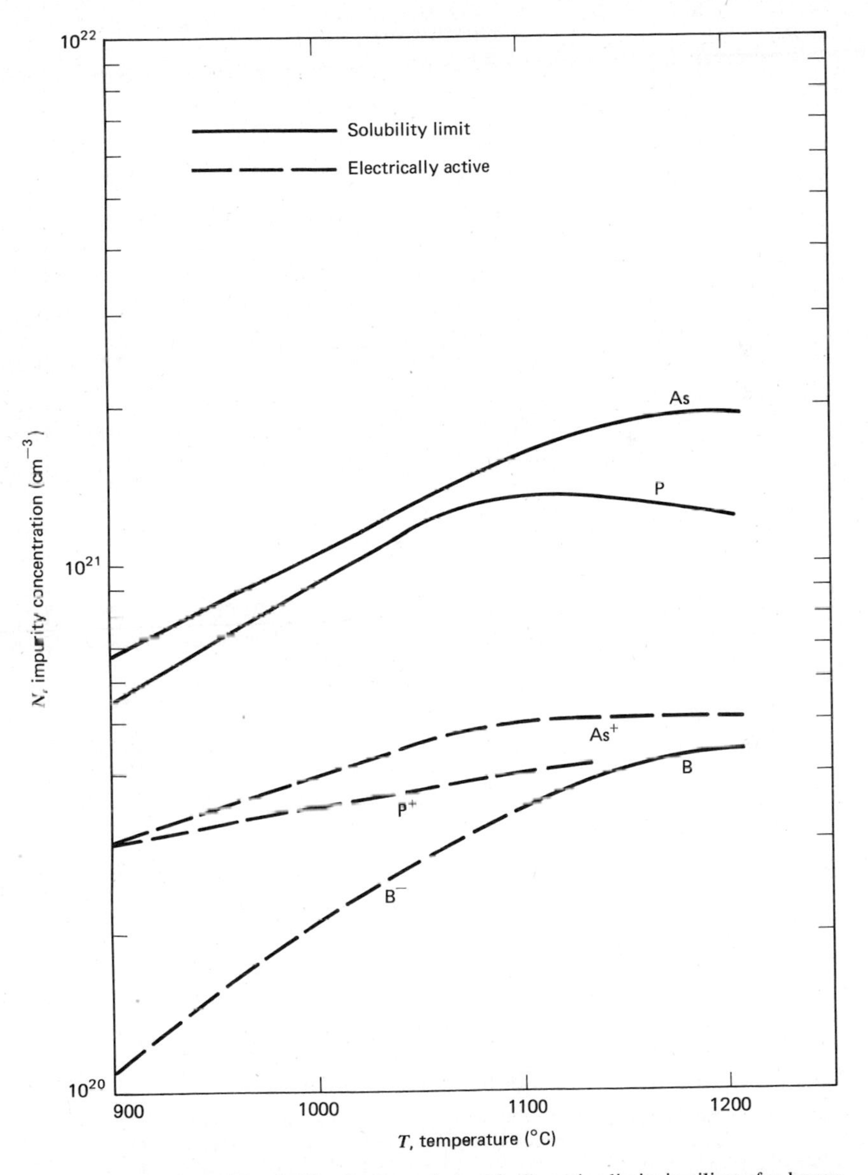

Figure 7-7. The solid solubility limits and electrically active limits in silicon for boron, phosphorus, and arsenic (1). This figure was originally presented at the Spring 1977 Meeting of The Electrochemical Society, Inc., held in Philadelphia, Pennsylvania.

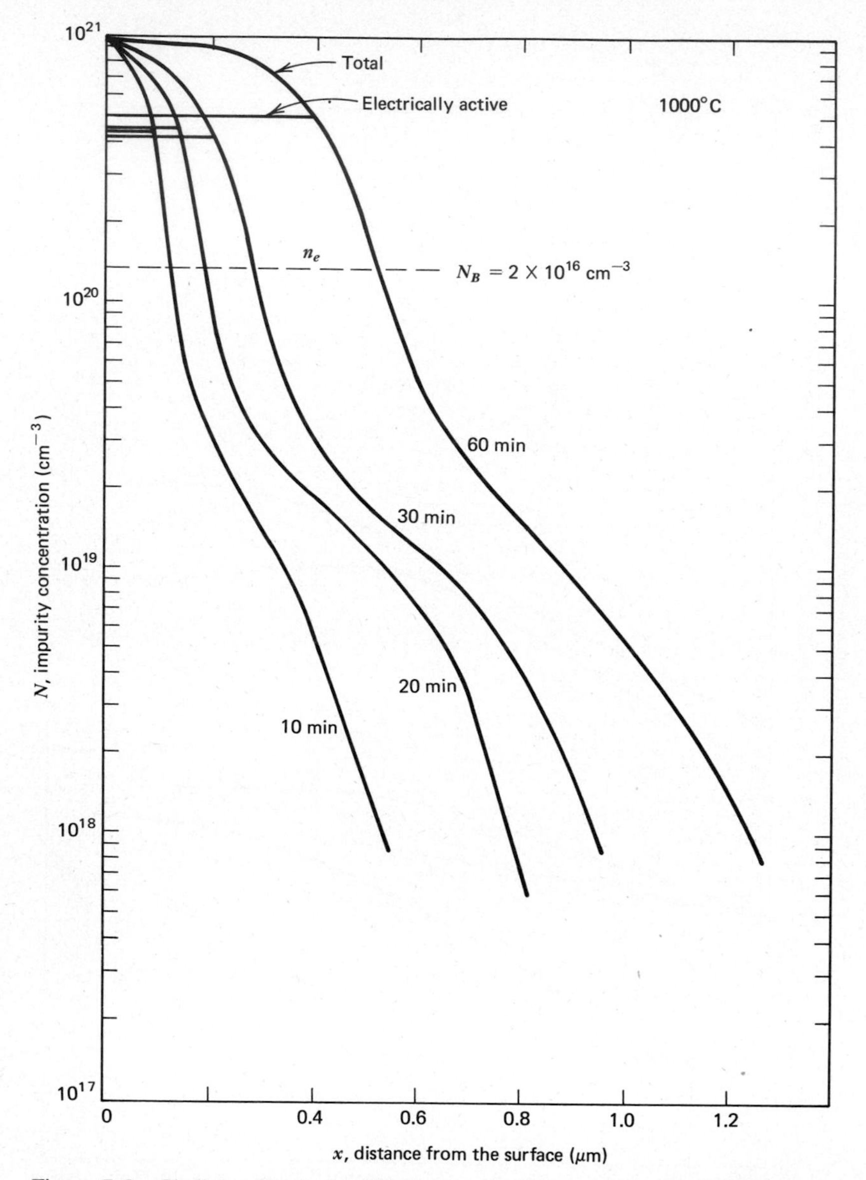

Figure 7-8. Shallow phosphorus diffusion profiles for constant source diffusions at 950°C (12). Copyright © 1969 IEEE. Reprinted from *Proceedings of the IEEE*, September 1969, p. 1503.

electron density, n_e, indicated on the figure. This is the electron density at which the Fermi level has dropped to 0.11eV below the conduction band edge. This quantity can be calculated from the expression (3)

$$n_e(T) = 4.65 \times 10^{21} \exp\left(-\frac{0.39}{kT}\right) \tag{7-21}$$

and is plotted in Figure 7-9. The significance of the energy level associated with n_e is that the phosphorus-doubly charged vacancy pairs dissociate at this energy, leaving an abundance of vacancies to enhance the diffusion coefficient for concentrations below n_e. This results in a "kink" in the diffusion profile and a significant increase in the diffusion coefficient. For electron concentrations above n_o, the diffusion coefficient for phosphorus is given by (4)

$$D = h\left[D_i^0 + D_i^=\left(\frac{n}{n_{ie}}\right)^2\right] \tag{7-22}$$

where n_{ie} is the effective intrinsic carrier concentration enhanced by bandgap narrowing. For electron concentrations below n_e, the diffusion coefficient, D_{TAIL}, is given by (4)

$$D_{TAIL} = D_i^0 + \left(\frac{n_s^3}{n_e^2 n_{ie}}\right)\left[1 + \exp\left(\frac{0.3}{kT}\right)\right]D_i^- \tag{7-23}$$

which is large compared to the diffusion coefficient for densities above n_e.

A model has been presented by Tsai (12) for calculating phosphorus diffusion profiles. The basic geometry for this model is shown in Figure 7-10. The region of constant impurity concentration, denoted by x_0, is assumed to increase linearly with time so that

$$x_0 = \alpha t \tag{7-24}$$

where α is the growth rate of the surface layer phase. The impurity concentration, $N(x, t)$, is given by

$$N(x, t) = N_s \qquad \text{for} \quad x \leqq x_0 \tag{7-25}$$

$$N(x, t) = N_1(x, t) + N_2(x, t) \qquad \text{for} \quad x > x_0 \tag{7-26}$$

where

$$N_1(x, t) = \frac{1}{2}\left(1 - \frac{N_2(x_0)}{N_s}\right)N_s \exp\left[-\left(\frac{\alpha}{2D_1}\right)(x - \alpha t)\right]F_1(x, t) \tag{7-27}$$

and

$$N_2(x, t) = \frac{N_2(x_0)}{2} \exp\left[-\left(\frac{\alpha}{2D_2}\right)(x - \alpha t)\right]F_2(x, t) \tag{7-28}$$

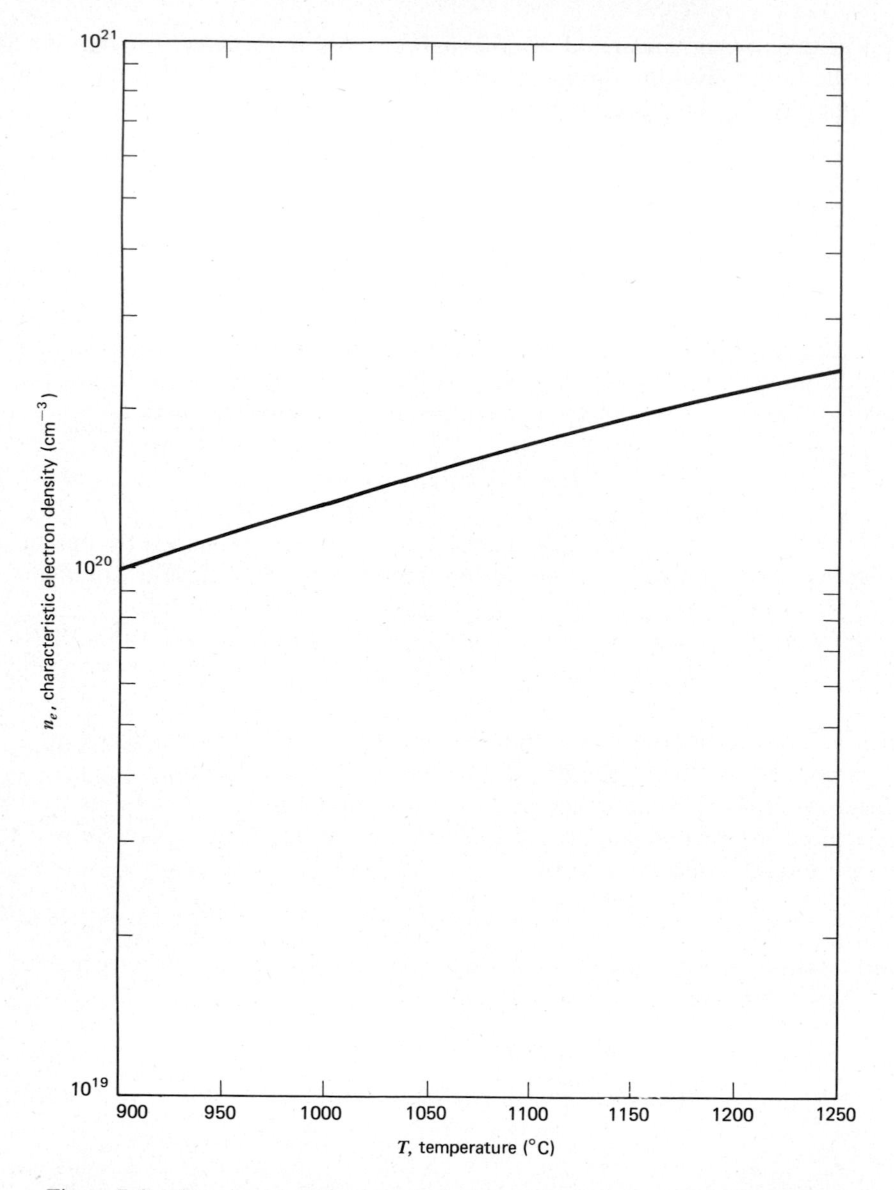

Figure 7-9. The characteristic electron density associated with the "kink" formation in phosphorus diffusions (3).

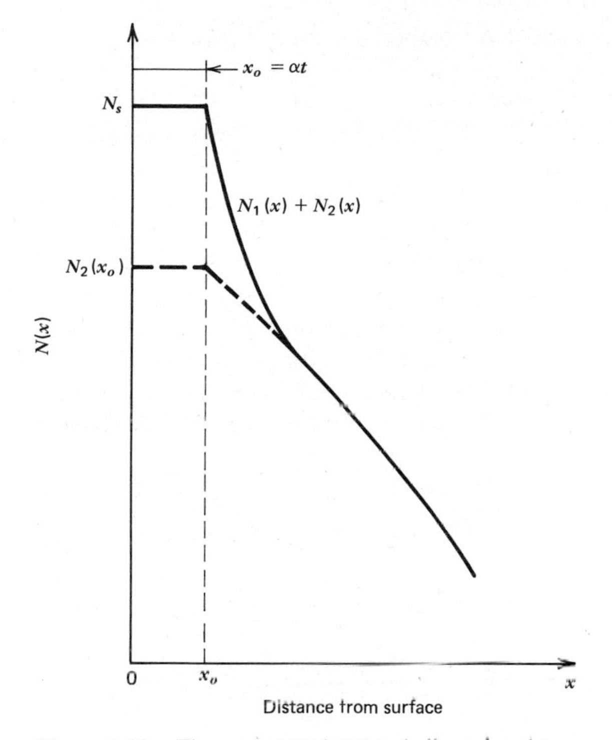

Figure 7-10. The geometry for the shallow phosphorus diffusion model (12). Copyright © 1969 IEEE. Reprinted from *Proceedings of the IEEE*, September 1969, p. 1500.

where

$$F_1(x, t) = \text{erfc}\left[\frac{(x+\alpha t)}{(2\sqrt{D_1 t})}\right] + \text{erfc}\left[\frac{(x-3\alpha t)}{(2\sqrt{D_1 t})}\right] \tag{7-29}$$

$$F_2(x) = \text{erfc}\left[\frac{(x+\alpha t}{(2\sqrt{D_2 t})}\right] + \text{erfc}\left[\frac{(x-3\alpha t}{(2\sqrt{D_2 t})}\right] \tag{7-30}$$

The quantities α, D_1, D_2, and $N_2(x_0)$ are assumed to be temperature dependent with the form

$$R = R_0 \exp\left(-\frac{E_A}{kT}\right) \tag{7-31}$$

where E_A is an activation energy. Experimentally determined values for these quantities are given in Table 7-2.

This model gives an accurate prediction of the phosphorus impurity profile. Since the location of the emitter-base junction is important in determining the frequency response characteristics of bipolar transistors, this model

***Table* 7-2. Phosphorus Profile Quantities**

Quantity	R_0	E_A
α	$0.18\ \text{cm} \cdot \text{s}^{-1}$	1.75 eV
D_1	$49.3\ \text{cm}^2 \cdot \text{s}^{-1}$	3.77 eV
D_2	$2.49 \times 10^{-5}\ \text{cm}^2 \cdot \text{s}^{-1}$	2.0 eV
$N_2(x_0)$	$3.95 \times 10^{23}\ \text{cm}^{-3}$	0.9 eV

should be used to estimate the emitter doping profile. Clearly this model does not lead to a simple method for estimating the diffusion coefficient. If, however, diffusions are limited to temperatures below 1000°C, where D_2 is the dominant diffusion coefficient, Equation 7-27 can be approximated in the vicinity of the junction by

$$N(x_j, t) \simeq \frac{N_2(x_0)}{2} \exp\left[-\left(\frac{\alpha}{2D_2}\right)\left(\frac{3x_j}{4}\right)\right] \operatorname{erfc}\left[\frac{x_j - 3\alpha t}{2(D_2 t)^{1/2}}\right] \tag{7-32}$$

which can be solved for t by finding z_1 from

$$\operatorname{erfc} z_1 = \frac{2N(x_j, t)}{N_2(x_0)} \exp\left[\left(\frac{\alpha}{2D_2}\right)\left(\frac{3x_j}{4}\right)\right] \tag{7-33}$$

and inserting it in

$$t \simeq \left(\frac{3\alpha x_j + 2D_2 z_1^2}{9\alpha^2}\right) + \left[\left(\frac{3\alpha x_j + 2D_2 z_1^2}{9\alpha^2}\right)^2 - \left(\frac{x_j^2}{9\alpha^2}\right)\right]^{1/2} \tag{7-34}$$

Phosphorus exhibits an enhancement of the diffusion coefficient under oxidation conditions, 1.8 for dry and 3.3 for steam. The orientation dependence is only present under oxidizing conditions with the (100) diffusion coefficient being 1.8 times that for (111) crystal orientation.

Interacting Diffusions. It has been observed that diffusions involving two or more impurity species are not independent of one another. In particular, the sequential diffusions for establishing the base and emitter regions of a bipolar transistor indicate that the base-collector junction under the emitter region is not at the same depth as the base-collector junction outside of the emitter region. For a phosphorus emitter and a boron base, the base-collector junction is deeper under the emitter. This is called enhanced diffusion under the emitter, emitter "push," or emitter "dip." For an arsenic emitter and a boron base, the base-collector junction is shallower under the emitter. This is called emitter retardation or emitter "pull."

Enhanced diffusion under the emitter is due to the same mechanism as the enhancement of the "tail" diffusion coefficient for phosphorus. The dissociation of phosphorus-doubly charged vacancy pairs releases vacancies for boron diffusion also. As a result, that portion of the boron diffusion which occurs during the phosphorus diffusion does so with a diffusion coefficient given by

$$D = D_{iB}^{0}\left(\frac{D_{TAIL}}{D_{iP}^{0}}\right) \tag{7-35}$$

Since this enhancement factor can be on the order of 100, this can be a significant effect. It has also been suggested (3) that a buried layer under the emitter should experience approximately the same enhancement.

Emitter retardation occurs because of the formation of arsenic-vacancy complexes, reducing the concentration of vacancies available for boron diffusion. This effect is much less than the enhancement due to phosphorus. Arsenic-vacancy complexes do not occur if ion implantation is used as the source for the diffusion, rather than the conventional chemical sources. Under these conditions, emitter retardation is not observed.

The Two-Step Diffusion

Junctions are formed at locations where the total concentration of donor atoms is equal to the total concentration of acceptor atoms within the semiconductor. If a p-type impurity with a concentration $N(x, t)$ is diffused into a substrate with a uniform n-type dopant with a background concentration N_B, the junction occurs when

$$N(x_j, t) = N_B \tag{7-36}$$

where x_j is the distance from the surface to the junction.

An important parameter of a diffused layer is the sheet resistance. Data have been obtained for Gaussian and complementary error function diffused layers for both p- and n-type diffusions with background concentration as a parameter. These data are shown in Figures 7-11 through 7-14. From these curves, the surface concentration for a particular sheet resistance-junction depth product can be obtained.

A popular technique for obtaining a desired impurity profile is the two-step diffusion. The first step is a very shallow constant source diffusion called a predeposition. The second step uses the result of the predeposition as the source for a limited-source diffusion, usually in an oxidizing atmosphere, called the drive-in.

During the predeposition, the quantity of impurity, Q, transported through the surface during the time interval t_1, is given by

$$Q = 2N_{01}\left(\frac{D_1 t_1}{\pi}\right)^{1/2} \tag{7-37}$$

where N_{01} is the surface concentration maintained during the constant source

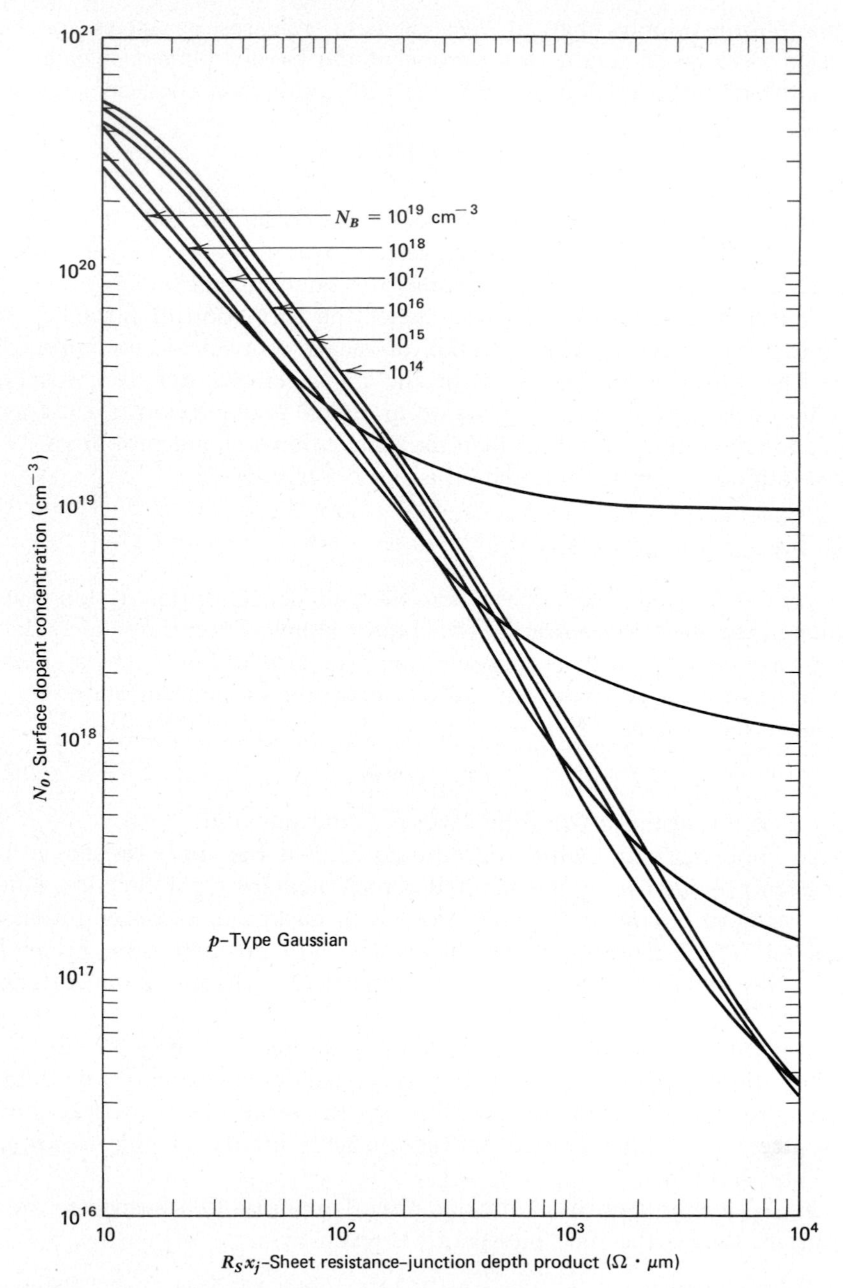

Figure 7-11. The surface dopant density of a p-type Gaussian diffusion in uniformly doped n-type silicon as a function of average resistivity at 300°K [adapted from Irvin (34)]. Copyright © 1962 American Telephone and Telegraph Company. Reprinted by permission from *The Bell System Technical Journal.*

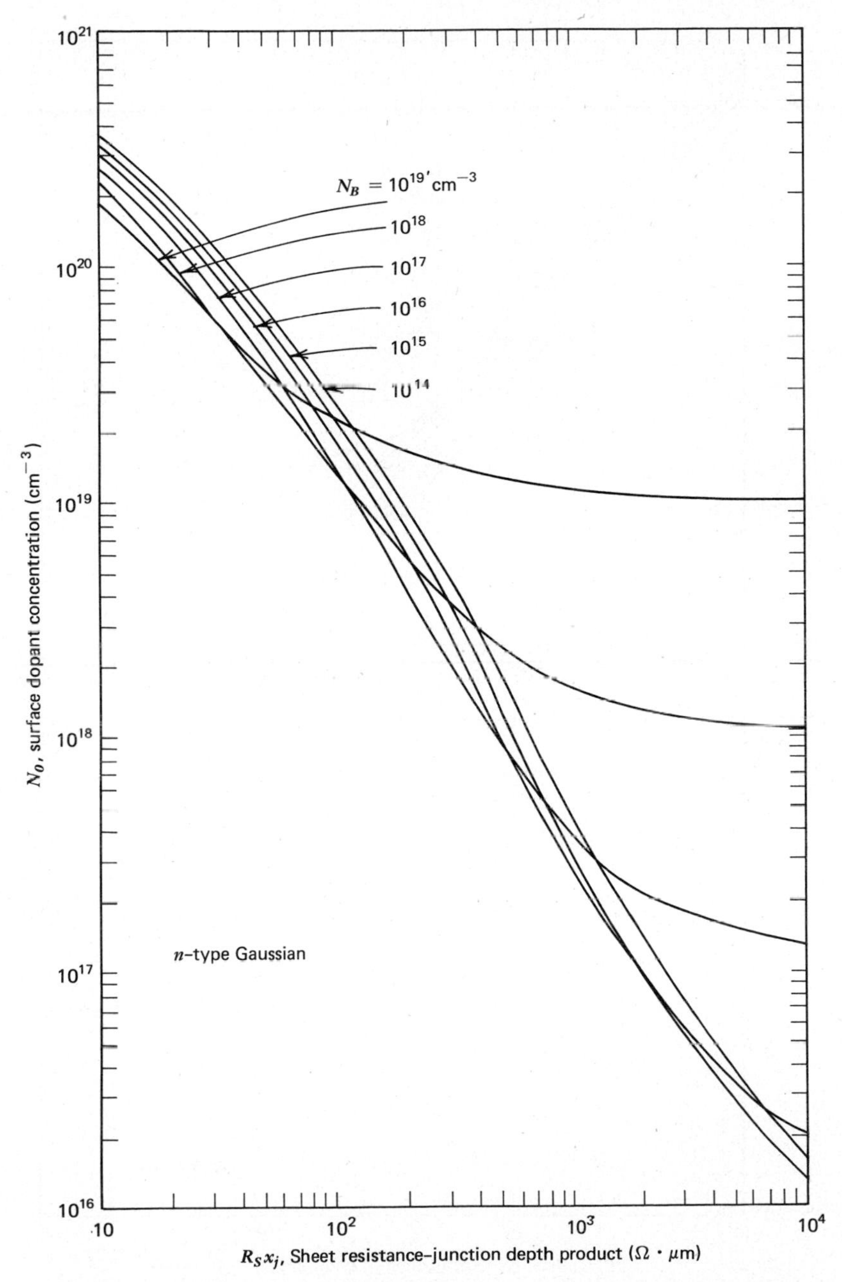

Figure 7-12. The surface dopant density of an n-type Gaussian diffusion in uniformly doped p-type silicon as a function of average resistivity at 300°K [adapted from Irvin (34)]. Copyright © 1962 American Telephone and Telegraph Company. Reprinted by permission from *The Bell System Technical Journal.*

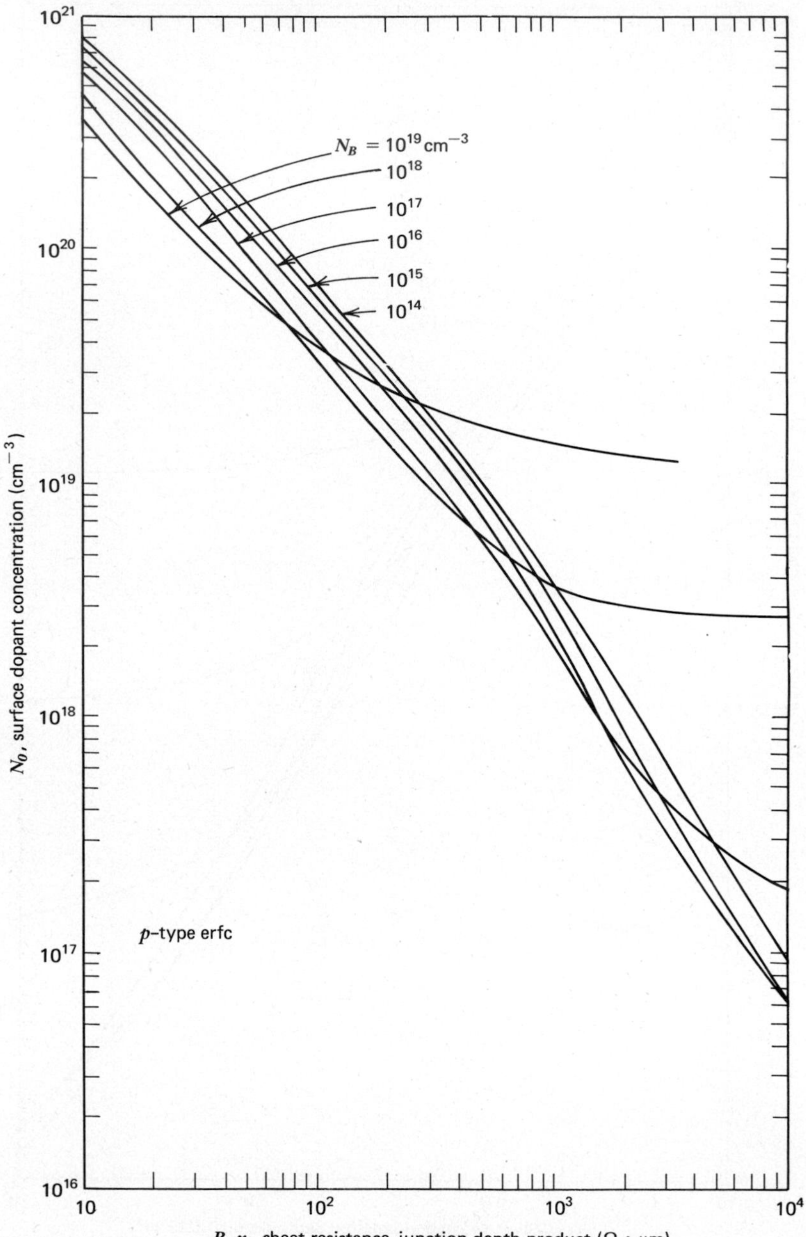

Figure 7-13. The surface dopant density of a p-type erfc diffusion in uniformly doped n-type silicon as a function of average resistivity at 300°K [adapted from Irvin (34)]. Copyright © 1962 American Telephone and Telegraph Company. Reprinted by permission from *The Bell System Technical Journal*.

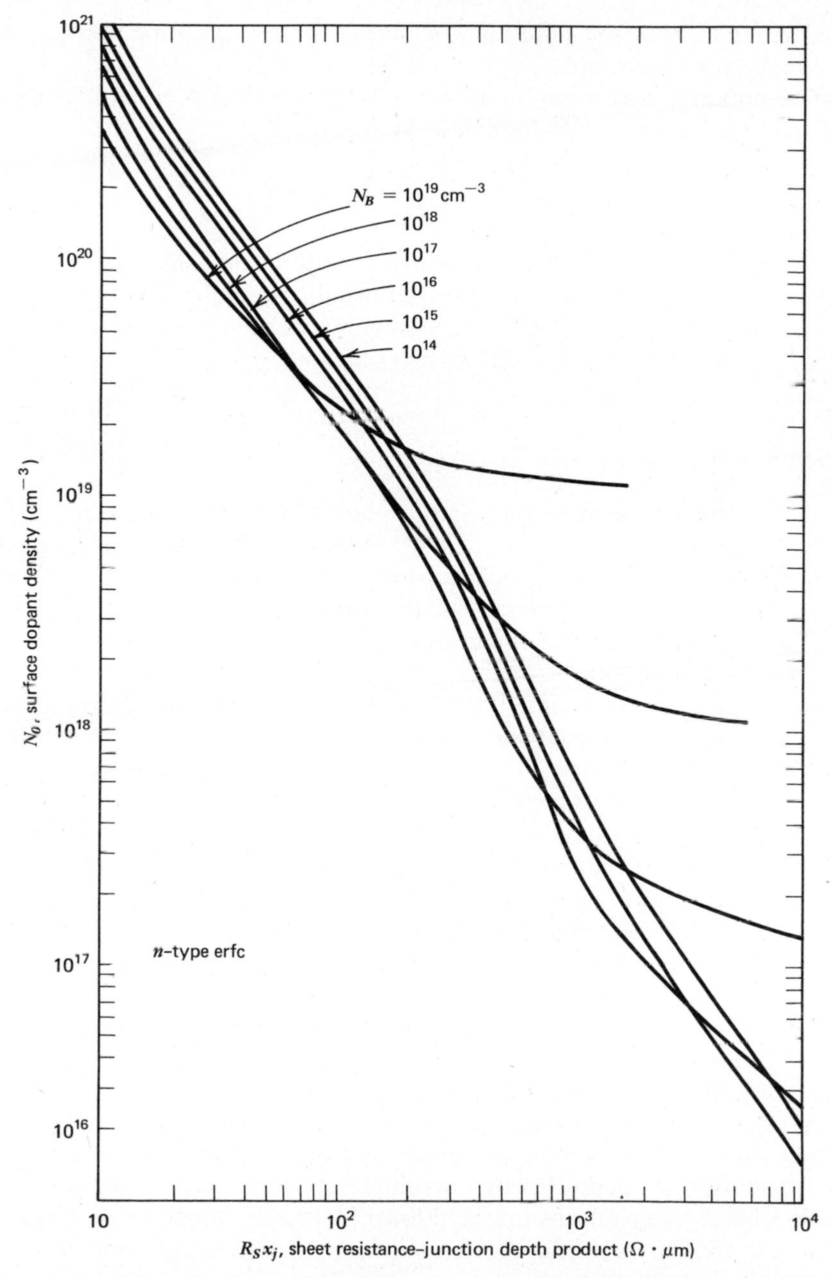

Figure 7-14. The surface dopant density of an n-type erfc diffusion in uniformly doped p-type silicon as a function of average resistivity at 300°K [adapted from Irvin (34)]. Copyright © 1962 American Telephone and Telegraph Company. Reprinted by permission from *The Bell System Technical Journal.*

diffusion, and D_1 is the diffusion coefficient for the diffusing species at the predeposition temperature.

The impurity concentration during the drive-in step is found by substituting Equation 7-37 into Equation 7-6. The result is

$$N(x, t_1, t_2) = \left(\frac{2N_{01}}{\pi}\right)\left(\frac{D_1 t_1}{D_2 t_2}\right)^{1/2} \exp\left(-\frac{x^2}{4D_2 t_2}\right) \tag{7-38}$$

where the subscript "2" refers to the drive-in diffusion. This equation is only applicable when $D_1 t_1 \ll D_2 t_2$, which is the usual case, since the predeposit step is usually performed at a much lower temperature than the drive-in step, and, therefore, with a much lower diffusion coefficient.

Surface Concentration Control

The most reliable method for establishing a surface concentration for a constant source diffusion is to provide an impurity concentration greater than the solid-solubility limit in the vicinity of the wafer so that the wafer will accept the solid-solubility concentration at the diffusion temperature. The solid-solubility limits for important impurities in silicon as a function of temperature are shown in Figure 3-6.

As indicated in the section on diffusion systems later in this chapter, most of the impurity sources used for diffusions are oxides of the impurity. These oxides may be in solid form, in which case, the temperature of the source can be controlled to produce a controlled concentration. In many cases, the oxide is formed in the diffusion tube by providing a controlled amount of oxygen in the gas stream along with a compound of the impurity. If the ratio of oxygen to impurity compound is varied, concentrations below the solid-solubility limit can be obtained. Mass flowmeters should be used for this process. A significant difficulty encountered with controlled impurity concentrations is the strong possibility of doping of the furnace tube. If precautions are not taken, the furnace tube will become an additional source of impurity, and the concentration will become solid-solubility limited. An HCl cleaning cycle can be used to reduce the effects of furnace tube contamination.

A more reliable technique for limiting the surface concentration for a constant source diffusion is to make use of a doped oxide as the source. The doped oxides can be deposited by low-temperature chemical vapor techniques, or from spin-on preparations in liquid form. A thin, undoped oxide layer, such as that formed by an oxidizing cleaning step in nitric acid, can be used to further reduce the surface concentration (5). Doped oxides have been used to obtain phosphorus surface concentrations controllable over the range from 5×10^{17} to 2×10^{20} cm^{-3}. The substitutional impurities in silicon are network formers in silicon dioxide. If the impurity concentration in the oxide is too large, approximately 10^{19} cm^{-3} for boron, the doped oxide will be converted to a glass at the diffusion temperature resulting in a different mechanism for impurity transfer from the layer into the silicon.

Ion implantation is sometimes used as an alternative to the constant-source step of a two-step diffusion. The concentration control afforded by ion implantation is better than that which can be obtained by diffusion.

Diffusion Masks

Integrated circuits are made by the selective doping of areas of a wafer to fabricate planar devices. This is made possible by the use of materials that act as masks for the deposition of impurities. The most popular of the masking materials is thermally grown silicon dioxide. Diffusion windows are created in the oxide by photoengraving. The thickness of the oxide necessary to provide an effective mask depends on the diffusing species, the temperature of the diffusion, and the time at the high temperature during the process.

The diffusion coefficients for boron, arsenic, and phosphorus in silicon dioxide are two to three orders of magnitude lower than the corresponding values in silicon. Other impurities, like gallium, indium, and aluminum have diffusion coefficients in silicon dioxide of the same order of magnitude as their corresponding values in silicon. Silicon dioxide is not an effective mask for these latter impurities, and care must be exercised to prevent process contamination from these relatively fast diffusing species.

The mechanism of diffusion of boron and arsenic through the amorphous thermal oxide on a silicon wafer is much different from the mechanism of diffusion in single-crystal silicon. If there is a sufficient concentration of impurity atoms present, approximately 10^{19} cm^{-3}, the oxide is converted into a doped glass. The conversion process starts at the surface of the oxide and gradually proceeds until the entire oxide is converted. At this point, the doped glass becomes a source for impurity diffusion into the silicon wafer, and is no longer a mask. The minimum oxide thicknesses to provide diffusion masks for boron and phosphorus are shown in Figure 7-15. The diffusion coefficient for arsenic in silicon dioxide is approximately an order of magnitude lower than that of boron. Thus, the same oxide thicknesses that will serve for masks for boron will also serve for arsenic. The diffusion coefficient for antimony in silicon dioxide shows a stronger temperature dependence than those of the other dopants. It is readily masked for temperatures up to 1050°C but requires a thicker masking oxide than phosphorus for temperatures above 1150°C.

Diffusion masks can also be formed using deposited layers of silicon dioxide or silicon nitride. Deposited oxides are usually less dense than thermal oxides and it is customary to use a thicker layer for a particular diffusion condition than would be required for a thermally grown oxide. Silicon nitride is an excellent diffusion mask, but the additional processing steps required for the

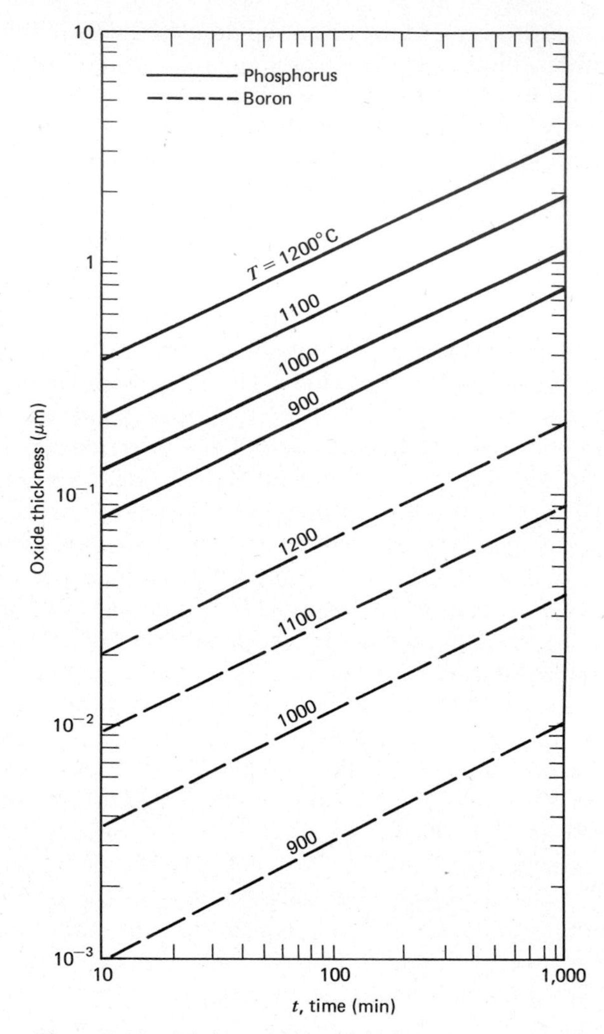

Figure 7-15. Minimum oxide thickness to serve as a diffusion mask for phosphorus and boron diffusions (35).

deposition and etching of the nitride layer have resulted in limited use of silicon nitride masks.

The mask thickness necessary for a constant source diffusion is significantly larger than that required for a two-step diffusion to achieve the same impurity profile. The mask must prevent diffusion during the time that the wafer is exposed to the impurity source, which is the entire process for a constant source diffusion, and only the predeposit step in the two-step diffusion. Unless a doped oxide is used as a source, both sides of the wafer are exposed to the impurity source. In some cases, it is undesirable to permit a diffusion into the backside of the wafer. Under these circumstances, it is necessary to have a protective oxide on the back of the wafer. This oxide is formed during the thermal oxidation cycle, but will be removed, unless it is protected, during the process of opening the diffusion windows on the front side of the wafer. The backside oxide can be protected by applying a photoresist or wax coating on the back after the pattern has been developed on the front.

The actual dimensions of the diffused region are slightly larger than the windows in the mask due to lateral diffusion. The boundary conditions for lateral diffusion are different from those for vertical diffusion, resulting in lateral junctions locations that are slightly less than those measured vertically from the surface. Figure 7-16 provides the information necessary to determine the extent of the lateral diffusion associated with a particular vertical diffusion.

Lateral diffusion has several effects on the electrical properties of the junctions formed by selective diffusions. The most important result of the lateral diffusion is that the intersection of the junction with the surface of the silicon occurs under a protective layer of silicon dioxide. This provides a means for forming junctions with reproducible reverse breakdown characteristics.

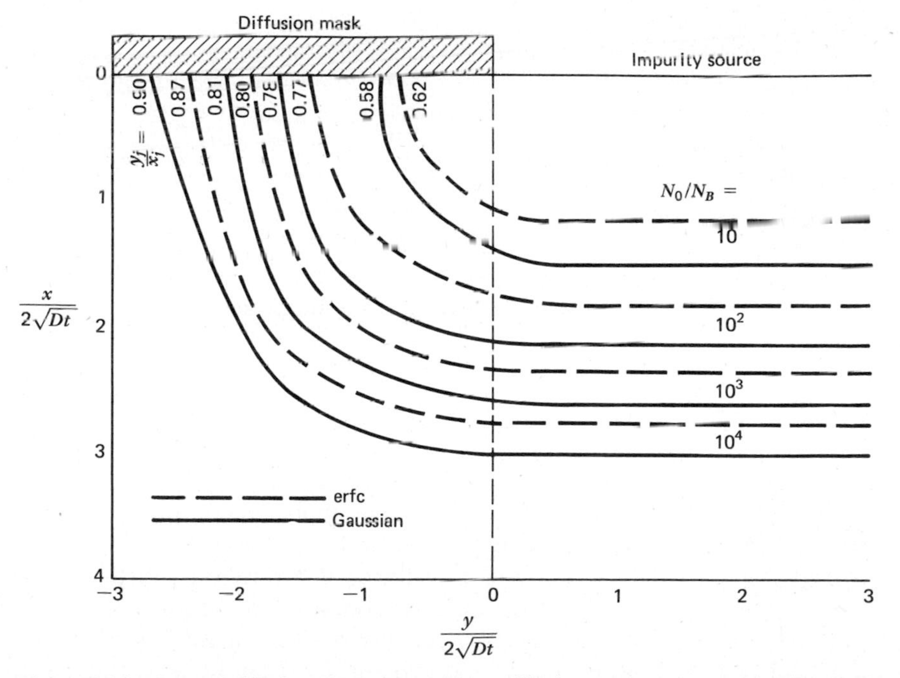

Figure 7-16. Lateral diffusion dimensions under a diffusion mask as compared to the depth of the diffusion (36). Copyright © 1965 by International Business Machines Corporation; reprinted with permission.

Unprotected junctions can have breakdown characteristics that are dominated by surface conditions. Another result of lateral diffusions is that the distance along the surface between diffused regions can be made smaller than the minimum dimensions that can be defined by the photoengraving process. This can be used to improve the characteristics of lateral *pnp* transistors. An undesirable side effect of lateral diffusions is the increase in the capacitance associated with the junction due to the increased area.

Oxidation During Diffusion (7, 9, and 10)

It is common practice to combine the drive-in step of a two-step diffusion with an oxidation. This serves the dual purpose of providing an oxide layer for the next photoengraving process and creating steps in the surface of the silicon wafer, since the oxide growth rate in the window area is much greater than that under the already oxidized regions, to permit alignment of patterns. The oxidation causes a redistribution of the impurity at the surface, and, thus, has an effect on the impurity profile. Boron is readily incorporated into the growing oxide, depleting the surface concentration, and phosphorus is rejected by the growing oxide, increasing the surface concentration.

If a steam oxidation is used for a boron drive-in, the rapid oxide growth can remove up to 80% of the boron from the predeposit. For this reason, it is common practice to use an inert atmosphere or dry oxygen during the early stages of boron drive-in diffusions, followed by a steam cycle to achieve the necessary oxide thickness during the latter stages of the process. A typical boron impurity profile after an oxidizing drive-in cycle is shown in Figure 7-17.

The rejection of phosphorus during oxidation results in a buildup of the surface concentration of the impurity. The result, called the "snow-plow" effect, is an apparent reduction of the activation energy for the diffusion from 3.6 to 2.5 eV, which represents a significant enhancement of the diffusion coefficient.

Impurity Redistribution During Epitaxial Growth (12, 13)

The impurity profile in an epitaxial layer depends on the intentionally added impurity during the growth process and the impurity concentration in the substrate. If the substrate is heavily doped, as in the case of discrete bipolar transistors where a lightly doped n-type layer is grown on a heavily doped n-type substrate, there is a significant autodoping from the substrate. The autodoping occurs from three sources; evaporation from the backside of the wafer which is incorporated in the gas stream and deposited in the growing layer, evaporation from the front side of the wafer which occurs during initial growth, and solid-state diffusion from the substrate into the growing layer.

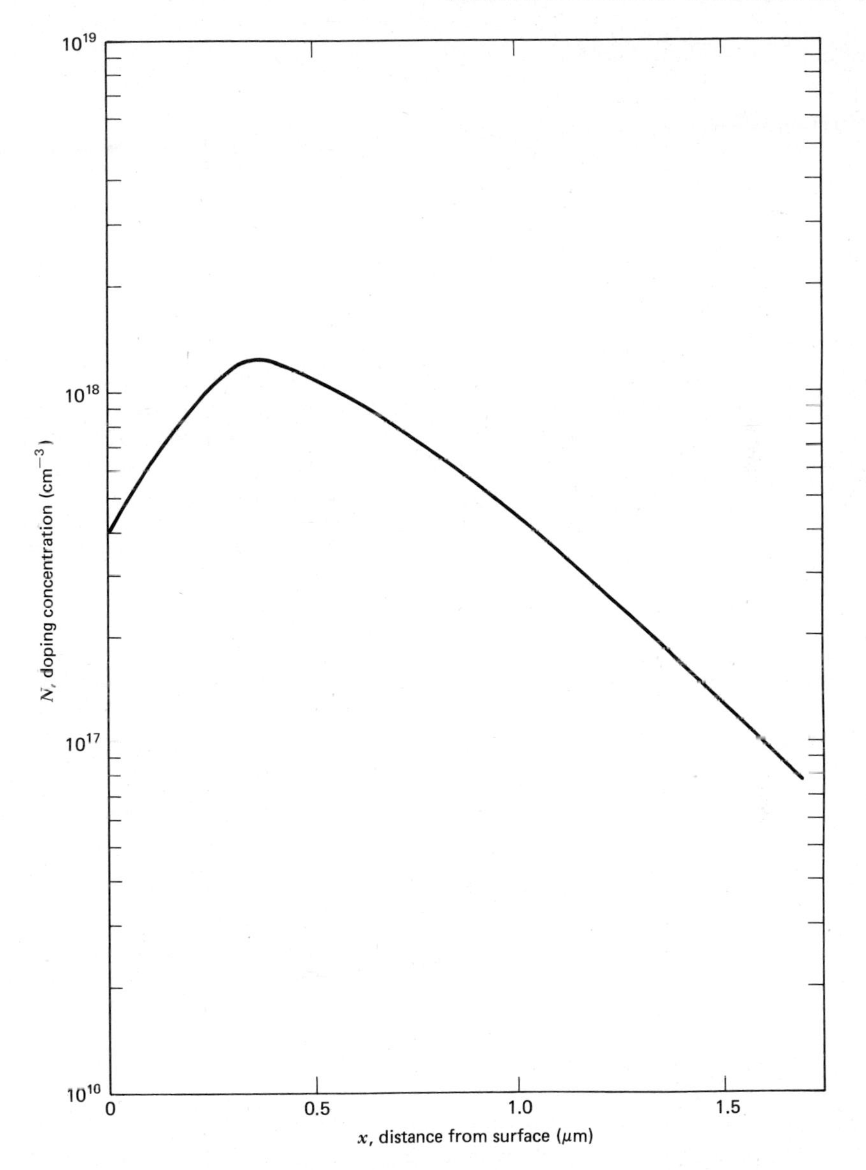

Figure 7-17. The impurity profile of a Gaussian boron diffusion showing the reduction in concentration at the surface due to drive-in in an oxidizing atmosphere (37). Reprinted by permission of the publisher, The Electrochemical Society, Inc.

These effects make it difficult to obtain an abrupt junction for this process. In the case of a junction-isolated bipolar integrated circuit, the substrate is lightly doped p-type and the epitaxial layer is lightly doped n-type, and, while the same effects are present, the doping concentration of the substrate is so much lower that the junction is very nearly abrupt. In most situations, however, heavily doped n-type buried layers are used under the collector regions of the bipolar transistors. These buried layers are formed by selectively diffusing an n-type impurity, usually arsenic or antimony because of their relatively low diffusion coefficients, into regions of the p-type substrate before the epitaxial layer is grown. The buried layer is partially incorporated into the epitaxial layer during the growth process, primarily due to diffusion.

The diffusion of the buried layer into the growing epitaxial layer can be modeled as a process with a moving boundary. The diffusion equation under this situation becomes

$$D\frac{\partial^2 N}{\partial x^2}=\frac{\partial N}{\partial t}+v\frac{\partial N}{\partial x} \tag{7-39}$$

where v represents the rate at which the layer thickness is increasing. The origin is the surface of the layer. This equation can be solved under the two conditions where (1) the impurity concentration at the surface of the layer is N_1 and the impurity concentration in the substrate is initially zero, and (2) the impurity concentration at the surface of the epitaxial layer is essentially zero and the substrate is initially uniformly doped to a uniform concentration N_2. The superposition of the two solutions is a reasonable approximation to the conditions for a buried layer diffusion into a growing epitaxial layer. This solution is

$$\begin{aligned}N(x, t) = \left(\frac{N_1}{2}\right)\Bigg[&\operatorname{erfc}\left(\frac{x-vt}{2\sqrt{D_1 t}}\right) \\ &+\exp\left(\frac{vx}{D_1}\right)\operatorname{erfc}\left(\frac{x+vt}{2\sqrt{D_1 t}}\right)\Bigg] \\ &+\left(\frac{N_2}{2}\right)\left[1+\operatorname{erf}\left(\frac{x-vt}{2\sqrt{D_2 t}}\right)\right]\end{aligned} \tag{7-40}$$

where D_1 and D_2 are the diffusion coefficients of the impurities represented by N_1 and N_2 respectively. Equation 7-40 is plotted for some typical situations in Figure 7-18. The location of the buried layer relative to the base-collector junction is an important consideration in determining the collector-base breakdown voltage for a bipolar transistor.

Diffusion Systems

The physical apparatus used for diffusions is essentially the same as that used for oxidation, as described in Chapter 5. Essentially all production diffusion systems are open-tube arrangements, although it is possible to perform

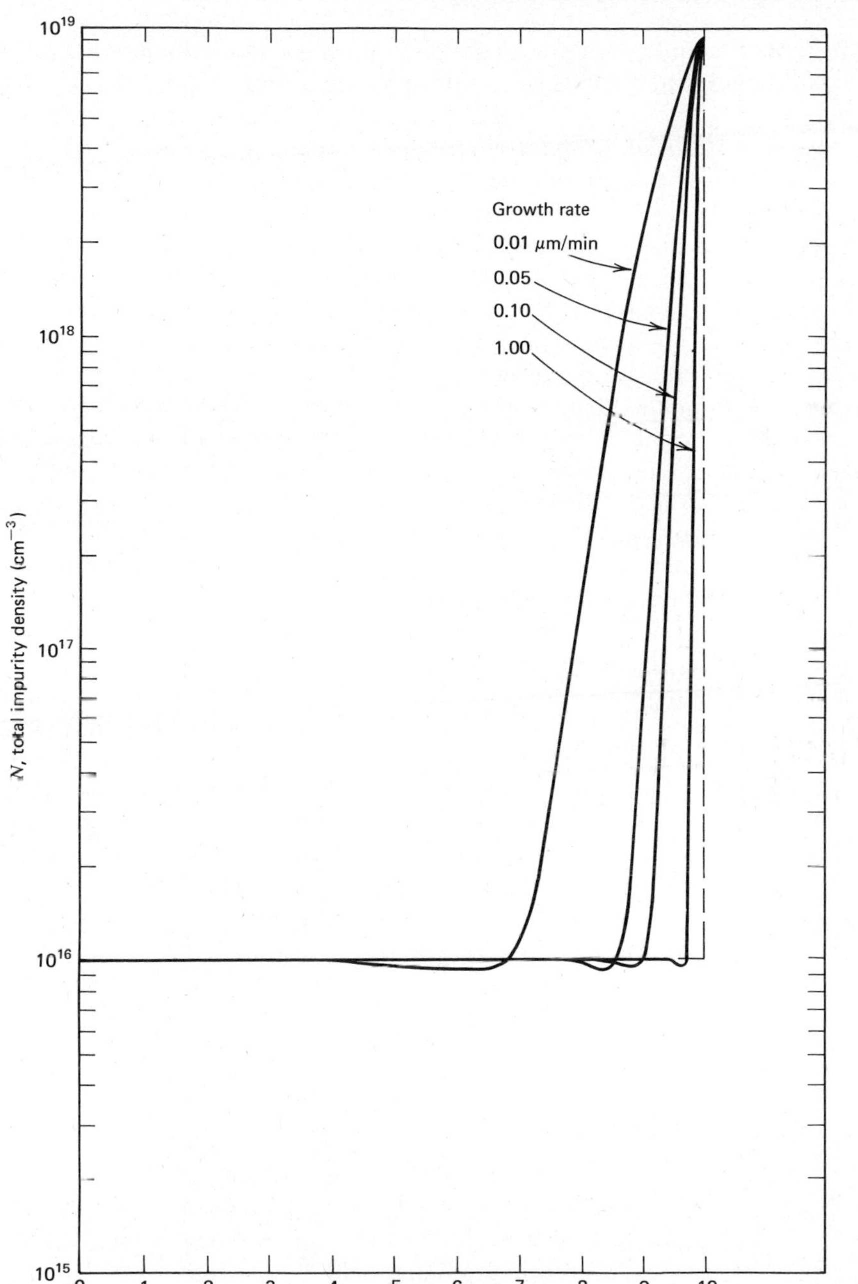

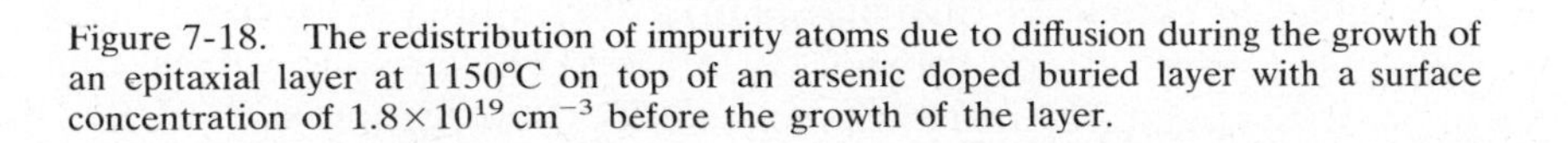
Figure 7-18. The redistribution of impurity atoms due to diffusion during the growth of an epitaxial layer at 1150°C on top of an arsenic doped buried layer with a surface concentration of 1.8×10^{19} cm^{-3} before the growth of the layer.

diffusions at reduced pressures in sealed fused silica tubes. Sources are available in solid, liquid, and gaseous forms for most impurities.

Phosphorus Diffusion Systems. The most popular phosphorus impurity systems use phosphorus oxychloride ($POCl_3$), a liquid, or phosphine (PH_3), a gas, as the impurity source. In both cases the impurity compound is reacted with oxygen to form phosphorus pentoxide (P_2O_5) in the furnace tube. Solid P_2O_5 has been used as a source, with nitrogen as a carrier gas, but it must be maintained at a temperature between 200 and 300°C in a source furnace, and there must be an increasing temperature gradient from the source to the wafers to prevent the deposition of the source material on the apparatus instead of the wafers. Planar solid sources consisting of ceramic wafers containing P_2O_5 and cadium oxide (CaO) have been developed. These wafers are placed in a fused silica carrier in every third slot with silicon wafers in the other slots with each silicon wafer facing a ceramic wafer. A carrier gas is used to transport some of the P_2O_5 to the silicon wafers.

When a liquid source, like $POCl_3$, is used, it is maintained at a constant temperature, typically 20°C, while dry nitrogen is bubbled through the liquid. The gas leaving the bubbler carries sufficient $POCl_3$ to provide the impurity concentration for the diffusion. This gas is mixed with additional nitrogen and a carefully controlled flow of oxygen. The concentration of P_2O_5 formed in the furnace tube depends on the oxygen flow. Most phosphorus diffusions are performed under solid-solubility limit conditions, but reducing the oxygen flow can be used to reduce the surface concentration.

Phosphine is a highly toxic gas, and is available in dilute mixtures with argon, nitrogen, helium, or hydrogen. For diffusion systems, argon is the usual diluent, with concentrations of 2000 to 3000 ppm of phosphine. This mixture is combined with more argon and oxygen before entering the furnace tube. An unfortunate by-product of the reaction is phosphoric acid (HPO_3), which deposits at the cold, exit end of the furnace.

Doped oxide sources, either deposited or spun-on, are diffusion sources that are directly on the wafer. These are processed in nitrogen or nitrogen/oxygen atmospheres.

No matter what the source, the actual phosphorus diffusion occurs from a layer of phosphosilicate glass in contact with the wafer. The concentration of phosphorus pentoxide in the glass depends on the temperature of formation, with a higher concentration occurring for glasses formed below 1030°C. At temperatures above 1120°C the layer becomes liquid. The phosphosilicate glass is rapidly etched in buffered hydrofluoric acid.

Boron Diffusion Systems. Like phosphorus, boron impurity sources are available in many forms. Boron trioxide (B_2O_3) is a popular solid source, that can be used at temperatures from 600 to 1200°C. Boron tribromide (BBr_3) is a liquid source, which is used like phosphorus oxychloride in a temperature controlled bubbler through which nitrogen is passed. Oxygen is reacted with the BBr_3 in the furnace tube to form B_2O_3. The most popular gaseous source for boron is diborane (B_2H_6), a highly toxic material. This is also reacted with oxygen in the furnace tube to form B_2O_3. Solid planar sources are more popular for boron than for phosphorus. One type uses oxidized wafers of hot pressed boron nitride (BN) to provide B_2O_3. Another type consists of glass-ceramic wafers containing B_2O_3, SiO_2, Al_2O_3, MgO, and BaO. Doped oxides, either in deposited or spun-on form are also used for boron diffusions.

As in the case of phosphorus, the actual source for boron diffusions is a layer of borosilicate glass in contact with the silicon surface. This layer is liquid at diffusion temperatures. At temperatures above 1200°C, boron reacts with silicon to form silicides, which are extremely difficult to remove. Pitting of the surface also occurs. Boron glasses are difficult to remove since they etch very slowly in buffered hydrofluoric acid. In some cases, a soak in fluoboric acid (HBF_4) prior to etching increases the etch rate. An alternative approach (17) uses a selective etchant called "R" etch (100 parts 70% HNO_3, 100 parts water, 1 part 49% HF), which etches boron glass very slowly, but at a rate three to six times that at which it etches silicon dioxide, depending on the boron concentration. In a two-step diffusion, the boron glass can be removed after the predeposition step without a photolithographic process using this etchant. It is common practice to perform a brief steam oxidation at the conclusion of a boron predeposit in order to increase the etch rate of the boron glass in buffered hydrofluoric acid.

Arsenic Diffusion Systems. Most compounds containing arsenic are highly toxic, and, for this reason, arsenic diffusions were largely avoided in the industry for a number of years. During this period, antimony was used for buried layers, even though it has a lower solubility in silicon than arsenic and, thus, resulted in higher sheet resistance. Improvements in the techniques for handling toxic gases has made arsine (AsH_3) a viable open-tube arsenic diffusion source. Doped oxides are also in wide use as arsenic sources.

ION IMPLANTATION (18–27)

Ion implantation is a low-temperature technique for the introduction of impurities into semiconductors. It provides a flexibility not available with diffusion. In bipolar transistors, for example, the base can be implanted through the emitter. In MOS transistors, an ion implantation can be used to accurately tailor the threshold voltage. Another interesting application of ion implantation is the formation of high valued resistors with very low-temperature coefficients of resistance.

The designer does not have complete control over the impurity profile from an ion implantation. The ion dose determines the total impurity concentration deposited in the silicon, but the shape and location of the impurity profile depends on the implant energy and the heat treatment subsequent to the implant. Damage to the material limits the energy of the implant such that the peak of the implanted distribution is usually less than 1 μm from the surface. Ion implantation can be used as a source for a diffusion redistribution cycle for deeper impurity profiles.

Implanted Profiles

When energetic ions enter a material, they lose energy by two principal mechanisms, nuclear collisions and electronic interactions. At high energies, electronic interactions predominate. As the ion energy decreases, nuclear collisions become more important, resulting in large deflection angles for incoming ions and displacements of the substrate atoms. If the substrate is crystalline, as is the case in semiconductor wafers, the angle of the entering ion beam, relative to the crystal lattice, has a significant effect on the depth of penetration of the ions. If the angle of the ion beam is within a few degrees of one of the principal crystal directions, a phenomena called "channeling" occurs. The average stopping distance for heavy ions, like arsenic, can be up to 50 times as large under channeling conditions, as compared to that into amorphous materials. For this reason, it is customary for ion implantations to be performed at an angle of 7° from the perpendicular. Under these conditions, the substrate appears amorphous.

When a monoenergetic ion beam enters an amorphous material there is a statistical distribution of the final resting places of the ions. The actual distance traveled by the individual ions is relatively unimportant since, due to deflections, the ions take "zig-zag" paths to their ultimate destinations. The important measure is the average distance traveled parallel to the direction of the beam. This is called the "projected range" and is designated R_p. The other quantities of interest are the "straggle" in the projected range, ΔR_p, and the peak concentration, N_p. The distribution of ions, $N(x)$, takes the form of a Gaussian and is given by

$$N(x) = N_p \exp\left[\frac{-(x-R_p)^2}{2\,\Delta R_p{}^2}\right] \tag{7-41}$$

where x is the distance into the substrate measured from the surface. The total dose of ions, N_s, in ions per centimeters squared, is given by

$$N_s = \int_{-\infty}^{\infty} N(x)\,dx \tag{7-42}$$

which yields

$$N_s = \Delta R_p N_p \sqrt{2\pi} \tag{7-43}$$

Table 7-3. **Properties of the Gaussian**

$N(x)/N_p$	1	0.5	0.1	0.01	0.001	0.0001
$x = R_p \pm$	0	$1.2\,\Delta R_p$	$2\,\Delta R_p$	$3\,\Delta p$	$3.7\,\Delta R_p$	$4.3\,\Delta R_p$

or

$$N_p = \frac{N_s}{\sqrt{2\pi}\,\Delta R_p} = \frac{0.4 N_s}{\Delta R_p} \tag{7-44}$$

From Equation 7-44, it is seen that the peak concentration depends on the total dose and the straggle. The straggle is a function of the ion species and the energy of the implant. The basic properties of the Gaussian are summarized in Table 7-3. The projected range and straggle for Group III and Group V dopants in silicon are shown in Figure 7-19.

The sheet resistance associated with a particular implant can be estimated in the following manner. The approximate carrier concentration, n_o (or p_o), is given by

$$n_o = \frac{N_s}{2.5\,\Delta R_p} \tag{7-45}$$

This value can be used to determine the mobility, μ, of the carriers from Figure 7-3. The sheet resistance, R_s, is calculated from

$$R_S = (qN_s\mu)^{-1} \tag{7-46}$$

The actual sheet resistance observed will be dependent on the annealing cycle. If the annealing temperature is sufficient to activate all of the implanted ions, the sheet resistance will be close to that indicated by Equation 7-46.

Ion Implantation Apparatus

The equipment used for ion implantation is massive and expensive. A typical system is shown in Figure 7-20. The ions are created by means of a confined electric discharge, which is sustained by a vapor of the material to be ionized. The ion beam is extracted from the source and accelerated. The beam passes through a mass analyzing magnet, which separates the desired ions from ions having different masses. The beam is then usually electrostatically deflected to provide a uniform dose to the substrates by a raster scan. Mechanical scanning by moving the substrates is also used, but with less uniform results. The ion beam is focused to a small diameter, which has a Gaussian spatial distribution, and must be scanned in a pattern in which the traces overlap. Overscan is used to improve uniformity. The substrate chamber is evacuated. Implants can be performed at room temperature or with heated substrates. Some heating of the substrates occurs during room temperature implants, depending on the energy deposited during the process.

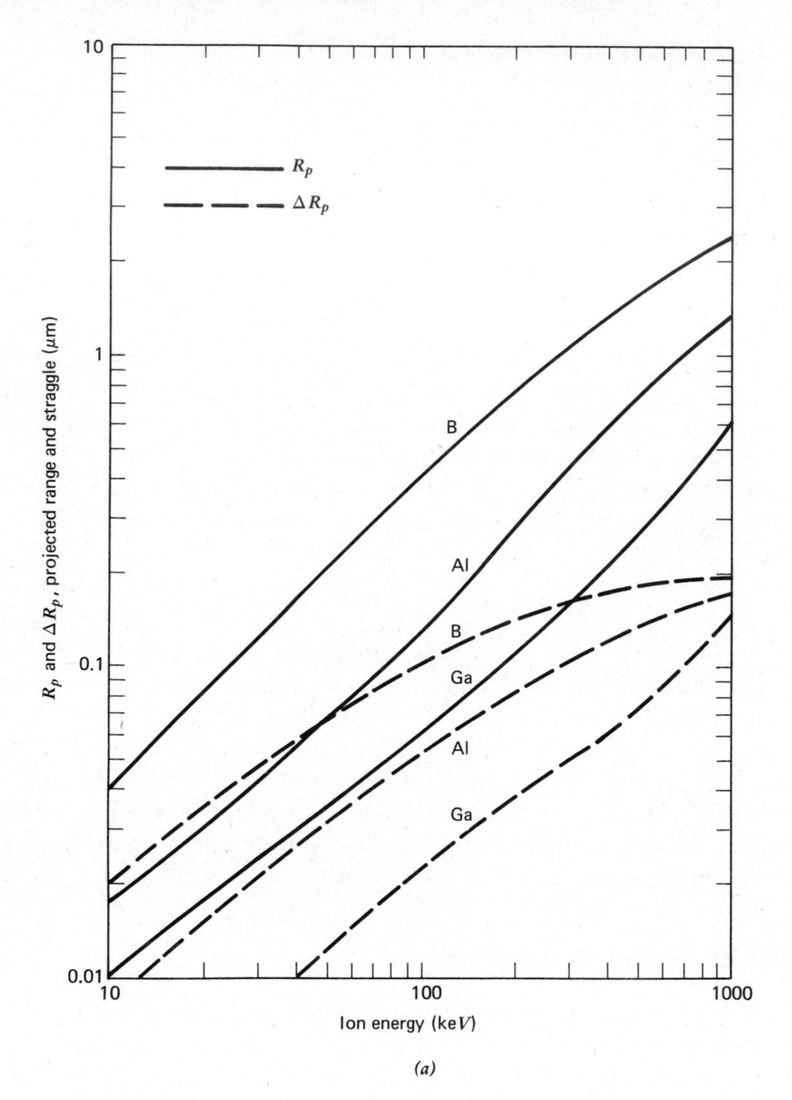

(a)

Figure 7-19. Projected range and straggle as a function of implant energy for (*a*) *p*-type dopants, and (*b*) *n*-type dopants in silicon (26). Reprinted by permission of the Atomic Energy Research Establishment.

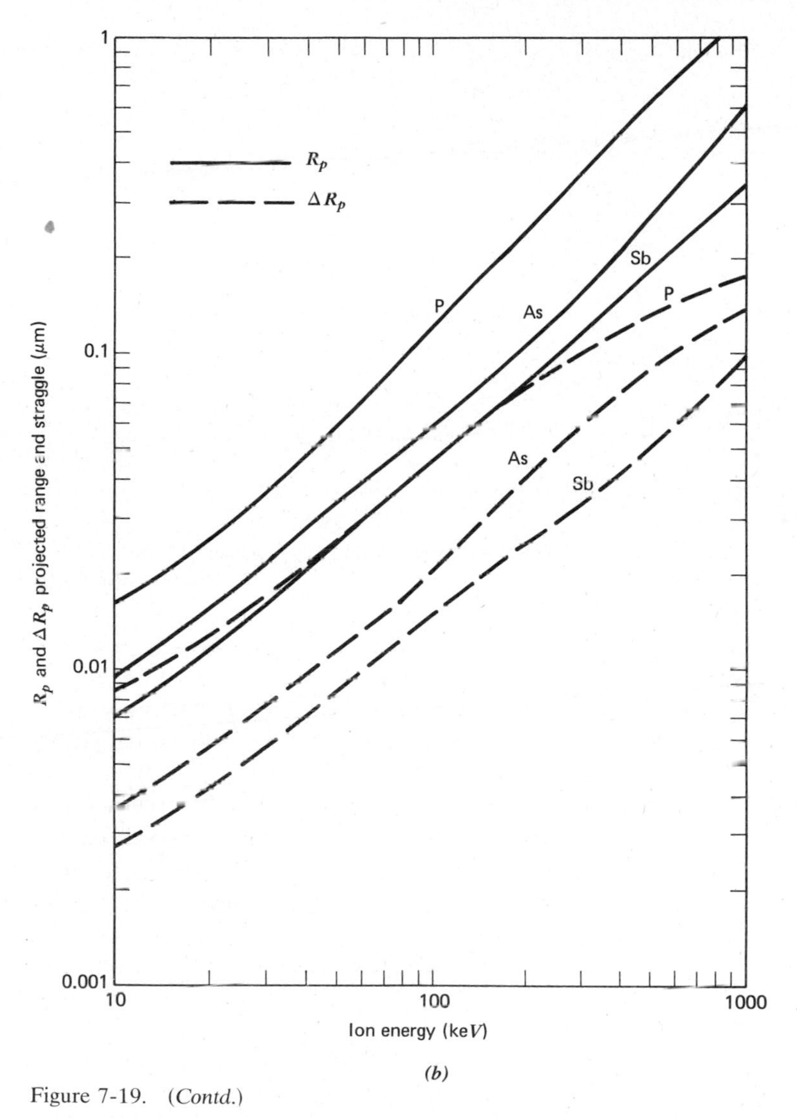

(b)

Figure 7-19. (*Contd.*)

Masking for Ion Implantation

Since ions are absorbed to a certain extent by all solids, a variety of materials can be used as masks for selective ion implantation. In order to act as a mask a layer must be thick enough to absorb essentially all of the ion dose impinging on the surface of the mask materials. Referring to Table 7-3, this implies that the minimum thickness for masking, t_m, is

$$t_m = R_p + 4.3\,\Delta R_p \tag{7-47}$$

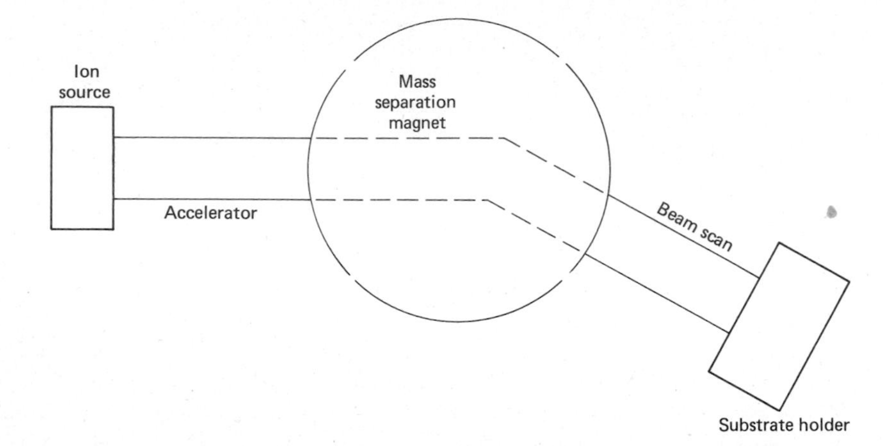

Figure 7-20. The essential components of an ion-implantation system.

where R_p and ΔR_p are for the masking material. Extensive data for a wide variety of material are available in tabular form (27) for the estimation of the masking thickness required. These data are based on the stopping power of a particular material for a particular ion, at a specific energy.

For low-energy (less than 100 keV) implants, common masking materials are silicon dioxide, silicon nitride, and photoresist. Using silicon as a reference, the relative stopping powers for these materials for boron or phosphorus implants at 40 keV are 1.25, 1.62, and 0.75, respectively. Silicon nitride is slightly better masking material, but silicon dioxide and photoresist are materials that are in common use for many of the other processes involved in device fabrication, and therefore are very popular as implant masks. If a deep implant is required, a heavy metal like gold, platinum, tungsten, or tantalum is deposited on top of a silicon dioxide layer and patterned by photolithography to form the mask. The relative stopping power of these materials is between 2 and 4 for 1-MeV boron and phosphorus implants. If photoresist is used as an implant mask, the ion current must be kept low enough to prevent excessive heating. If this is not done, the photoresist becomes very difficult to remove after the implantation. In some cases, it is desirable to locate the peak of the implant profile near the silicon surface. This can be accomplished by implanting through an oxide layer so that a significant part of the implanted dose is left in the oxide.

Annealing of Implanted Profiles

After an ion implantation, the silicon is left in a damaged condition and the implanted ions are not necessarily in substitutional sites. It is, therefore, necessary to anneal the material to remove the crystalline damage and activate the implanted impurities. This can be achieved by a heat treatment at temperatures of 950°C or higher. It should be recognized that only one annealing cycle is necessary, even though several implants may have taken place. It is not necessary to have a specific anneal cycle since any high-temperature process, like an oxidation or a diffusion, can serve a dual purpose. If the anneal is to be performed in nitrogen, it is recommended that a deposited oxide layer be placed on the wafer to reduce junction leakage currents.

During the annealing process, there will be some redistribution of the implanted profile, due to diffusion. Implanted arsenic emitters exhibit abnormally large emitter-base reverse leakage currents unless the impurity distribution is allowed to diffuse away from the damaged region. The diffusion coefficient for implanted impurities is enhanced in the damaged region, but, if an implanted distribution is to be used as the source for a deep diffusion, a normal diffusion coefficient can be expected during most of the process. The implant can also have an effect on the diffusion coefficient of impurities already in the substrate prior to the implant. An example of this is the use of ion implanted protons to selectively increase the diffusion coefficient of boron under the implanted areas.

It is also possible to anneal ion implantation damage by scanning the wafer with a laser or electron beam. The energy necessary to permit the restructuring of the crystal lattice in the damaged region is absorbed near the surface when this technology is used, thus eliminating the need for a high-temperature process for the entire wafer. This reduces the diffusion effects associated with the annealing process and provides additional flexibility for processing.

CHARACTERIZATION OF IMPURITY PROFILES (11, 14, 28, 29)

After an impurity distribution has been introduced into a silicon wafer, it is desirable to be able to determine the sheet resistance, junction depth, and profile of the doped region. Some of this information, in particular, sheet resistance, can be determined by including test patterns on the wafer, but, in most cases, control wafers, which have been through all of the process steps except photolithography, are included in the production run for the purpose of evaluation.

Sheet Resistance Measurements

The most common technique for measuring sheet resistance makes use of an in-line four-point probe. The outer two probes are used to pass current through the layer to be evaluated, and the inner two probes are used to measure the voltage in a potentiometer arrangement. The probes are spring loaded and pass through sapphire bearings to maintain equal spacing.

The sheet resistance of an infinitely extending sheet is given by

$$R_S = \left(\frac{V}{I}\right)\left(\frac{\pi}{\ln 2}\right) = 4.5324\left(\frac{V}{I}\right) \tag{7-48}$$

In a practical situation with finite dimensions, a correction factor based on the geometry replaces the constant in the calculation. The correction factor for circular wafers is shown in Figure 7-21. If the impurity is introduced by a diffusion from all sides of the wafer, the back and edges of the sample contribute to the measurement. If the wafer is circular, the effect of the conducting surface on the back is to make the wafer appear like an infinite sheet, and Equation 7-48 should be used. If the wafer is noncircular, the backside diffusion should be removed by etching or lapping and the appropriate correction factor should be used.

Junction Depth Measurements

The accurate determination of junction depths, particularly for shallow junctions, is difficult. The most common techniques are the angle-lap-and-stain and groove-and-stain. Angle lapping is performed by mounting the wafer on a special fixture so that the wafer can be lapped at a very small angle, typically 1

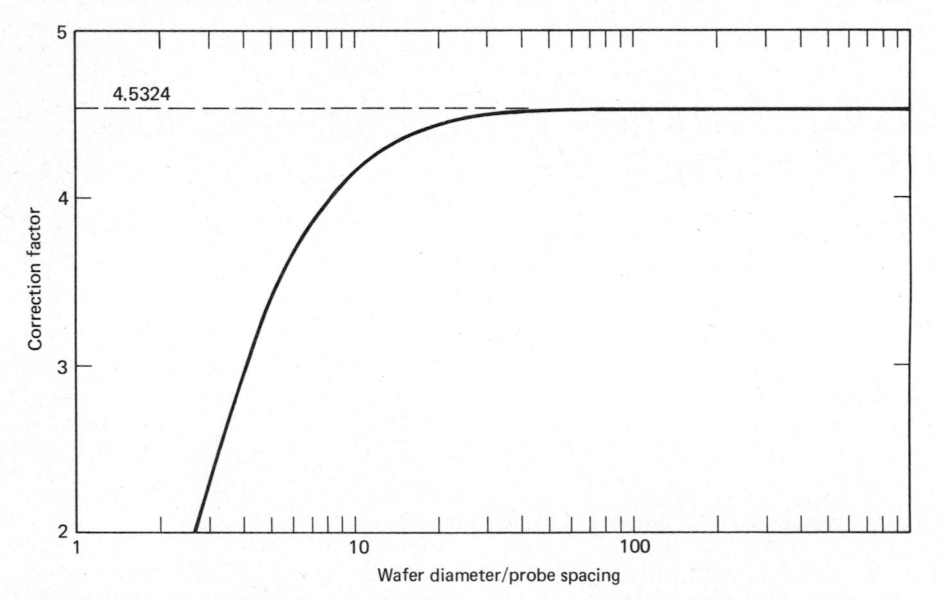

Figure 7-21. The correction curve for four-point probe resistivity measurements on round wafers (35).

to 5°, as shown in Figure 7-22. The junction depth is magnified by this arrangement, so that the distance measured on the lapped surface, d, is related to the junction depth, x_j, by

$$d = \frac{x_j}{\sin \theta} \tag{7-49}$$

where θ is the angle of the fixture. The junction can be delineated by a chemical etchant that selectively stains either n-type or p-type material. An effective stain for delineating junctions is concentrated hydrofluoric acid with 0.1 to 0.5% nitric acid. High-intensity light improves the staining process. The p-type region turns dark. The groove-and-stain technique is similar, except that a cylindrical groove is machined in the surface of the wafer instead of angle-lapping. After the junction has been delineated, the junction depth is measured by an interference microscope in a manner similar to the measurement of oxide thickness.

A different approach to junction delineation makes use of a scanning electron microscope. For this method electrical contact must be made to both sides of the junction. The junction is exposed, either by angle-lapping or sectioning. The electron beam is used to produce electron-hole pairs in the silicon. These carriers produce no current in the external circuit unless they are

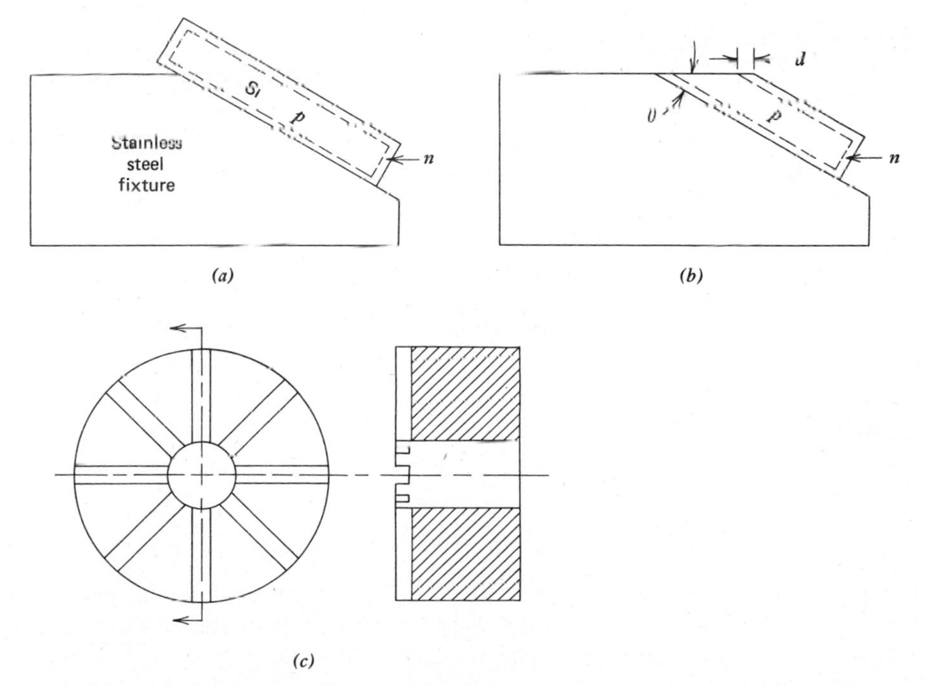

Figure 7-22. A fixture for angle-lapping samples in preparation for the stain-etch technique for determining junction depth. (a) Before polishing. (b) After polishing. (c) Holder for staining fixture.

created in the presence of an electric field. In a *pn* junction, there is a built-in electric field that peaks at the location of the metallurgical junction. As the scanning beam traverses the junction, the external current peaks. By use of a timed scan, the junction depth can be determined.

Doping Profile Measurements

Impurity profiles can be determined by several techniques. Most of these are called "differential" methods, which implies that a measurement is made, then a thin layer of silicon is removed, followed by another measurement and the removal of another layer, ad infinitum. The information of interest is determined from the rate of change of the data. Perhaps the most accurate method for determining impurity profiles is the use of radioactive isotopes in a differential technique. Suitable isotopes exist for phosphorus and arsenic, but, unfortunately, not for boron. The result of this type of measurement is the total impurity concentration as a function of depth, but of more interest is the concentration of electrically active impurities as a function of depth. In order to obtain this information, an electrical test is employed, usually the differential Hall effect or the differential conductivity measurement. In order to perform these tests, very thin layers of silicon must be removed in a controlled manner. This can be done by etching, using a slow isotropic etchant, but in most cases, anodization of the silicon into silicon dioxide, followed by oxide removal in hydrofluoric acid, is employed. A common electrolyte is tetrahydrofurfuryl alcohol with a small quantity of potassium nitrite (KNO_2). An anodization voltage of 100 V is applied until the current drops to a low value. This removes approximately 200 Å of silicon. The actual amount of silicon removed can be determined by an interference microscope or by determining the weight loss during the process. These differential processes are obviously tedious and destructive to the wafer.

An electrical technique has been developed for estimating the profiles of impurity distributions if the profile is not excessively steep. This is based on the change in small signal capacitance of a reverse biased Schottky barrier diode as a function of bias voltage. The carrier concentration for such a structure is given approximately by (29)

$$N(x) = \frac{-C^3}{q\varepsilon_s A^2}\left(\frac{dC}{dV}\right)^{-1} \tag{7-50}$$

where C is the small signal capacitance of the structure, q is the electronic charge ($1.6 \times 10^{-19} C$), ε_s is the total dielectric constant of the semiconductor, and A is the area of the diode. The distance from the surface, x, is given by

$$x = \frac{\varepsilon_s A}{C} \tag{7-51}$$

This technique is limited to the range where x is greater than the value obtained for zero bias, and is useless for cases where the impurity distribution changes so rapidly that the approximation

$$\Delta V \cong x\, \Delta \mathscr{E} \qquad (7\text{-}52)$$

is no longer valid, where $\mathscr{E}$ represents the electric field intensity. The Schottky barriers are usually formed by evaporating circular patterns through a metal shadow mask. This method is relatively fast (automatic equipment is available for this measurement), but, because its accuracy is limited, it should be compared with one of the differential methods to establish its validity for a particular situation.

SUMMARY

In this chapter methods of introducing impurities into semiconductor wafers have been discussed. Diffusion is the primary technique for forming *pn* junctions. The most popular approach makes use of a two-step process, consisting of a constant source predeposit that serves as a limited source drive-in redistribution step. At least part of the drive-in step is performed in an oxidizing atmosphere to provide an oxide layer for the next photoengraving step. Ion implantation is a versatile technique for introducing impurities into silicon, but is limited to impurity profiles very near the surface, unless the implant is followed by a diffusion. Methods have been developed for determining the sheet resistance and junction depth for diffused or implanted junctions, but accurately determining the impurity profile is a tedious and destructive process.

REFERENCES

1. R. B. Fair, *Semiconductor Silicon 1977*, H. R. Huff and E. Sirtl, eds., The Electrochemical Society, Princeton, 1977.
2. R. B. Fair, *Journal of the Electrochemical Society* 122, 6 (1975).
3. D. A. Antoniadis et al., *SUPREM II—A Program for IC Process Modeling and Simulation*, Technical Report No. 5019-2, Standard Electronics Lab., June 1978.
4. R. B. Fair and J. C. C. Tsai, *Journal of the Electrochemical Society* 124, 7 (1977).
5. G. Masetti et al., *Solid State Electronics* 19, 6 (1976).
6. J. Middelhoek and J. Holleman, *Journal of the Electrochemical Society* 121, 1 (1974).

7. R. K. Jain and R. J. Van Overstraeten, *IEEE Transactions on Electron Devices* ED-21, 2 (1974).
8. D. J. Hamilton and W. G. Howard, *Basic Integrated Circuit Engineering* (New York: McGraw-Hill, 1975).
9. H. F. Wolf, *Silicon Semiconductor Data* (Oxford: Pergamon Press, 1969).
10. L. A. Hall and H. Guckel, *Solid State Electronics* 18, 10 (1975).
11. G. Masetti et al., *Solid State Electronics* 16, 12 (1973).
12. J. C. C. Tsai, *Proceedings of the IEEE* 57, 9 (1969).
13. P. H. Langer and J. I. Goldstein, *Journal of the Electrochemical Society* 121, 4 (1974).
14. W. R. Runyan, *Silicon Semiconductor Technology* (New York: McGraw-Hill, 1965).
15. R. M. Burger and R. P. Donovan, *Fundamentals of Silicon Integrated Device Technology, Vol. I* (Englewood Cliffs N.J.: Prentice-Hall, 1967).
16. J. J. Steslow et al., *Solid State Technology* 18, 1 (1975).
17. S. N. Ghosh Dastidar, *Solid State Technology* 18, 11 (1975).
18. L. R. Planger, *Journal of the Electrochemical Society* 120, 10 (1973).
19. S. Prussin and A. M. Fern, *Journal of the Electrochemical Society* 122, 6 (1975).
20. J. L. Stone and J. C. Plunkett, *Solid State Technology* 19, 6 (1976).
21. J. George and J. Chruma, *Solid State Technology* 16, 11 (1973).
22. A. Axmann, *Solid State Technology* 17, 11 (1974).
23. J. Sansbury, *Solid State Technology* 19, 11 (1976).
24. G. Carter and W. A. Grant, *Ion Implantation of Semiconductors* (New York: Wiley, 1976).
25. J. W. Mayer et al., *Ion Implantation in Semiconductors* (New York: Academic Press, 1970).
26. B. J. Smith, *Ion Implantation Range Data for Silicon and Germanium Device Technologies* (Bristol, England: Adam Hilger Ltd., 1978).
27. R. R. Troutman, *IEEE Transactions on Electron Devices* ED-24, 3 (1977).
28. J. F. Gibbons et al., *Projected Range Statistics, Semiconductors and Related Materials, 2nd ed.* (Strousbourg, Pa.: Dowden, Hutchinson, and Ross, 1975).
29. G. V. Lukianoff, *Solid State Technology* 16, 3 (1971).
30. R. S. Muller and T. I. Kamins, *Device Electronics for Integrated Circuits* (New York: Wiley, 1977).

31. E. M. Conwell, *Proceedings of the IRE* 46, 6 (1958).
32. S. M. Sze and J. C. Irvin, *Solid State Electronics* 11, 6 (1968).
33. B. Ross and J. R. Madigan, *Physical Review* 108, 6 (1957).
34. J. C. Irwin, *Bell System Technical Journal* 41, 2 (1962).
35. S. K. Ghandhi, *The Theory and Practice of Microelectronics* (New York: Wiley, 1968). Figures reprinted by permission of the publisher.
36. D. P. Kennedy and R. R. O'Brien, *IBM Journal of Research and Development* 9, 3 (1965).
37. J. S. Huang and W. C. Welliver, *Journal of the Electrochemical Society* 117, 12 (1964).

PROBLEMS

7-1. The electrical current density, J_E, for positively charged particles in a one-dimensional semiconductor is

$$J_E = qN\mu\mathscr{E} - qD\frac{\partial N}{\partial x}$$

where N is the particle density, q is the charge per particle, μ is the mobility, $\mathscr{E}$ is the electric field intensity, and D is the diffusion coefficient. Using this equation and Equation 7-2, determine a general partial differential equation for positively charged particles in a semiconductor.

7-2. Create a table of diffusion coefficients for boron and phosphorus for various conditions at 900, 950, 1000, 1050, 1100, 1150, 1200, and 1250°C. List values for D_i and D for $N = 5 \times 10^{17}$, 10^{18}, 5×10^{18}, 10^{19}, 5×10^{19} cm^{-3} and the solid-solubility limit, assuming a background concentration of 10^{16} for boron and 10^{17} for phosphorus.

7-3. Find the time necessary to form a *pn* junction 2 μm from the surface by performing a constant source, solid-solubility limited boron diffusion into a background phosphorus concentration of 10^{16} cm^{-3} at 900, 1000, and 1100°C. Determine the sheet resistance for each case.

7-4. A *pn* junction is to be formed 2 μm from the surface in n-type material with a doping concentration of 10^{16} phosphorus atoms per cubic centimeter. The sheet resistance is to be 150 $\Omega/\square$. What is the surface concentration of boron necessary to satisfy these conditions if a constant source diffusion is to be used? Can this be a solid-solubility limited diffusion?

7-5. The junction for Problem 7-4 is to be formed by a two-step diffusion. The predeposit is to be solid-solubility limited at 900°C and the drive-in is to be performed at 1100°C. What is the appropriate diffusion schedule?

7-6. Assume that the diffusion of Problem 7-5 has been performed, and that the boron distribution does not change during a subsequent phosphorus diffusion (these changes can be taken care of by reducing the boron drive-in time so that the total D_2t_2 product remains the same). The phosphorus diffusion is a solid-solubility limited constant source diffusion at 1050°C to produce an *np* junction 1.5 μm from the surface. Find the time for the phosphorus diffusion and the resulting sheet resistance.

7-7. It is desired to form an *np* junction by a shallow phosphorus diffusion from a solid-solubility limited constant source for 5 min at 1000°C into a boron doped silicon wafer with a background concentration of 5×10^{18} cm^{-3}. How deep is the junction?

7-8. Find the impurity profile for an epitaxial layer grown at 1100°C for 10 min at 1 μm/min with an impurity concentration of 10^{16} phosphorus atoms per cubic centimeter on top of an arsenic buried layer with a concentration of 1.8×10^{21} cm^{-3}.

7-9. Determine the energy and dose necessary to produce a boron implant with a peak located 0.4 μm from the surface of a silicon wafer. The sheet resistance of the implant is to be 500 $\Omega/\square$. Assume that the background impurity concentration is 10^{15} cm^{-3}.

7-10. A boron implant is to be made into a coated silicon wafer at 40 keV with the peak of the implant to occur at the interface between the coating and the silicon wafer. How thick should the coating be, if the material is (a) silicon dioxide, (b) silicon nitride, and (c) photoresist? How thick should these coatings be in order to essentially mask the implant?

7-11. Derive Equation 7-50, stating assumptions.

Chapter 8
Assembly and Packaging

The final stages in the fabrication of microelectronic circuits are the series of processes which transform the circuits from the wafer form which facilitates mass production to the rugged individual circuits ready to withstand severe environments and provide for electrical connection to the outside world. The assembly and packaging processes range in sophistication from the conventional soldering techniques, which were well developed before microcircuits appeared on the scene, to the highly specialized techniques for bonding gold or aluminum wires, which are less than one-third the diameter of a human hair. Package types include hermetically sealed metal units with glass-to-metal seals around the leads, hermetically sealed ceramic dual-in-line packages (DIP's), injection molded plastic structures, and virtually no package at all, in certain types of hybrid microcircuits (it can be argued that the hybrid assembly process is a packaging process). Various aspects of the assembly and packaging processes have been the subjects of intensive development work to provide for automatic operations. The complete automation of assembly and packaging will further reduce the already very low cost of producing microcircuits.

MONOLITHIC IC ASSEMBLY

In this section, the steps used in the assembly and packaging of monolithic integrated circuits are described. These include wafer probing, die separation, die bonding, wire bonding, and package sealing.

Wafer Probing

Before monolithic integrated circuits are separated into individual chips, it is essential that nonfunctional circuits be identified. This is accomplished by probing all of the circuits on a wafer and performing a series of dc tests. Circuits failing to meet the specifications of these tests are marked with an ink dot. In some cases, when a circuit may be sold in more than one performance category, such as military, industrial, commercial, or high-frequency, different colored inks are used to identify those circuits which do not qualify for the highest category but still meet minimum specifications. In a prototype facility, manual probe stations containing individually controlled micromanipulator probe heads are used. Probe cards with prepositioned probes are used in a production situation, with automatic x-y indexing, and a minicomputer programmed series of tests.

Die Separation (1)

The separation of an integrated circuit into individual dice is accomplished by a scribe-and-break technique. Scribing streets, devoid of oxide and metal, are provided on the wafer. These streets are typically 75 μm wide, and are oriented along crystallographic cleavage planes to facilitate fracturing. The orientation is accomplished by orienting the initial photolithographic pattern parallel to the orientation flat on the wafer. For (111) oriented wafers, the natural cleavage planes parallel to the orientation flat are perpendicular to the surface, but those at right angles to the flat are not, resulting in the possibility of a nonrectangular cross section in that direction. This can result in chipping of the edges during die bonding. The optimum shape, from a fracturing viewpoint, for circuits on (111) oriented wafers is that of an equilateral triangle, and some small transistors are fabricated in this shape, but this is impractical for an integrated circuit. This problem does not exist for (100) oriented wafers, since a natural cleavage plane perpendicular to the surface exists at right angles to the flat, as well as parallel to the flat.

The wafer can be scribed by one of three techniques. For many years, the standard scribing tool was the corner of an industrial diamond. The tool was drawn across the wafer with a light pressure. A significant amount of compressive stress remains in the wafer after the scribing process. The yield from the diamond scribe-and-break process was frequently considerably less than 100%. This led to the development of alternative scribing techniques. A pulsed laser can be used to create a scribe line by localized melting of the silicon. The third technique, which has become the standard method for scribing large-scale integrated circuits, makes use of a diamond impregnated saw blade to cut the scribe lines. The yield after breaking for saw-scribed or laser-scribed wafers is higher than that for diamond-scribed wafers. In all three techniques, the scribe lines penetrate less than halfway through the wafer, to avoid unintentional breaking during postscribing cleaning.

The breaking process consists of placing the wafer, scribed side down, on a soft, flexible support, covering the wafer with a thin plastic sheet, and passing a roller over the wafer under pressure. In some cases, an automatic wafer breaking machine is used. For this process, the wafer is attached to a plastic backing sheet and held between two flexible metal belts which are passed over a stationary roller such that each row of dice is broken from the following row which is held fixed. The wafer is then rotated 90° and the process is repeated.

In many cases, it is necessary to thin the wafers by back-lapping before the scribing process. Wafers with diameters of 75 to 100 mm are typically 500 to 600 μm thick. For proper breaking and to fit into many packages, wafers must be thinned to 200 μm or less.

One special type of integrated circuit requires an entirely different type of die separation process. This type of integrated circuit is designed for attachment to hybrid circuits face down by electroplated gold beam leads. Before die separation, the beam-lead processed wafers are mounted face down on sapphire wafers, and lapped to thicknesses of 75 to 100 μm. Using infrared mask alignment, the wafers are photolithographically patterned for die separation. Beam-lead circuits are fabricated on (100) oriented wafers, so that a preferential etch, like potassium hydroxide or hydrazine, can be used to separate the dice with a minimum of surface area to be removed between the circuits. The result is a set of dice with beam-leads protruding from the edges, ready for facedown bonding. Beam-lead bonding is discussed in the section on die attachment for hybrid microcircuits.

Die Attachment (2–5)

The attachment of integrated circuit dice to packages serves as the means for thermal contact between the circuits and the packages, and, in some cases, also provides for electrical contact between the substrate of the integrated circuit and one pin on the package. If electrical contact is not required, an electrically insulating material like glass frit or thermally conducting epoxy can be used. If electrical contact is necessary, gold or silver filled epoxy or a metallic phase alloy is required for die attachment.

One of the most effective techniques for die attachment makes use of the gold-silicon eutectic alloy. This alloy consists of 3.6 atomic percent silicon and 96.4 atomic percent gold. It melts at 380°C. The source of the gold in the alloy can be plating on a metal package, thick-film ink on a ceramic package, plating on the backside of the die, or a thin alloy preform placed between the die and the package. If a preform is used, the preferred alloy contains 2 atomic percent silicon, which wets better than the eutectic alloy. The eutectic die bonding processs makes use of mechanical scrubbing motion to form the bond. The scrubbing action breaks down the thin native oxide layer on the back of the silicon die, making silicon available for alloy formation, and reduces the possibility of voids under the die. The bonding stage is maintained at a temperature between 400 and 420°C, and is flooded by nitrogen or forming gas to prevent the formation of an oxide. The mechanical scrubbing is provided either manually, with the die held by tweezers, or ultrasonically, with the die held in a special die collet with a vacuum pickup. The collet is designed with sloping sides so that contact to the die is along the upper edges, as indicated in Figure 8-1.

Epoxy die bonding has become a popular die attachment technique, particularly for devices which are not subjected to temperatures above 300°C. Epoxy compositions are available in unfilled and filled forms. Thermally conducting, but electrically insulating, epoxies contain alumina or similar high-thermal conductivity powder. Electrically conducting epoxies contain either gold or silver. The epoxy is usually dispensed by an automatic system, which places a controlled quantity of epoxy on the package base. The die is picked up by a collet and mechanically scrubbed into position on the base, which has been heated to between 150 and 200°C, depending on the epoxy composition. The mechanical scrubbing reduces the formation of voids under the die.

Wire Bonding (6–11)

Electrical connections have been a dominant cause for failures in virtually all electrical systems. A significant driving force behind the development of microelectronics was the expected increase in reliability due to the reduction in the number of interconnections. This goal has obviously been accomplished. Try to imagine building a microprocessor with 3000 discrete transistors and the associated 9000 solder joints, instead of the 40 solder joints required for the integrated circuit version. Unfortunately, there is still a requirement for some

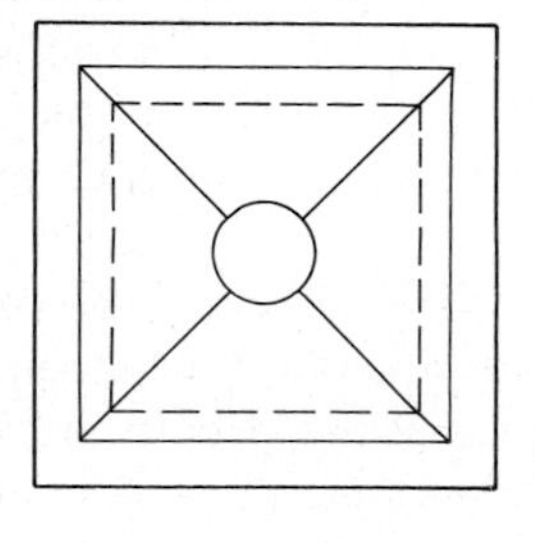

(a)

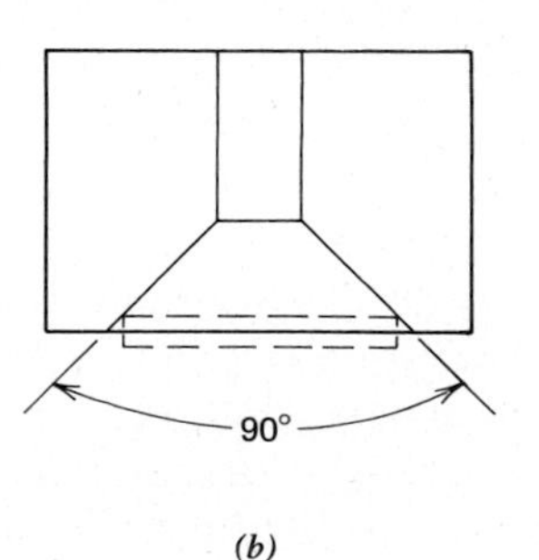

(b)

Figure 8-1. A die collect for die bonding. (a) Bottom view. (b) Cross section of side view (courtesy of Gaiser Tool).

interconnections. The current levels used in microelectronic circuits are low enough that wires between 20 and 40 μm in diameter are sufficient for most cases. The reliable attachment of these wires to both the integrated circuit bonding pads and the package leads is the subject of this section.

A vast majority of package leads are gold plated, at least in the area where wire bonding is to take place. In ceramic packages, the gold is frequently screen printed. For most other packages, the leads are gold plated Kovar®, an iron-nickel-cobalt alloy with the same thermal coefficient of expansion as glass, which was originally developed for glass-to-metal seals in vacuum tubes.

The most popular metal system for intraconnections for monolithic integrated circuits is aluminum, or predominantly aluminum alloys containing a small percentage of silicon and/or copper. Aluminum is a good conductor, adheres well to silicon dioxide, and penetrates the thin layer of native silicon dioxide that covers the contact holes after etching. It forms an ohmic contact with heavily doped p- and n-type silicon, and a Schottky barrier rectifier with lightly doped n-type silicon. The addition of a small percentage of silicon to aluminum reduces the tendency for aluminum-silicon alloy spikes to penetrate through shallow junctions in high-frequency transistors. The addition of copper to the aluminum reduces the electromigration effect. Electromigration is the transport of metal atoms in the presence of electrical current. This effect is most evident under conditions of high-current densities and high temperatures. These conditions can occur in integrated circuits where the metalization crosses an oxide step, creating a thinner layer of metal at the corner that has a higher resistance than the rest of the metal path. This high-resistance region is at a higher temperature than the rest of the metal and current density is increased due to the reduction in cross section. If electromigration occurs, it compounds the problem by further reducing the thickness of the metal in this localized area. The addition of copper and a well-designed deposition system (to assure proper step coverage) are used to alleviate this effect. Another shortcoming of aluminum is its low resistance to corrosion. If moisture enters the package due to a nonhermetic seal, narrow aluminum stripes will eventually fail. This can be avoided by depositing a low-temperature glass over the metal layer with openings etched only over the bonding pads. Another possible drawback to aluminum metalization is incompatibility with gold bonding wire, as discussed below. In spite of all these possible problems, aluminum is a highly successful metal for integrated circuit intraconnections.

An alternative metal system used in some integrated circuits requires four depositions. Ohmic contact is accomplished by depositing platinum on the wafer and converting it to platinum silicide in the regions where the platinum contacts the silicon by heating in air. The platinum is then removed by an etchant that does not attack the silicide. Then layers of titanium, platinum, and gold are deposited on the wafer, and patterned photolithographically. The titanium adheres well to silicon dioxide, but is a relatively poor electrical conductor and has poor wire bonding characteristics. Gold is an excellent conductor, but does not adhere well to silicon dioxide. It also diffuses into titanium, thus necessitating a platinum diffusion barrier. The etching of the trimetal system presents some difficulties, usually requiring sputter etching of the titanium and platinum.

There are two basic methods for attaching fine wires to integrated circuit bonding pads and package leads. The first of these is called thermocompression, ball, or nail head wire bonding. This works well with gold wire. The second technique makes use of ultrasonic energy to form the bonds, and is used primarily with aluminum wire.

Thermocompression wire bonding, as the name implies, makes use of a combination of temperature and pressure to form welds that would not form at the temperature or pressure alone. The gold wire is passed through a capillary which may or may not be heated. A small hydrogen torch or an electric spark is used to melt the end of the wire to form a ball with a diameter approximately twice that of the wire. The ball is brought into contact with the bonding pad on the integrated circuit, which has been heated to 220 to 250°C. The capillary deforms the ball into the shape of a nail head by applying pressure to form the bond. The capillary is raised, allowing the bond to pull the wire away from the reel. The wire forms a loop as the capillary is repositioned over the package lead. The second bond is formed by deforming the wire with the capillary. The capillary is then raised, unreeling a small length of wire, before a magnetic wire clamp is actuated. The wire breaks at the point weakened by the bond, leaving a tail-less bond. A new ball is formed and the clamp is released, leaving the bonder ready for the next bonding operation. The sequence is shown in Figure 8-2. The bond formed by this process is a strong, reliable weld. The best results are obtained for gold–gold bonds. The gold–aluminum bonds formed when the integrated circuit bonding pads are made from aluminum are not as reliable as the bonds to the gold plated package leads. In the presence of heat, gold and aluminum can react to form a number of intermetalic compounds. One of these compounds is purple in color and is often referred to as the "purple plague." The purple compound is a relatively good conductor, but its presence is usually accompanied by a tan compound that is brittle and a poor conductor. Bonds that include these compounds can fail after thermal cycling. The "plague" can be avoided by limiting the temperature of the thermocompression bonding process.

Thermocompression pulse bonding is used when it is desirable to keep the bonding area at a lower temperature. The substrate is maintained at 150°C and the bonding tip is pulsed to 450°C during the bonding process.

Ultrasonic bonding is a low-temperature process. The heat generated in this process is due to the friction between the wire and the bonding area. Aluminum wire with a small percentage of silicon (typically 1%) is the most popular wire for this type of bonding, but some success is obtained with gold

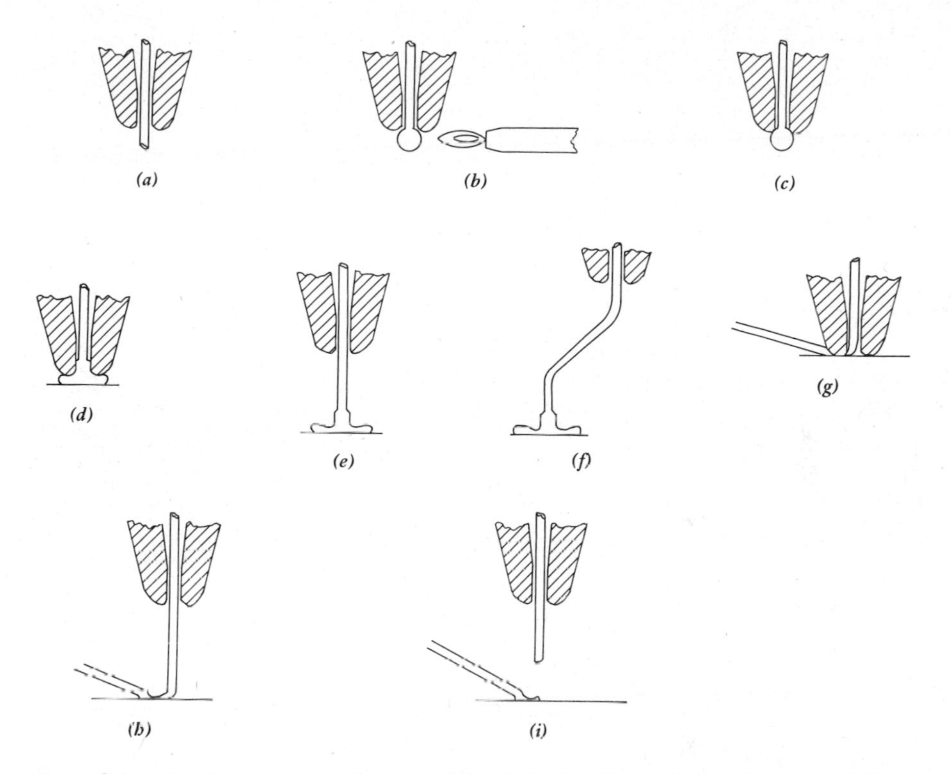

Figure 8-2. The thermocompression or nail-head wire bonding technique (courtesy of Gaiser Tool). (*a*) Tip of capillary with 1-mil gold wire through hole. (*b*) Ball melted on end of wire from flame off torch. (*c*) Wire drag holds ball against tip. (*d*) Ball bond on bonding pad surface. (*e*) Capillary and bonding arm raising toward loop position. (*f*) Capillary at loop height and substrate moving. (*g*) Stretch bond on substrate lead. (*h*) Capillary rising and ready for tail pulling. (*i*) Wire clamp is on and tail is pulled.

wire. A typical ultrasonic bonding tool is shown in Figure 8-3. The wire passes through a magnetic wire clamp and through the hole in the back of the tool. With the wire clamp closed and the wire protruding slightly beyond the front of the tool, the bonding wedge is positioned above the bonding area. The tool is then brought down vertically, and, with the application of ultrasonic energy and pressure, the wire is deformed as the bond is formed. The wire clamp is opened as the tool is raised, allowing the wire to be unreeled. A loop is formed in the wire as the bonding tool is positioned over the second bonding area. The bonding tool is brought down at a slight angle from the vertical so that the back edge of the bonding foot is lower than the front edge. As the bond is formed, the back edge of the bonding foot deforms the wire into a weakened condition. The wire clamp is closed as the tool is lifted, breaking the wire at the weakened point. The wire clamp then moves forward slightly, to position the

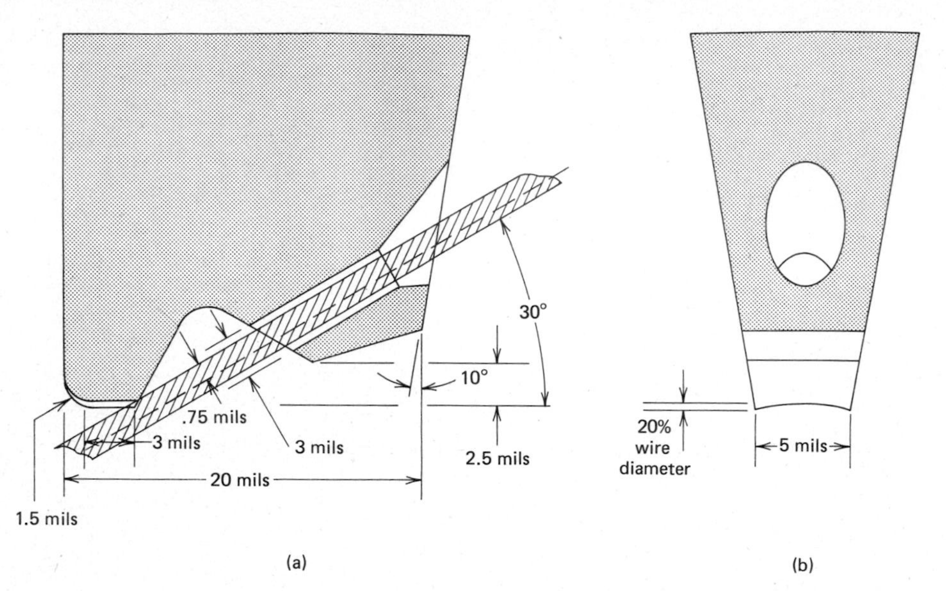

Figure 8-3. An ultrasonic bonding tool. (*a*) Cross section of side view. (*b*) Back view showing threading hole (courtesy of Gaiser Tool).

wire under the bonding foot for the next bond. Ultrasonic bonds are reliable but not as strong as thermocompression bonds.

A third type of bond combines the two techniques. It is called thermosonic bonding. The bonding process is similar to thermocompression bonding, except that the bonding area is maintained at 150°C and ultrasonic energy is added to form the bond. The temperature at which thermosonic bonding takes place is below the threshold for the formation of Au–Al intermetallics and does not disturb epoxy die bonds.

The wire bonding process is very expensive compared to most of the other fabrication processes. Up to, and including, the die separation process, each wafer contains anywhere from a hundred to several thousand individual circuits, depending on the die size, which are processed simultaneously. The die bonding process is applied to each die individually. The wire bonding process, however, is used a number of times for each circuit, ranging from 6 bonds for a basic voltage regulator to 128 bonds for complex, very large-scale integrated circuits. Obviously, this is a very expensive part of the fabrication process, particularly if an operator must manually position the bonding tool for each bond. Automatic bonding equipment is available which is programmed to perform a series of bonds after an operator positions a small circle of light, called a "spotlight," on the first bonding pad. Even automatic bonding is time consuming, and considerable effort has been expended for the development of techniques for simultaneously performing all of the bonding operations. The most promising of these "gang" bonding methods is the tape carrier system. In this process, thick gold bumps are plated on the integrated circuit bonding pads. Gold or tin plated copper lead patterns are formed on a plastic tape. The lead pattern is thermocompression bonded to the gold bumps in a single operation. The circuit can then bonded to a package lead frame or a hybrid substrate in a single operation. This process is discussed in more detail in the hybrid assembly section later in this chapter.

Packages and Package Sealing (3, 12–14)

The selection of a package for an integrated circuit depends on the die size, number of bonding pads, power dissipation, and anticipated environment. In general, metal packages are superior for heat transfer, but are more expensive than ceramic or plastic packages. For commercial applications, where an hermetic seal is not required, injection molded plastic packages, usually in dual-in-line configuration, are very popular.

The major consideration, from the point of view of power dissipation, is to prevent the temperature of the circuit from reaching the level at which the devices become inoperable. For a typical bipolar transistor, this temperature is approximately 350°C. It should be recognized that this is the internal "hot spot" temperature, not the ambient temperature. For each package type there is a parameter called the thermal resistance, which has the units of degrees celsius per watt and is used to estimate the temperature difference between the "hot spot" and the ambient for a particular power dissipation. Values for thermal resistances for selected standard packages are listed in Table 8-1. The thermal resistance is related to the die bonding process, and is lowest for eutectic die attachment. The values given in the table are for free air conditions.

Table 8-1. **Thermal Resistances for Standard Integrated Circuit Packages**

Package Type	Thermal Resistance (Junction to Ambient) °C/W
TO-5 (metal, 8-10-12 pins)	150–300
TO-86 (Ceramic flat pack, 14 leads)	250–350
TO-116 (Plastic dual-in-line, 14 leads)	200–300

The addition of fins, fan cooling, or liquid cooling can be used to increase the power handling capability of a particular package. Many commercial integrated circuits are available in more than one package configuration.

If the anticipated environment for the integrated circuit is severe, like in military, space, or automotive applications, an hermetic seal is required. The sealing process is performed in an inert, dry atmosphere. The sealing materials are usually gold-tin eutectic solder (20% tin) or a solder glass like PbO–ZnO–B_2O_3. The solder glasses are used primarily for sealing ceramic packages and require processing temperatures in excess of 450°C. The gold-tin solder is used in the form of stamped preforms to seal metal or metalized ceramic packages. The solder melts at 280°C.

HYBRID ASSEMBLY TECHNIQUES (2, 4, 14–20)

Hybrid microcircuits are assembled in a wide variety of forms. Some hybrids are assembled by techniques very similar to those used for the assembly of monolithic integrated circuits. Others are assembled by methods that are used to assemble circuits on printed circuit boards. In many cases, hybrid microcircuit technology is a form of packaging technology for monolithic integrated circuits.

Passive Add-On Components for Hybrid Microcircuits

The increased production of hybrid microcircuits has resulted in the availability of a variety of miniature electronic components prepared specially for use in hybrids. These include all of the electronic components which cannot be economically fabricated by thick- or thin-film techniques. In this section, capacitors, resistors, inductors, and transformers for hybrid circuits are described.

Capacitors. Multilayer ceramic chip capacitors are frequently used in both thick- and thin-film hybrid microcircuits, because the capacitance density available in chip form far exceeds that obtainable by conventional deposition techniques. Chip capacitors usually are terminated with thick-film conductor inks, as shown in Figure 8-4, so that the components can be mechanically attached and electrically connected by reflow soldering. In some thin-film applications, soldering is undesirable, and chip capacitors are attached by nonconducting epoxy and electrically connected by wirebonding. The use of aluminum ribbon instead of the conventional round wire in ultrasonic bonding has proved effective in this application.

Another type of chip capacitor is the thin-film capacitor deposited on an oxide coated silicon substrate. This capacitor has both terminations on the top

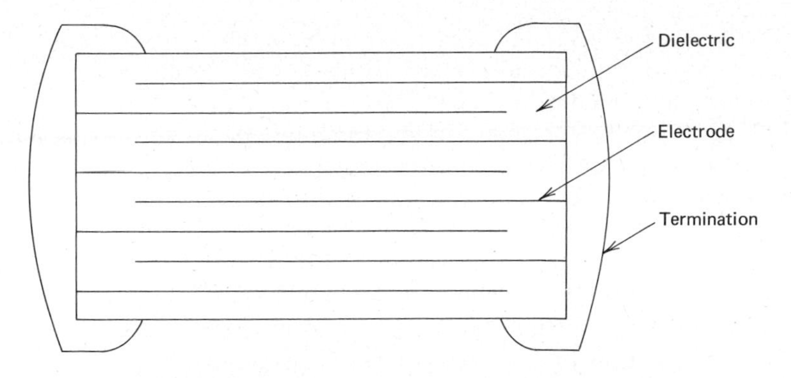

Figure 8-4. A cross-sectional view of a ceramic multilayer chip capacitor (Johanson).

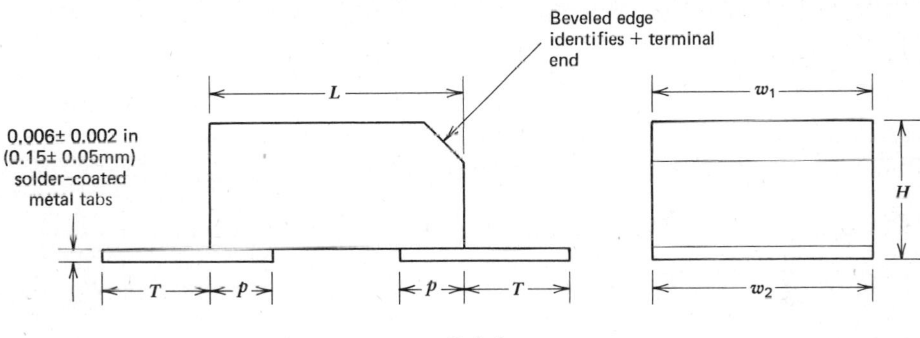

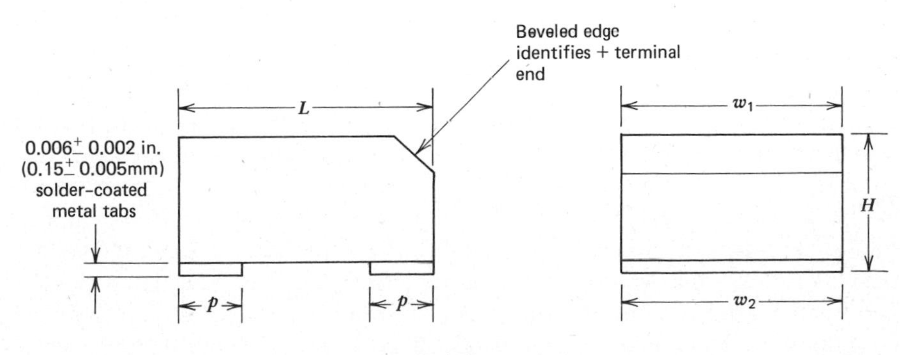

Figure 8-5. The Domino capacitor package for solid tantalum electrolytic capacitors. Copyright © 1973, Sprague Electric Co.

for wire bonding into the circuit. The chip can be attached by eutectic or epoxy die bonding. Some of these devices are available with beam leads for inverted thermocompression bonding.

For large valued capacitors, solid tantalum electrolytic components in molded Domino® packages are popular. This package type is shown in Figure 8-5. These capacitors are usually attached by reflow soldering.

Resistors. In some cases, it is more economical to add resistors in chip form instead of forming them by the film deposition technique. In thick-film circuits, it may be possible to print all of the resistors except one by using one paste, and then add the other resistor as a chip. In thin-film circuits, high- or low-valued resistors may take up an undesirably large area if formed by the conventional deposition process, thus indicating the use of a chip resistor fabricated in a different material.

Thick-film chip resistors are printed on ceramic substrates with thick-film conductor terminations. They can be attached with epoxy and wire bonded into the circuit, or by reflow soldering.

Thin-film chip resistors are available on ceramic or silicon substrates. These can be attached by eutectic or epoxy die bonding and wire bonded into the circuit. Some of these devices are also available with beam leads for inverted thermocompression bonding.

Inductors and Transformers. Where possible, inductors and transformers are avoided in hybrid microcircuit design. There are a very limited number of subminiature devices of this type available, usually wound on toroidal cores. They are usually attached with epoxy and wire bonded into the circuit.

Diodes, Transistors, and Integrated Circuits for Hybrid Microcircuits

Semiconductor devices and integrated circuits are attached to thick- and thin-film substrates by a variety of techniques. In many cases, thick-film hybrids are assembled so that no additional packaging, other than, perhaps, a conformal dip coating, is required. Under these circumstances, the semiconductor components are attached in packaged form. It is common practice to use standard package configurations, like flat packs, dual-in-line, TO-5, and glass diode packages, for this type of assembly. In some cases, special packages like the Microtab® and Minibloc® transistor configurations have been developed for hybrid applications. At the other extreme, due to the unused volume contained in most packages, significant reductions in the total volume occupied by a particular circuit can be achieved by using bare silicon chips, which are wire bonded directly into a thin-film circuit, and placing the entire hybrid into a final package. In many cases, the entire circuit can be included in a package that is the same size or smaller than the package normally used to contain one of the semiconductor devices used in the circuit. In between these extremes are a number of special configurations, which have been developed primarily to enable the assembly operations to be automated.

Hybrid Assembly with Packaged Devices. The assembly of hybrid microcircuits with packaged add-on components provides some advantages over the assembly of printed circuit boards to accomplish the same function. There is a reduction of size and weight, but only by a factor of two or three, depending on the circuit configuration. There is an economic advantage if a large number of identical circuits are required. Perhaps the most important advantage is due to the use of resistor trimming. Production resistor trimming routinely achieves ±1% resistors, but, a more significant technique, active trimming of operating circuits, eliminates the need for bulky and expensive potentiometers in many circuits.

The thick-film hybrid microcircuit configuration lends itself particularly well to assembly with packaged components. The most popular device attachment technique under these circumstances is reflow soldering. The solder can be applied in the form of a screen printed paste, which is tacky enough to support the preformed leads of the components. The entire substrate is then heated until the solder flows, bonding the components both electrically and mechanically to the circuit.

Hybrid Assembly with Bare Chips. The assembly of hybrid microcircuits with bare semiconductor chips is similar to the packaging of an integrated circuit. There are some problems, however, which are unique to the hybrid assembly process. If eutectic die bonding is to be used for more than one die, it is necessary to provide a means for concentrating the heat to a small area of the substrate. This prevents damage to the other components and the possible de-bonding of other die. Several techniques have been developed to localize the substrate heating, including focused infrared sources, miniature hydrogen torches, and an electrical heating technique in which a pair of fingers on either side of the die collet are used to pass a large-current pulse through the gold metalized bonding area. Because of these difficulties, epoxy die bonding has become popular in hybrid microcircuit assembly. For thick-film circuits, the epoxy is frequently screen printed, which provides one of the most repeatable methods for dispensing the bonding agent. Wire bonding can be performed by pulse-tip thermocompression or ultrasonic techniques. Conventional thermocompression wire bonding softens epoxy die bonds, due to the high temperature. The bonding characteristics of the various conductor materials determine the best bonding process for the application.

Once a hybrid microcircuit has been assembled by the chip-and-wire technique, it must be placed in a package to protect the semiconductor devices from the environment and the delicate wires from mechanical damage. There is a wide selection of hybrid packages available, including metal, ceramic, and plastic types, in many shapes and sizes. In some cases, the package is shaped to fit a particular application, like the automobile voltage regulator that is designed to bolt to the alternator. A typical hybrid packaging operation consists of attaching the substrate to the package with epoxy and wire or ribbon bonding the circuit leads to the package leads. This is followed by a conventional package sealing cycle.

Special Device Configurations for Automated Assembly. One difficulty encountered with the chip-and-wire assembly technique for hybrid microcircuit assembly is the inability to test the individual dice before assembly. There are also a significant number of manual operations associated with this assembly technique. As a result of this difficulty, several special configurations have been developed for hybrid assembly, which lie somewhere between bare chip and conventional packaging. Some of these configurations are described in the paragraphs below.

The beam-lead sealed junction device and integrated circuit manufacturing process was developed by Bell Telephone Laboratories. In this process, the junctions are protected by a silicon nitride passivation layer, which provides a level of protection similar to that of a hermetic package. The metal intraconnection system is a platinum silicide, titanium-platinum-gold system with 2-μm-thick gold plated beams that protrude beyond the edges of the die, as shown in Figure 8-6. The die is attached in a facedown position by thermocompression bonding of the gold beam leads to a gold metalization pattern on the substrates. The bonding tool "wobbles" around the outside of the circuit, bonding all of the leads in a single operation. Electrical tests on the individual dice can be performed before the bonding operation. Essentially all of the heat transfer from the die to the substrate occurs through the leads. This system was originally developed for thin-film hybrids, but advances in fine-line screen printing have made this a viable process for thick-film hybrid microcircuits also. This process is applicable to chip capacitors and resistors, making it possible to assemble an entire hybrid microcircuit by this technique.

The flip-chip bonding process was developed by IBM for the Series 360 computer. In this process, solder bumps are attached to the device or integrated circuit bonding pads. The die is inverted and mounted on the substrate metal pattern by a "controlled collapse" soldering process. This is shown in Figure 8-7.

The ceramic chip carrier has evolved from the ceramic channel mounted device and the leadless-inverted-device (LID). The basic forms of these special configurations are shown in Figure 8-8. The channel mount consists of a two-level alumina ceramic with deposited gold in the channel and on the two upper level areas. The device is eutectically die bonded in the channel and wire

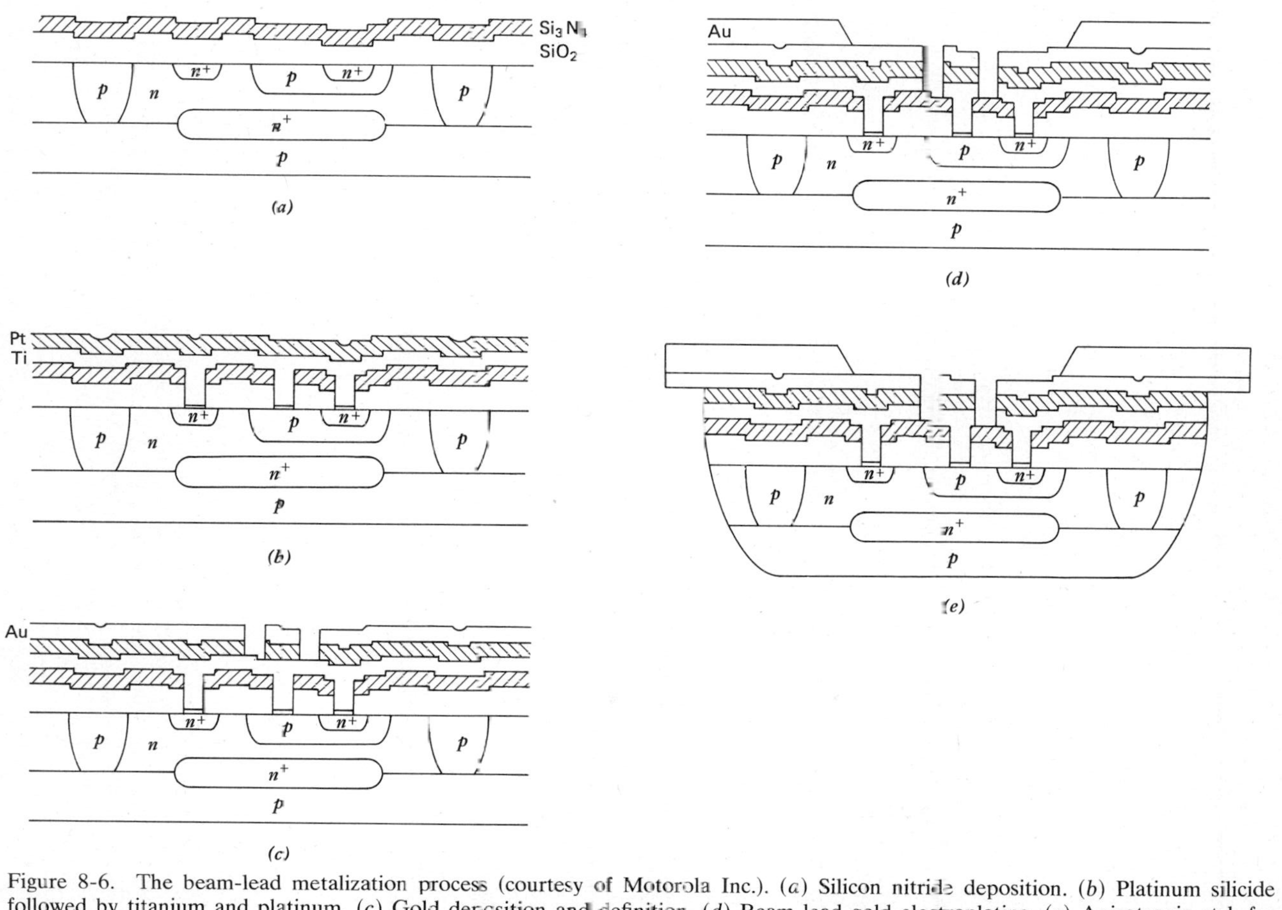

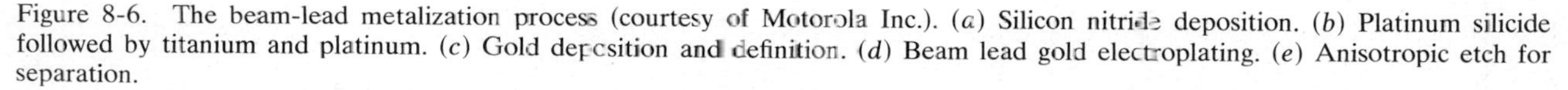

Figure 8-6. The beam-lead metalization process (courtesy of Motorola Inc.). (*a*) Silicon nitride deposition. (*b*) Platinum silicide followed by titanium and platinum. (*c*) Gold deposition and definition. (*d*) Beam lead gold electroplating. (*e*) Anisotropic etch for separation.

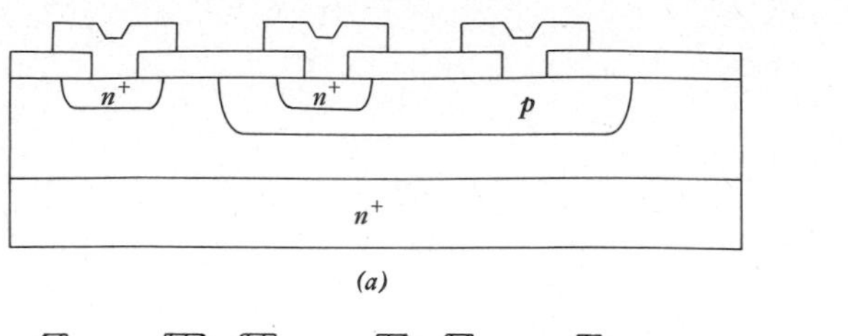

(a)

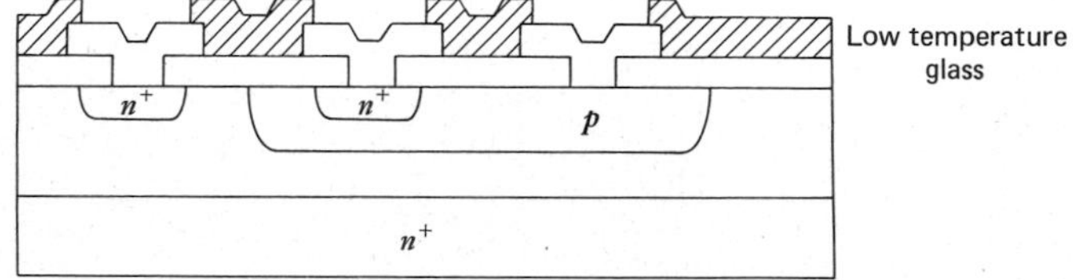

(b)

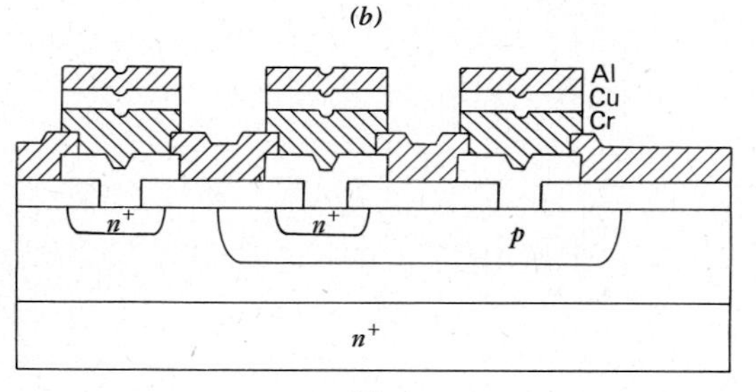

(c)

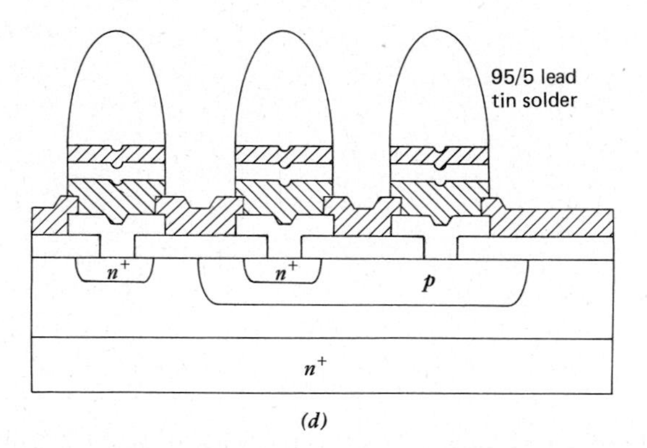

(d)

Figure 8-7. One type of flip-chip solder ball process (courtesy of Motorola Inc.). (*a*) Typical transitor process with aluminum metallization. (*b*) Deposit low-temperature gas. (*c*) Deposit chromium, copper, and aluminum and pattern. (*d*) Deposit lead-tin solder through metal mask and heat to reflow.

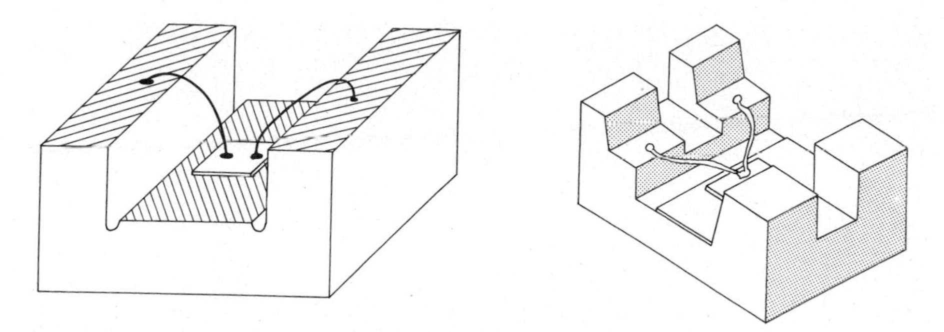

Figure 8-8. Two forms of ceramic chip mounting structures, the channel mount and the leadless inverted device (LID) (Amperex).

bonded to the upper levels. This structure is epoxied, face up, to the substrate and wire bonds are made to the three gold areas to complete the circuit. The LID is a multilevel alumina ceramic structure similar to the channel mount, except that there is an additional level of gold plated feet. Once the device is mounted and wire bonded in place, epoxy is used to protect it from the environment. The LID structure is attached to the circuit by inverting the unit and soldering the feet to the substrate metal pattern. The LID concept has been extended to cover a family of integrated circuit configurations in dual-in-line structures up to 24 leads and a square structure with 32 leads. A most promising form of ceramic premounted chip configuration is the hermetic chip carrier, as shown in Figure 8-9. This unit is essentially the portion of a ceramic package which contains the integrated circuit. The circuit is mounted in the carrier, wire bonded, and hermetically sealed, as in a conventional package. The closely spaced leads are then reflow soldered to the substrate metal pattern, resulting in the possibility of a densely packed configuration of hermetically sealed units.

In the structures described above, except for beam-lead devices, individual wire bonds are required at some point in the assembly process. The automatic tape carrier bonding configuration makes use of gang bonding to form both the inner lead bonds to the semiconductor die and the outer lead bonds to the metal pattern on the substrate. To prepare the semiconductor dice for tape carrier bonding, gold bumps are plated over the bonding pads, before die separation. If the integrated circuit has an aluminum intraconnect pattern, a diffusion barrier, like titanium-palladium-gold is deposited over the wafer surface. The pattern for the bumps is then defined using a dry film photoresist. After the bumps have been electroplated, the photoresist is removed and the barrier layer is selectively etched away. The tape carrier is a polyimide plastic into which have been punched sprocket holes for advancing the tape and openings for the circuit dice. A copper foil is laminated to the tape with an

Figure 8-9. Ceramic chip carriers for integrated circuits (Photo Courtesy of 3M Company).

epoxy adhesive and etched to form the proper lead configuration. The copper is then coated with nickel, followed by gold, to provide for reliable thermocompression bonding, or tin for gold-tin eutectic bonding, to the gold bumps on the die. After the inner lead bond, the integrated circuits can be individually tested. The outer lead bond is formed by punching the circuit out of the tape over a die and making a thermocompression bond with the metal pattern on the substrate in a single operation. Since the same process can be used to bond a circuit to a package or lead frame, circuits fabricated on tape carriers can be sold in packaged form or for automated hybrid assembly without any special processing.

SUMMARY

The assembly and packaging of monolithic integrated circuits include a number of steps that involve operators working with individual dice, like die bonding and wire bonding. This is in great contrast with the fabrication steps leading to the final assembly, where many circuits are fabricated simultaneously on each wafer. Assembly is time consuming, and adds significantly to the overall fabrication cost. Packaging an inoperable circuit is an expensive waste.

Hybrid circuit assembly takes on many forms. In some cases, thick-film hybrids are assembled by soldering, like printed circuit boards. In other cases, the same type of operations are used as those for the assembly and packaging of integrated circuits. Automated device handling techniques are being applied to the production of hybrid microcircuits, which have resulted in reduced cost and increased reliability.

REFERENCES

1. J. F. Marshall, *Solid State Technology* 17, 9 (1974).
2. D. N. Bull, *Solid State Technology* 17, 9 (1974).
3. C. E. T. White and H. C. Sohl, *Solid State Technology* 18, 9 (1975).
4. R. F. S. David, *Solid State Technology* 18, 9 (1975).
5. J. Kimball, *Solid State Technology* 16, 19 (1973).
6. T. H. Ramsey, *Solid State Technology* 16, 10 (1973).
7. *Thermocompression Bonding Manual*, Gaiser Tool Company, 1976.
8. K. I. Johnson et al., *Solid State Technology* 20, 3 and 4 (1977).
9. G. L. Schnable and R. S. Keen, *Proceedings of the IEEE* 57, 9 (1969).
10. L. E. Terry and R. W. Wilson, *Proceedings of the IEEE* 57, 9 (1969).
11. J. R. Black, *Proceedings of the IEEE* 57, 9 (1969).
12. T. H. Ramsey, Jr., *Solid State Technology* 17, 9 (1974).
13. R. W. Thomas and D. E. Meyer, *Solid State Technology* 17, 9 (1974).
14. C. A. Harper, *Handbook of Thick Film Hybrid Microelectronics* (New York: McGraw-Hill, 1974).
15. R. G. Oswald et al., *Solid State Technology* 21, 3 (1978).

16. A. S. Rose et al., *Solid State Technology* 21, 3 (1978).
17. R. L. Cain, *Solid State Technology* 21, 3 (1978).
18. A. Keizer and D. Brown, *Solid State Technology* 21, 3 (1978).
19. C. J. Dawes, *Solid State Technology* 19, 3 (1976).
20. D. W. Hamer and J. V. Biggers, *Thick Film Hybrid Microcircuit Technology* (New York: Wiley-Interscience, 1972).

Chapter 9
Devices for Bipolar Integrated Circuits

The principal devices in monolithic bipolar integrated circuits are the junction diode and the bipolar junction transistor. By specifying the surface geometry and doping profile, the designer has the option of using these basic device structures to form resistors, capacitors, diodes, junction field effect transistors, and bipolar junction transistors. This impressive array of electronic components is available on all monolithic bipolar integrated circuits without any special processing (except in cases where junction field effect transistor performance is to be optimized), since the most complex device, from a processing viewpoint, is the bipolar junction transistor. From an economics viewpoint, designers are free (essentially obligated) to use as many transistors as possible in their designs, since there is no increase in the cost of the circuit to do so. In fact, since the "real estate" occupied by a transistor can be considerably smaller than that occupied by a capacitor or a resistor, it is less expensive to use a transistor than one of these other components. Thus, the economic pressure on the integrated circuit designer is to use as many transistors as possible, while economic considerations for the designer of a discrete component or hybrid circuit is to use as few transistors as possible. Another interesting consequence of integrated circuit processing is that the transistors within an integrated circuit tend to have closely matched characteristics. In addition, since the devices are on the same substrate, the temperature characteristics are intimately coupled, providing thermal tracking. While the thermal interactions are frequently beneficial, there are other interactions within the integrated circuit, in the form of parasitic circuit elements resulting from processing constraints, which can degrade the anticipated performance of the circuit. Inadequate simulation of these parasitic effects makes it difficult to transfer a discrete component circuit to an integrated circuit with similar electrical characteristics. Computer aided design systems contain device models including the parasitic elements that permit accurate simulation of circuit operation.

The major emphasis in this chapter is the design of components that are compatible with the fabrication of a junction-isolated double-diffused epitaxial bipolar transistor with a buried layer. This device is shown in cross section, approximately to scale, in Figure 9-1. Components that require deviations from this "standard process" will add to the fabrication cost of the circuit, and, thus, the enhanced performance of the circuit due to these "special" components must justify the additional cost.

As mentioned in the *Preface*, it is assumed that the reader has been introduced to semiconductor device physics at a level equivalent to the text, *Device Electronics for Integrated Circuits*, by R. S. Muller and T. I. Kamins (Wiley, 1977). A brief treatment of basic device physics is presented in the *Appendix* to this book to provide a review in a notation consistent with that used in this chapter and Chapter 10. The device physics presented is that which is necessary to extend the basic principles so that the characteristics of practical devices can be predicted, as influenced by the surface geometry and doping profiles.

In this chapter, the basic transistor structure is thoroughly investigated with the objective of establishing a "standard" process for integrated circuits.

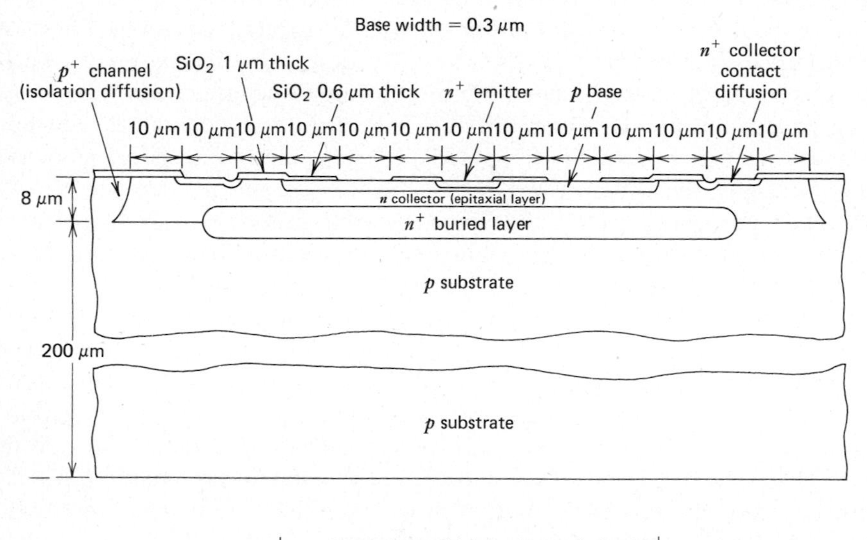

Figure 9-1. A cross section of an integrated circuit bipolar transistor drawn to scale (2). Reprinted by permission from Motorola, Inc.

The characteristics of diodes, capacitors, and resistors fabricated by the standard process are then examined. A section on dielectric isolation techniques and other nonstandard processes concludes the chapter.

THE BIPOLAR JUNCTION TRANSISTOR

The electrical characteristics of the bipolar junction transistors formed in monolithic integrated circuits are influenced by a number of factors, some of which are under the control of the designer. Much of the design is based on compromise, so that the fabrication process is compatible with the desired characteristics of the other components in the circuit. In this section, the reasons behind these compromises are explored.

It is important to recognize that each integrated circuit manufacturer has well-developed bipolar processes and associated libraries of transistors with established electrical characteristics, which are used in the design of new integrated circuit types. In addition, computer models of the processes, like Stanford University's SUPREM (1), have been developed, which permit much more accurate calculations of the impurity profiles than the relatively crude approximations used in this book.

In order to gain an insight into the device design process, it is assumed that the reader has no predesigned transistor structures, and is interested in developing both the geometrical layout and process schedule for a particular bipolar transistor structure to satisfy specific operating criteria. The fabrication process will impose certain constraints on the operating characteristics so that the designer is prohibited from specifying all of the parameters before initiating the design cycle. To determine which parameters may be specified and which are dependent, it is necessary to develop an appropriate model for the bipolar transistor.

Bipolar Transistor Model

There are a wide variety of transistor models based on physical principles and circuit properties. It is important to recognize that the simplest model that results in adequate agreement with experiment should be used. This principle is applied in this section.

dc Current Gain. The dc common-base forward current transfer ratio, with the transistor biased in the forward-active region of operation (base-emitter junction forward biased, base-collector junction reverse biased), is designated α_F. It is defined by

$$\alpha_F \equiv \frac{-I_C}{I_E} \tag{9-1}$$

where the current direction are as indicated in Figure 9-2, and I_C is the collector current, I_E is the emitter current, and I_B is the base current. Since

$$I_C + I_E + I_B = 0 \tag{9-2}$$

the ratio of I_C to I_B is given by

$$\frac{I_C}{I_B} = \frac{\alpha_F}{(1-\alpha_F)} \equiv \beta_F \tag{9-3}$$

where β_F is the common-emitter forward current transfer ratio.

To determine the effects of doping profiles on β_F, it is instructive to consider the factors that determine α_F in a one-dimensional physical model. It is convenient to divide the collector and emitter currents into their electron (designated by a subscript n) and hole (designated by a subscript p) components. Thus

$$I_C = I_{nC} + I_{pC} \tag{9-4}$$

and

$$I_E = I_{nE} + I_{pE}$$

Then, α_F can be written

$$\alpha_F = \frac{-I_C}{I_E} = \frac{-I_C}{I_{nC}} \cdot \frac{I_{nC}}{I_{nE}} \cdot \frac{I_{nE}}{I_E} \equiv M\alpha_T\gamma \tag{9-5}$$

for an *npn* transistor. The first factor, M, is the ratio of the total collector current to the electron current entering the collector-base junction space-charge region from the base, and is called the *collector multiplication* factor. The second factor, α_T, is the ratio of the electron current leaving the base at

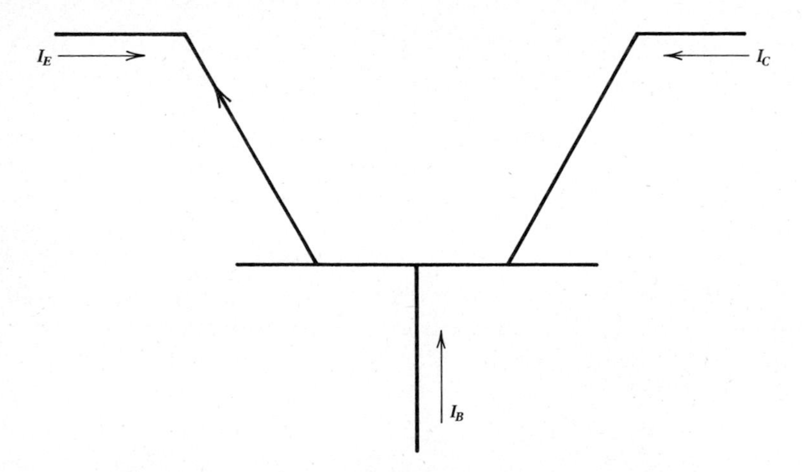

Figure 9-2. The current directions in the transistor model.

the edge of the collector-base junction space-charge region to the electron current entering the base from the emitter-base junction space-charge region, and is called the *base transport* factor. The third factor, γ, is the ratio of the electron current entering the base from the emitter-base junction space-charge region to the total emitter current, and is called the *emitter efficiency*.

The collector multiplication factor is given by the empirical relationship

$$M = \frac{1}{1-(V_{CB}/BV_{CBo})^n} \tag{9-6}$$

where n is between 2 and 4 for silicon, BV_{CBo} is the collector-base breakdown voltage with the emitter terminal open, and V_{CB} is the collector-base voltage. The factor, M, is always greater than unity, but operation at relatively low collector-base voltages significantly reduces the effect.

In the case of the uniformly doped base under low injection conditions, the base transport factor, α_T, is given by the relation

$$\alpha_T = \operatorname{sech}\frac{x_B}{L_n} \approx \frac{1}{1+x_B^{\,2}/2L_n^{\,2}} \tag{9-7}$$

where x_B is the width of the base between the base edge of the base-emitter space-charge region and the base edge of the base-collector space-charge region, and L_n is the diffusion length of electrons in the base region, which is defined by

$$L_n = (D_n\tau_n)^{1/2} \tag{9-8}$$

where D_n and τ_n are the diffusion coefficient and excess carrier lifetime for electrons in the base. The approximation is based on the condition that the diffusion length of minority carriers is much greater than the width of the base, which is satisfied in all modern bipolar transistor structures. The base transport factor is a measure of the probability that an electron injected into the base from the emitter will traverse the base region and be swept into the collector region, without recombining with a hole. The magnitude of α_T is less than, but very close to, unity for a modern transistor structure.

The effect of the nonuniform doping in the base of a double-diffused epitaxial transistor is to produce electric fields that, in essence, balance the tendency for diffusion of the majority carriers (holes), but can have a significant effect on the transport of minority carriers (electrons) across this region. A typical transistor doping profile is shown in Figure 9-3. Note that there is a region of positive doping gradient very near the emitter-base space-charge region, followed by a longer region of negative doping gradient. These two regions result in electric fields that are in opposite directions, the former in a direction such as to oppose the transport of electrons from the emitter to the collector, and the latter in a direction that aids this transport. Another consequence of nonuniform doping in the base is that the diffusion coefficient

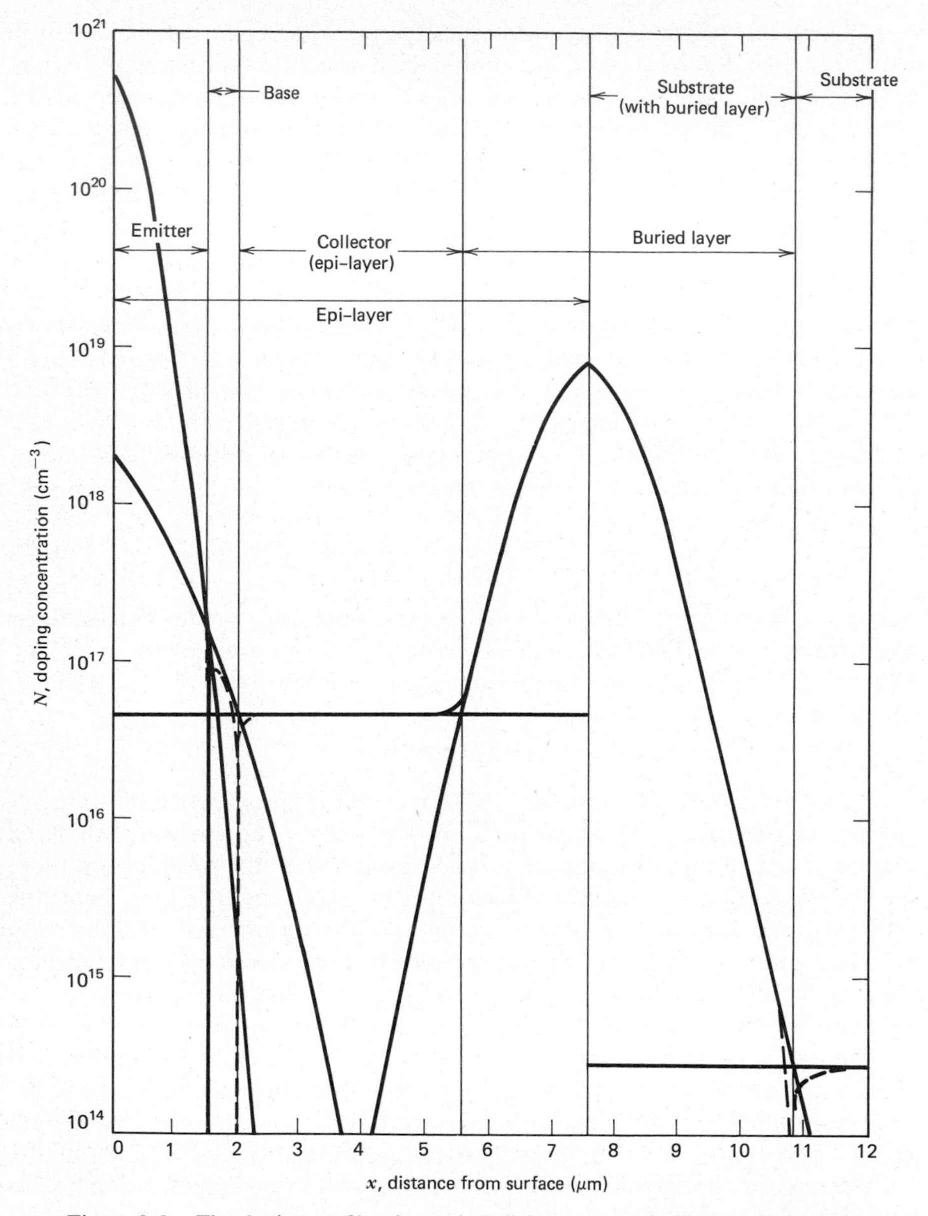

Figure 9-3. The doping profile of a typical digital integrated circuit transistor (3).

for minority carriers is a function of position. If an average value for the diffusion coefficient is used to calculate the base transport factor, the overall effect of the two regions of electric field in the base is essentially one of cancellation. This is in general agreement with a more detailed calculation considering the variation in diffusion coefficient with position across the base (6).

The level of injected electrons has an effect on the base transport factor. For injected levels at least an order of magnitude less than the majority carrier equilibrium concentration, the effect is negligible. As the injection level is increased above this "low injection" level, the majority carriers show a significant redistribution, since the excess hole concentration, p', must be approximately equal to the excess electron concentration, n', on a point-by-point basis. This results in an electric field to balance the diffusion of majority carriers and aid the flow of minority carriers across the base. This has the effect of enhancing the minority carrier diffusion coefficient. The enhancement factor is position dependent, since the minority carrier concentration decreases to a low value near the collector-base junction under the forward-active bias conditions. The maximum value of this enhancement factor, in the limit of very high injection, for a uniformly doped base, is two.

The emitter efficiency, γ, is a measure of useful emitter current for amplification, since it represents the electron current injected into the base from the emitter, as compared to the total emitter current. It is given by

$$\gamma = \frac{I_{nE}}{I_{nE} + I_{pE}} = \frac{1}{1 + I_{pE}/I_{nE}} \tag{9-9}$$

For the simple model with uniform doping in both the emitter and base, this can be shown to be

$$\gamma \approx \frac{1}{1 + x_B N_{aB} D_{pE} / x_E N_{dE} D_{nB}} \tag{9-10}$$

where N_{aB} and N_{dE} are the net doping concentrations in the base and emitter, respectively, D_{nB} and D_{pE} are the minority carrier diffusion coefficients in the base and emitter, and x_B and x_E are the widths of the quasineutral regions in the base and emitter. It is assumed, in this relationship, that both the base and emitter widths are short compared to their respective minority carrier diffusion lengths. This equation indicates that γ is increased toward a maximum value of unity if

$$N_{dE} \gg N_{aB} \tag{9-11}$$

or the donor concentration in the emitter is made many times larger than the acceptor concentration in the base. Unfortunately, the simplified model is inadequate to describe the emitter efficiency for a double-diffused transistor structure, particularly when the emitter doping concentration is high (6–13). For an integrated circuit bipolar transistor, it is more appropriate to write the emitter efficiency as

$$\gamma = \frac{1}{1+G_B/G_E} \tag{9-12}$$

where G_B is called the Gummel number and is calculated from

$$G_B = \int_{\text{base}} \left(\frac{N_{aB}}{D_{nB}}\right) dx \tag{9-13}$$

and G_E is an analogous quantity for the emitter which is given by

$$G_E = \int_{\text{emitter}} \left(\frac{N_{dE}}{D_{pE}}\right) dx \tag{9-14}$$

Equation 9-14 must be modified to include the effects of heavy emitter doping to complete the model. Note that some references define the Gummel number as $\int_{\text{base}} N_{aB}\, dx$.

It is possible for one of the factors, M, α_T, or γ, to overshadow the other two factors in the determination of the dc common-emitter current gain. If this is the case, the design criterion can be related to this factor alone. To investigate this possibility, consider that

$$\beta_F = \frac{\alpha_F}{1-\alpha_F} = \frac{1}{1/\alpha_F - 1} \tag{9-15}$$

but

$$\frac{1}{\alpha_F} = \frac{1}{M} \cdot \frac{1}{\alpha_T} \cdot \frac{1}{\gamma} \tag{9-16}$$

and, using Equations 9-6, 9-7, and 9-12,

$$\frac{1}{\alpha_F} = \left[1 - \left(\frac{V_{CB}}{BV_{CBo}}\right)^n\right]\left(1 + \frac{x_B{}^2}{2L_n{}^2}\right)\left(1 + \frac{G_B}{G_E}\right) \tag{9-17}$$

Then, since all three factors are very close to unity, neglecting second order terms

$$\beta_F \approx \frac{1}{G_B/G_E + x_B{}^2/2L_n{}^2 - (V_{CB}/BV_{CBo})^n} \tag{9-18}$$

and, if one of the three terms in the denominator is significantly larger than the other two, it will dominate the determination of β_F for the device. For operation so that $V_{CB} \ll BV_{CBo}$, the larger of two positive terms will be the most significant. Since L_n is on the order of 30 μm, as indicated in Figure 9-4,

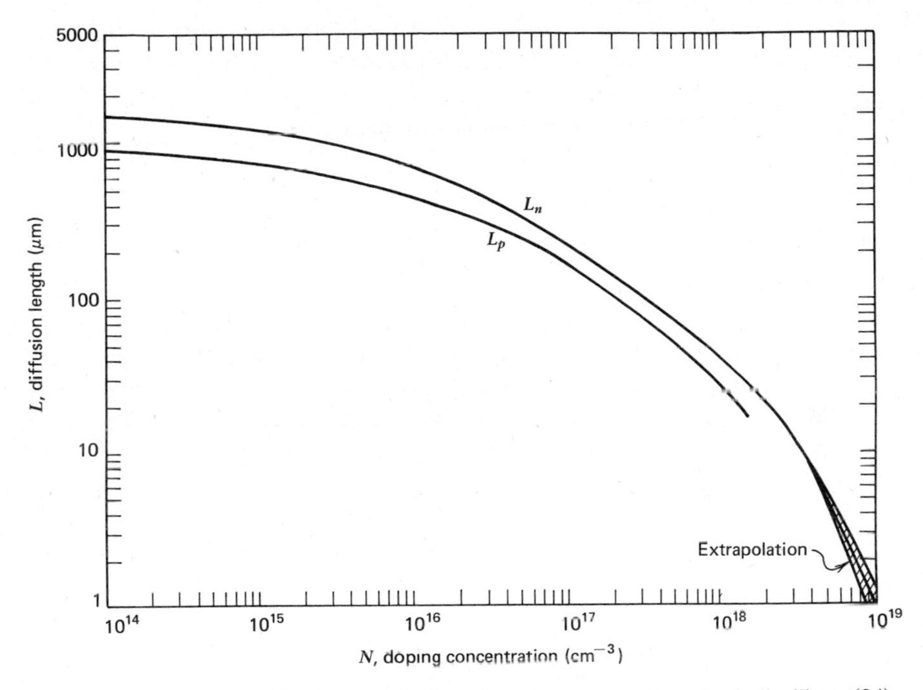

Figure 9-4. Diffusion length as a function of doping concentration for bulk silicon (24).

and typical values for x_B range from 0.2 to 1 μm, the term due to α_T is usually very small. The value of G_E is relatively fixed quantity (as described below), permitting the designer to vary G_B to exert control over the dc common-emitter current gain. For most modern *npn* transistors, the emitter efficiency will dominate the determination of β_F and the width of the base will primarily affect the high-frequency characteristics of the device.

Calculation of the Emitter Efficiency. The emitter efficiency can be readily calculated once the Gummel number, G_B, and the emitter figure of merit, G_E, are determined. The Gummel number is relatively easy to calculate, but the emitter figure of merit is very complicated.

The Gummel number is usually found by extracting the position dependent diffusion coefficient from the integral, and replacing it by an effective or average diffusion coefficient, $\tilde{D}_{nB}$, resulting in

$$G_B = \frac{1}{\tilde{D}_{nB}} \int_{\text{base}} N_{aB}(x)\, dx = \frac{Q_{Bo}}{q\tilde{D}_{nB}} \qquad (9\text{-}19)$$

where Q_{Bo} is called the built-in charge, and q is the electronic charge (1.602×10^{-19} C). Note that the units of Q_{Bo}/q are atoms per centimeters squared, the same units as the total dose of an ion implant. Since ion implantation can be carefully controlled, a "Gummel number implant" can be used to establish the doping in the part of the base under the emitter (the "active" part of the base). If the base is formed by a limited source diffusion, resulting in a Gaussian impurity profile with a final surface concentration of N_{0B}, the net acceptor concentration in the base is given by

$$N_{aB}(x) \approx N_{0B} \exp\left(\frac{-x^2}{4Dt}\right) - N_{dC} \tag{9-20}$$

where N_{dC} is the doping concentration in the uniformly doped collector. This equation is based on Equation 7-6. The approximation is due to the neglect of the donor concentration of the emitter, which overlaps the base very near the emitter-base junction. This is a reasonable approximation, since the emitter doping profile has a large gradient in this region. It is then possible to calculate Q_{Bo}/q using

$$\frac{Q_{Bo}}{q} = \int_{x_{EB}}^{x_{CB}} N_{aB}(x)\,dx = \int_{x_{EB}}^{x_{CB}} \left[N_{0B} \exp\left(\frac{-x^2}{4Dt}\right) - N_{dC}\right] dx \tag{9-21}$$

This integral can be evaluated by using the tabulated Area of the Normal Curve of Error (14). This function is given by

$$\text{area} = \int_0^t \left(\frac{1}{2\pi}\right)^{1/2} \exp\left(\frac{-y^2}{2}\right) dy$$

which is plotted in Figure 9-5. Changing variables, Equation 9-21 can be rewritten

$$\frac{Q_{Bo}}{q} = 2(\pi Dt)^{1/2} N_{0B}\left[\int_0^{t_2} \left(\frac{1}{2\pi}\right)^{1/2} \exp\left(\frac{-y^2}{2}\right) dy - \int_0^{t_1} \left(\frac{1}{2\pi}\right)^{1/2} \exp\left(\frac{-y^2}{2}\right) dy\right] - \int_{x_{EB}}^{x_{CB}} N_{dC}\,dx \tag{9-22}$$

where

$$t_1 = \frac{x_{EB}}{(2Dt)^{1/2}} \quad \text{and} \quad t_2 = \frac{x_{CB}}{(2Dt)^{1/2}}$$

The quantities, x_{EB} and x_{CB}, are the distances from the surface to the base edges of the emitter-base and collector-base space-charge regions, respectively, as shown in Figure 9-6. The difference, $x_{CB} - x_{EB}$, represents the active width of the base, x_B, in contrast to the difference, $x_{jC} - x_{jE}$, which is the metallurgical width of the base. It is difficult to determine the width of the space-charge regions of diffused junctions, but they can be estimated in the following manner. The base-collector junction is typically a Gaussian diffusion into a

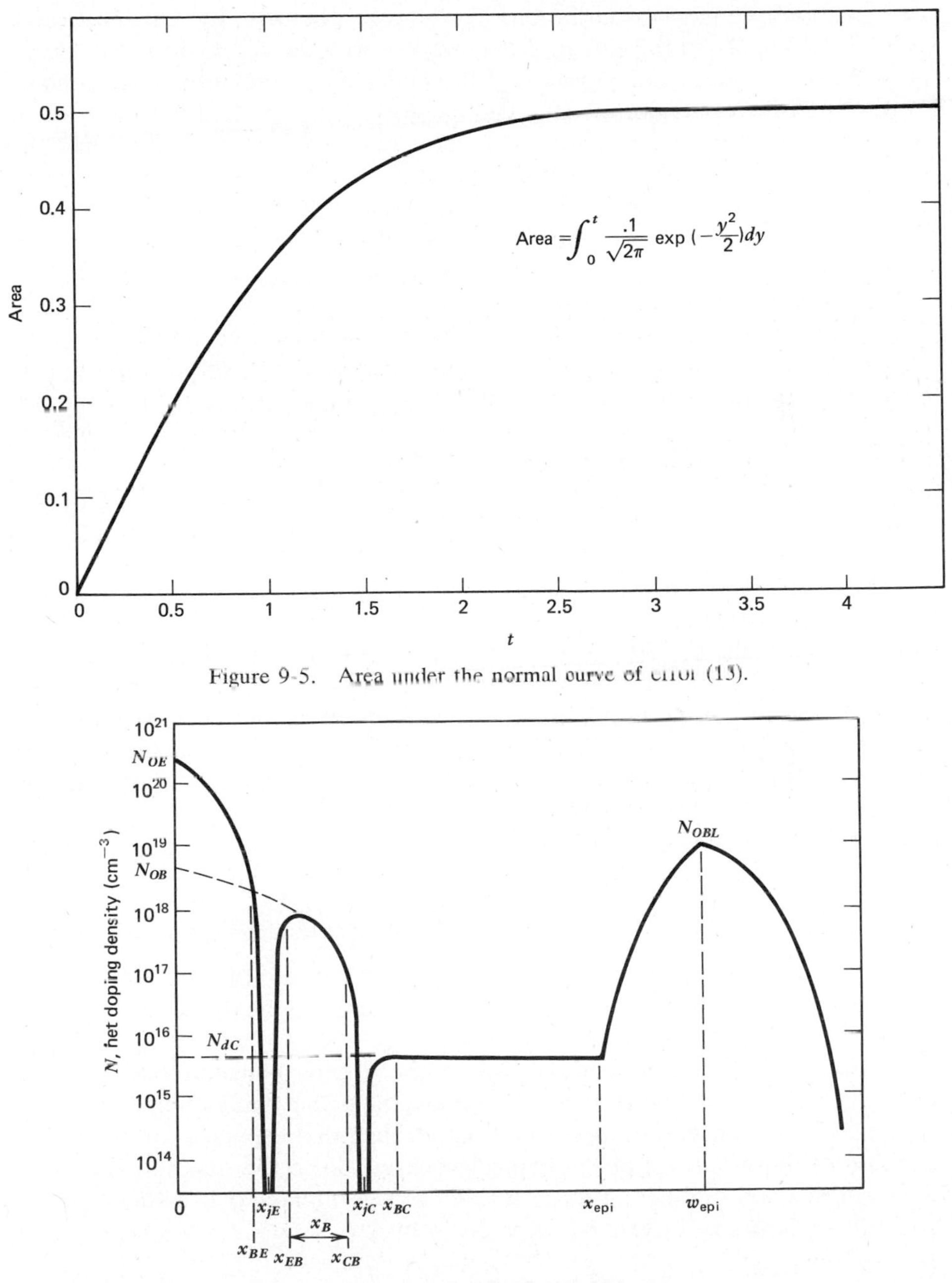

Figure 9-5. Area under the normal curve of error (13).

Figure 9-6. A transistor doping profile defining the distances used in the analysis of the device.

constant background concentration. The curves obtained by Lawrence and Warner (16) apply to this situation. Figure 9-7 was derived from these curves for a ratio of surface concentration of the Gaussian to background concentration in the collector region of 10^3. The quantity, V, in the figure is the sum of the built-in voltage and the applied reverse bias voltage across the junction. For zero bias conditions, $V \simeq 0.7$ V, and, for $x_j = 2\ \mu$m, $N_{dC} = 5\times10^{15}\ \text{cm}^{-3}$, and $N_{0B} = 5\times10^{18}\ \text{cm}^{-3}$, from Figure 9-7*a*, the total space-charge region width is 0.6 μm. Using Figure 9-7*b*, it is seen that 36% of this, or 0.22 μm is in the more heavily doped base side of the space-charge region. If the collector-base voltage is 15 V, the space charge region widens to 1.35 μm, with 25% of this, or 0.33 μm in the base side. Since both the base and emitter diffusion profiles vary with position, the Lawrence and Warner curves do not apply, but they can be used as an approximation, which indicates that $x_{EB} - x_{jE}$ is on the order of 0.1 μm and varies little with forward bias. For design purposes, it will be assumed that the active base width, x_B, will be

$$x_B = x_{jC} - x_{jE} - 0.4 \qquad (\text{in } \mu\text{m}) \tag{9-23}$$

for most situations.

The value selected for $\tilde{D}_{nB}$ in Equation 9-19 must represent the diffusion coefficient for electrons, averaged over the base region. As a first-order approximation, the average $\tilde{D}_{nB}$ will be that value from Figure 7-3 corresponding to a doping concentration of $N_{aB}(x_{EB})/2$.

The theoretical model for the emitter region does not agree as well with experiment as the corresponding model for the base region. As a consequence, the calculation of the emitter figure of merit, G_E, is not as accurate as the calculation of the Gummel number. Following the same basic procedure as that applied to the base region, G_E can be written as

$$G_E = \frac{1}{\tilde{D}_{pE}} \int_{\text{emitter}} N_{dE}(x)\, dx = \frac{Q_{Eo}}{q\tilde{D}_{pE}} \tag{9-24}$$

where $\tilde{D}_{pE}$ is the effective diffusion coefficient for holes in the emitter, and Q_{Eo} is the built-in emitter charge. The quantity, N_{dE}, is the effective emitter doping concentration.

For a heavily doped emitter, where the doping concentration is large enough to form an impurity band that overlaps the conduction band, the intrinsic carrier concentration, $n_{ie}(x)$, is larger than that expected for silicon, n_i, because of the reduction of the bandgap energy. Since the doping concentration is a function of position for a diffused or ion-implanted structure, n_{ie} is a function of position. In general, n_i is given by

$$n_i = (N_c N_v)^{1/2} \exp\left(\frac{-E_g}{kT}\right) \tag{9-25}$$

where N_c and N_v are effective densities of states in the conduction and valence

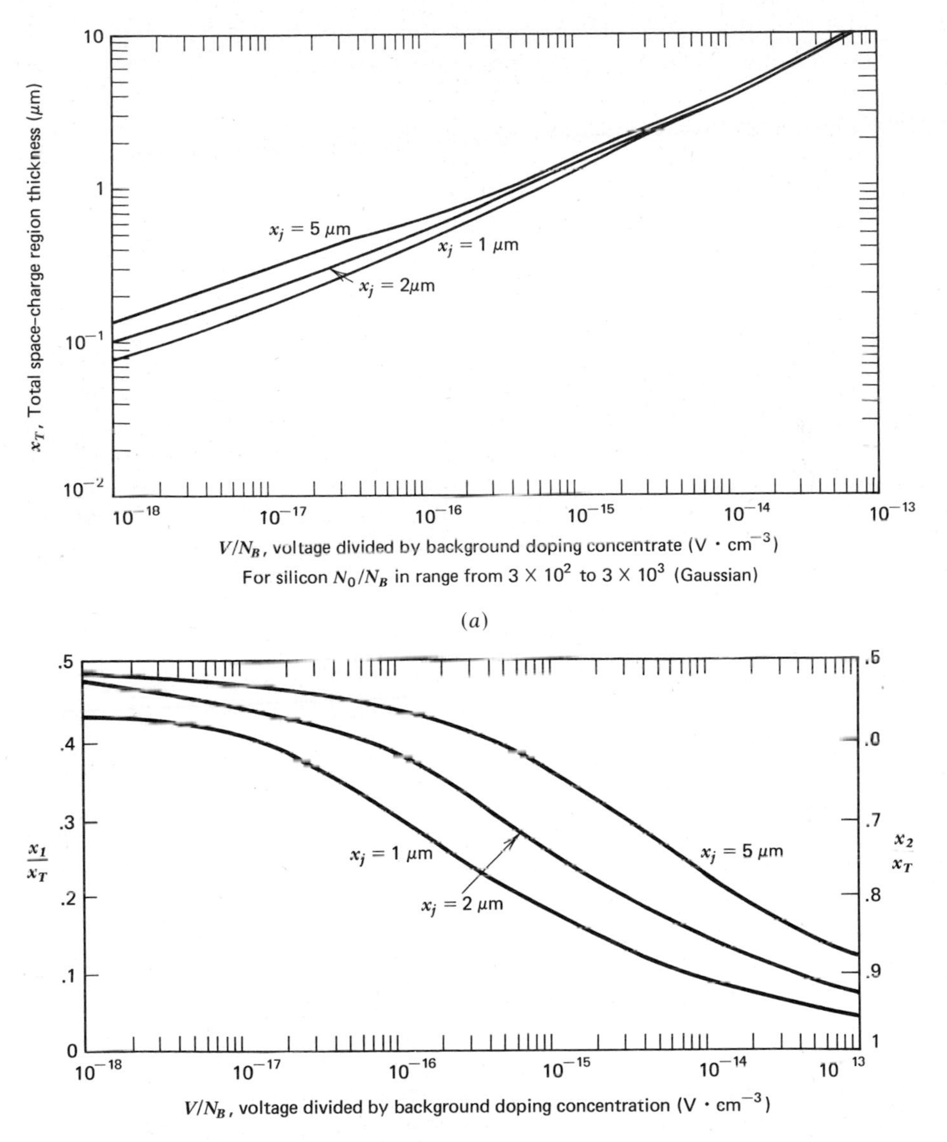

Figure 9-7. The space-charge region thickness as a function of voltage for a *pn* junction formed by a Gaussian diffusion into a constant background concentration (range of N_0/N_B from 3×10^2 to 3×10^3). (*a*) Total width, x_T. (*b*) Ratio of x_1/x_T and x_2/x_T where x_1 is the portion in the heavier doped side and x_2 is the portion in the lighter doped side [adapted from (16)]. Copyright © 1960 American Telephone and Telegraph Company. Reprinted by permission from *The Bell System Technical Journal.*

bands, and E_g is the energy associated with the forbidden gap. The densities of states are also altered by heavy doping. Since these effects begin to appear at doping densities between $10^{18}\ \text{cm}^{-3}$ and $10^{19}\ \text{cm}^{-3}$ and typical transistors have emitter doping surface concentrations on the order of $10^{21}\ \text{cm}^{-3}$ and base doping surface concentrations near $10^{19}\ \text{cm}^{-3}$, the heavy doping effects must be considered for both the valence and conduction bands, although the effects on the conduction band are more pronounced.

The effect of the bandgap narrowing on the effective emitter doping can be introduced in the following way. The effective emitter doping concentration can be written

$$N_{dE} = [N_E(x) - N_{aB}(x)]\left[\frac{n_i}{n_{ie}(x)}\right]^2 \tag{9-26}$$

where $N_E(x)$ is the emitter doping profile. A plot of n_{ie} as a function of net donor density with acceptor density as a parameter is shown in Figure 9-8. Another way of interpreting the effects of bandgap narrowing is to consider the effects on the minority carrier density. Since $n_o p_o = n_{ie}^2$, an increase in n_{ie} implies an increase in the equilibrium minority carrier density in the region. This results in a reduction in the emitter efficiency, since an increase in the equilibrium minority carrier density implies an increase in the injected minority carrier density at the edge of the space-charge region in the emitter.

There are two other effects due to heavy doping which also act to reduce the emitter efficiency. The first of these is a built-in electric field, which, due to the effective reduction of the doping density, can be opposite in direction to what would be expected if heavy-doping effects were not present. If this electric field is present, it can aid the flow of minority carriers into the emitter. The second effect is a significant decrease in excess carrier lifetime in the heavily doped region. The primary mechanism for recombination in lightly to moderately doped silicon is through recombination centers with energy levels near the middle of the energy gap. For heavily doped materials, the probability of collisions between carriers is substantially increased. This increases the importance of the three-body collision (Auger) mechanism for recombination. The excess carrier lifetime is very low, resulting in a diffusion length for holes in the emitter less than $0.3\ \mu\text{m}$, and, in some cases, as small as $0.01\ \mu\text{m}$.

The limits on the integral for the calculation of Q_{Eo}/q (Equation 9-24) are not as well defined as those for the calculation Q_{Bo}/q (Equation 9-19). The original expression for emitter efficiency (Equation 9-12) was developed on the premise that G_B/G_E was the ratio of the hole current injected into the emitter to the electron current injected into the base. In a modern transistor structure, the width of the base is much less than the diffusion length for electrons in the base. Therefore, the base region looks like a "short" diode, and the integration for Q_{Bo}/q is carried out over the quasineutral base width. In the emitter, however, the diffusion length for holes is very small, and it is common for the

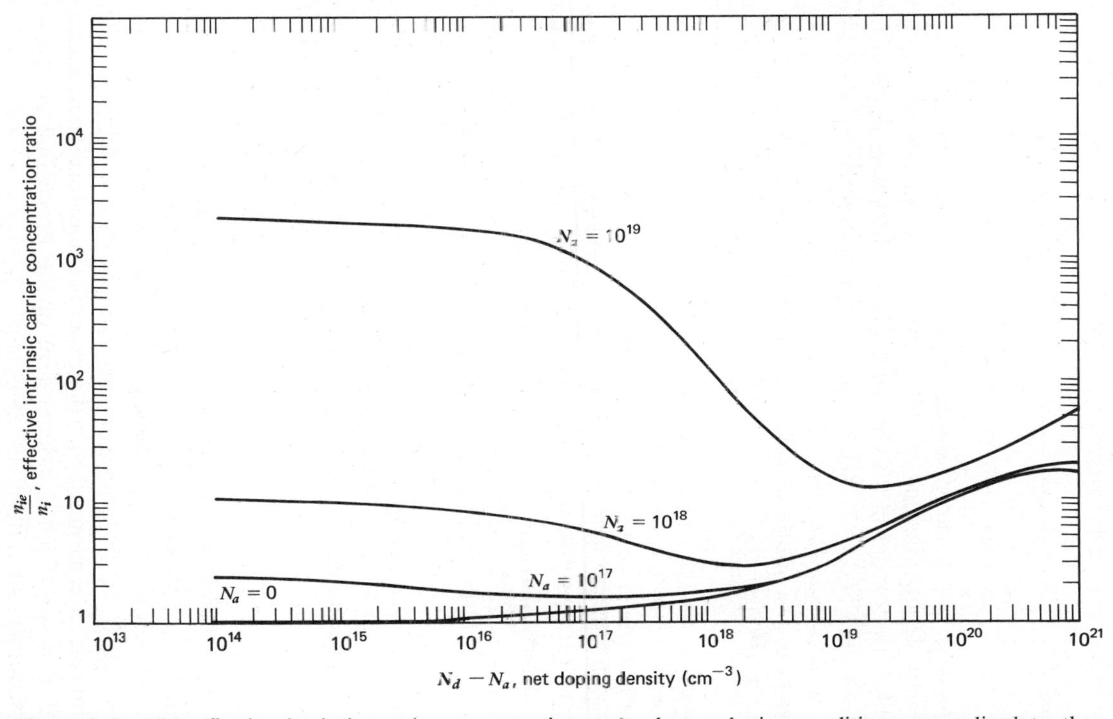

Figure 9-8. The effective intrinsic carrier concentration under heavy doping conditions, normalized to the intrinsic carrier concentration for no bandgap narrowing, as a function of net doping density with acceptor density as a parameter (10). Reprinted with permission from *Solid-State Electronics*, Vol. 16, No. 11, M. S. Mock, "Transport Equations in Heavily Doped Silicon, and the Current Gain of a Bipolar Transistor," Copyright © 1973, Pergamon Press, Ltd.

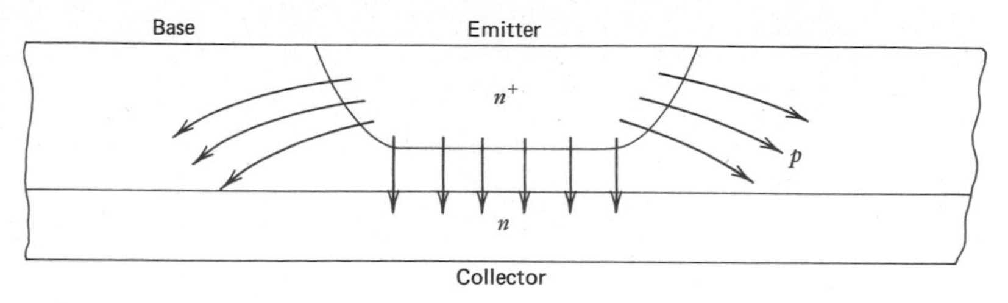

Figure 9-9. Emitter sidewall injection.

emitter region to look like a "long" diode. In this case the integral is carried out over a diffusion length, instead of the width of the quasineutral emitter region. This will be true except in the case of the very shallow emitters formed in microwave transistors and very high-speed logic circuits such as those found in the emitter-coupled logic families.

The final effect of the emitter profile on the emitter efficiency is related to injection of electrons from the sidewalls of the emitter into the base. These electrons are injected into regions of the base where the distance from the emitter-base junction to the collector-base junction is long compared to a diffusion length for electrons, as shown in Figure 9-9. These electrons, particularly those near the surface, are very likely to recombine with holes before they have the opportunity to reach the collector, thus, reducing the emitter efficiency. In many structures, this is a second order effect. The most important result of the emitter sidewalls is to increase the area of the emitter-base junction, and, thus, increase the transition region capacitance of this junction, which has a significant effect on the frequency response of the device.

The combined result of the contributions of the effects discussed in the preceding paragraphs for a heavily doped emitter structure is an emitter figure of merit, G_E, which is essentially independent of the process used to form the emitter. The values obtained for G_E range from 2×10^{13} to $9\times10^{13}\ \text{cm}^{-4}\cdot\text{s}$ with a typical design value of $5\times10^{13}\ \text{cm}^{-4}\cdot\text{s}$. To develop a process in which G_E has a controllable value, a structure with a lightly doped emitter must be used. One such device, the low-emitter-concentration (LEC) transistor, is shown in Figure 9-10. Common-emitter current gains in excess of 1000 can be obtained with this type of structure. The heavily doped region in the emitter of the LEC is needed to ensure that an ohmic contact can be made to the emitter. Since the vast majority of integrated circuit transistors are fabricated with heavily doped emitters, it is assumed that the process to be developed will have an emitter figure of merit which is independent of the processing parameters, and, thus, not within the control of the designer.

The Cutoff Frequency of the Intrinsic Transistor. If the width of the base is small compared to a diffusion length, the major effect of the base width on the transistor characteristics is related to the alpha cutoff frequency, ω_α, of the

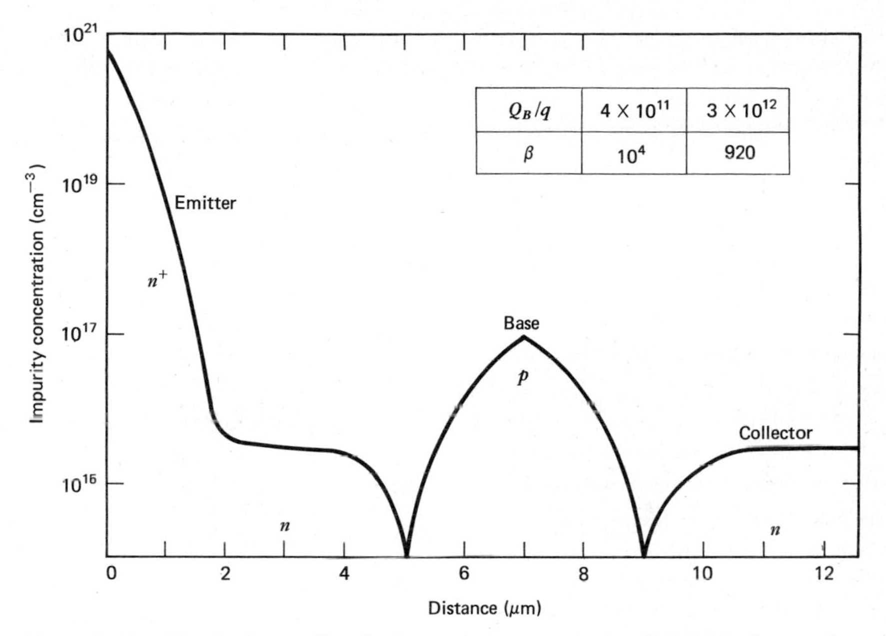

Figure 9-10. The doping profile of a low emitter concentration (LEC) bipolar transistor (25) Copyright © 1974 IEEE. Reprinted from *Technical Digest 1974 International Electron Devices Meeting*, December 9–11, 1974, Washington, D. C., p. 262.

device. This is the frequency at which the magnitude of the common-base current gain is reduced to 0.707 of its low-frequency value, in the one-dimensional model. This model is called the intrinsic transistor model, since it ignores the effects of the parasitic elements like the resistance of the inactive base and the transition region capacitances of the base-emitter and base-collector junctions. The values of the parasitic elements are mainly related to the geometrical layout of the device. If the alpha cutoff frequency is sufficiently high, the frequency response of the device will be dominated by the parasitic elements. The design approach is to ensure that the above condition is satisfied.

The alpha cutoff frequency is related to the transit time for minority carriers across the base. For a uniformly doped base,

$$\omega_\alpha = \frac{\pi^2}{4} \frac{D_{nB}}{x_B^2} \tag{9-27}$$

and

$$\tau_B = \frac{x_B^2}{2D_{nB}} \tag{9-28}$$

where τ_B is the transit time across the base. Then,

$$\omega_\alpha = \frac{\pi^2}{8\tau_B} \tag{9-29}$$

This relationship between alpha cutoff frequency and base transit time is a good approximation for all bipolar transistor structures. In general, the transit time can be written

$$\tau_B = \frac{x_B^{\,2}}{\nu \tilde{D}_n} \tag{9-30}$$

where ν ranges from 2 to 20, depending on the impurity gradient in the base and the injection level.

Junction Breakdown Voltages (15–17). Another set of important characteristics of junction transistors that depends on the doping profiles is the breakdown voltages of the base-emitter and base-collector junctions, and the resulting collector-emitter breakdown voltage. These breakdown voltages determine the range of power supply voltages which can be applied to the circuit for safe operation. In most cases, the breakdown voltage of a junction is determined primarily by the doping concentration of the more lightly doped side. In order to increase the breakdown voltage, it is necessary to reduce the doping concentration, which may not be consistent with other needs for circuit operation.

There are two basic mechanisms for junction breakdown, the Zener effect and avalanche multiplication. The Zener effect is a form of field-induced tunneling. The electric field due to the applied reverse bias results in a band condition such that electrons in the valence band are opposite empty states in the conduction band at the same energy level, separated by a distance that is small enough to make tunneling highly probable. The field necessary to produce Zener breakdown is $10^6\ \text{V} \cdot \text{cm}^{-1}$. The conditions that lead to the Zener effect can only occur if both sides of the junction are relatively heavily doped. If at least one side of the junction is lightly doped, the electric field in the space-charge region under reverse bias will become large enough to accelerate carriers to energies sufficient to ionize the silicon atoms in the space charge region. The electron-hole pairs produced in this process are also accelerated by the field, resulting in the formation of more electron-hole pairs, and so forth. This avalanche multiplication process occurs at electric fields smaller than that required to produce the conditions for the Zener effect in junctions with at least one side lightly doped. Since the voltage across a junction under breakdown remains essentially constant, the electric field does not reach the magnitude necessary to produce Zener breakdown in an avalanching junction. Zener breakdown is the predominant effect in junctions with breakdown voltages less than 5 V, and avalanche multiplication dominates in junctions with breakdown voltages greater than 8 V. Junctions that breakdown between 5 and 8 V exhibit a combination of the two breakdown mechanisms.

The base-emitter junction breakdown in a typical double-diffused transistor structure is usually within the mixed mechanism range. Since both diffusions have their maximum concentrations at or near the surface, the breakdown usually occurs near the surface. Figure 9-11 is a plot of the breakdown voltage of a double-diffused junction as a function of base-diffusion surface concentration, N_{0B}, with the base-emitter junction depth, x_{jE}, as a parameter. This figure is based on empirical values for the ionization coefficient associated with avalanche multiplication. The breakdown voltage for most base-emitter junctions is approximately 6 V. It is common practice to use a reverse-biased base-emitter junction as a reference diode, when such a device is required.

The breakdown voltage of the base-collector junction is usually higher than that of the base-emitter junction. Avalanche breakdown is the primary breakdown mechanism. When the transistor is connected in the common-emitter configuration, there are three possible ways in which breakdown can occur. The width of the space-charge region of the base-collector junction is a function of the magnitude of the reverse bias voltage. Since the base region is usually more heavily doped than the collector region (see (Figure 9-4), the space-charge region expands farther into the collector than into the base with

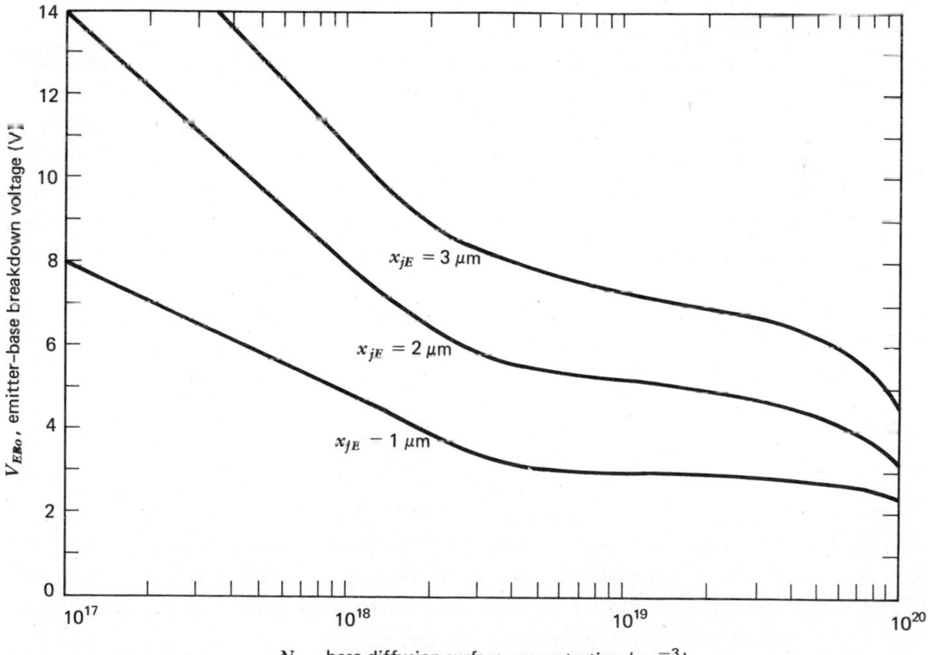

Figure 9-11. Emitter-base breakdown voltage as a function of base diffusion final surface concentration with emitter-base junction depth as a parameter [adapted from (17)]. Reprinted with permission from *Solid-State Electronics*, Vol. 17, No. 5, P. R. Wilson, "The Emitter-Base Breakdown Voltage of Planar Transistors," copyright © 1974, Pergamon Press, Ltd.

increasing voltage. If the conditions for avalanche breakdown are satisfied at a voltage below that necessary to widen the space-charge region to either the n^+ buried layer or the base edge of the base-emitter space-charge region, normal avalanche breakdown occurs. For a planar junction, such as that found in a typical integrated circuit, this breakdown is influenced by the curvature of the junction at the corners between the plane portion of the junction and the portion curving toward the surface. The effect of these corners is to reduce the voltage at which breakdown occurs to a value below that which would be expected for a plane junction. For this reason, the depth of the junction is a parameter in determining the avalanche breakdown voltage. If the space-charge region of the base-collector junction meets the n^+ buried layer at a voltage insufficient to produce breakdown, further widening of the space-charge region into the collector region is impossible. The voltage necessary to produce this effect is called the reach-through voltage. The results of this are an increase of the electric field in the collector region and an increase in the width of the space-charge region in the base. If the field in the collector region reaches a magnitude sufficient for avalanche multiplication before the space-charge region in the base reaches the space-charge region of the emitter-base junction, breakdown occurs at a voltage lower than that which would cause breakdown in the absence of a buried layer. The third type of breakdown occurs if the space-charge region in the base from the base-collector junction meets the space-charge region of the emitter-base junction at a voltage below that required for avalanche breakdown. A further increase in base-collector voltage results in large collector currents, since carriers are injected from the emitter directly into the base-collector space-charge region. This effect is called "punch-through," and is enhanced by the presence of a buried layer. The doping profile of the structure determines which type of breakdown occurs in a particular transistor.

The breakdown voltage of the base-collector junction with the emitter terminal open, BV_{CBo}, is a function of the impurity concentration in the collector, N_{dC}, the doping profile of the base, the depth of the base-collector junction, x_{jC}, and the distance from the surface to the buried layer, x_{epi}. In Figure 9-12, for a Gaussian base diffusion profile with a surface impurity concentration, N_{0B}, in the range from $3\times10^2\ N_{dC}$ to $3\times10^3\ N_{dC}$, the avalance breakdown voltage is plotted as a function of N_{dC} with x_{jC} as a parameter. Also included in this figure are curves representing reach-through conditions with $x_{epi}-x_{BC}$ as a parameter. The quantity, x_{BC}, is found by adding the collector portion of the base-collector transition region, as determined from Figure 9-7, to the base-collector junction depth, x_{jC}.

The common-emitter collector junction breakdown voltage usually depends on the electrical connection of the base. The most severe condition is the open base situation, and BV_{CEo} represents the lowest collector-emitter breakdown voltage. For this situation, the collector current equals the emitter

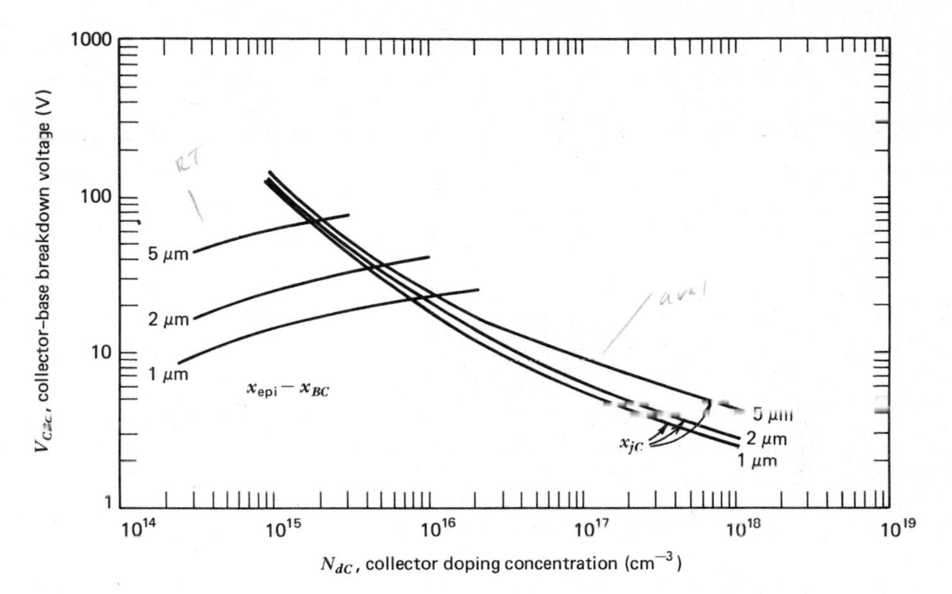

Figure 9-12. Collector-base breakdown voltage as function of collector doping concentration with collector-base junction depth as a parameter, modified by reach-through conditions [adapted from (18) and (26)]. Reprinted by permission of the publishers, Pergamon Press Ltd., and The Electrochemical Society, Inc.

current, resulting in

$$I_C = \frac{I_{Co}}{1-\alpha_F} = \beta_F I_{Co} \tag{9-31}$$

where I_{Co} is the collector-base leakage current with the emitter open. Since, from the definition of β_F, breakdown occurs when α_F equals unity, using Equations 9-5 and 9-6, it can be shown that

$$BV_{CEo} \simeq \frac{BV_{CBo}}{(\beta_F)^{1/n}} \tag{9-32}$$

where, as before, n is between 2 and 4 for silicon, if avalanche multiplication is the breakdown mechanism. The current-voltage characteristic of an open base transistor is shown in Figure 9-13. Note that BV_{CEo} is specified, typically at 5 mA, instead of 10 or 100 μA like BV_{CBo} or BV_{EBo}. This is because of the "snap-back" or negative resistance portion of the curve, eliminating the possibility of a higher voltage than that which the device can sustain from being specified for BV_{CEo}. The snap-back effect is due to the variation of β_F with collector current (discussed later in this chapter) and Equation 9-32. For very low-current levels, β_F is low, but, an increase in collector current due to the start of breakdown (V_1 in the figure) results in a lower voltage for breakdown and a higher current, until an equilibrium sustaining breakdown occurs.

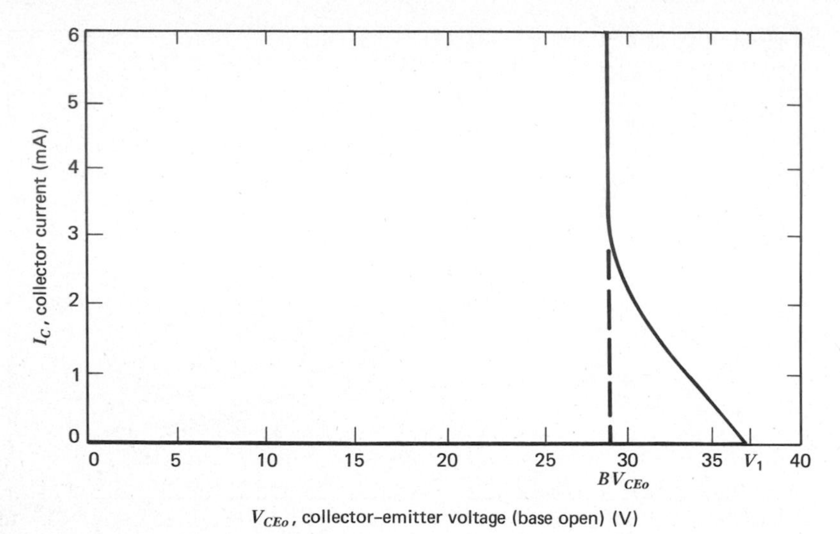

Figure 9-13. The open-base collector-emitter breakdown voltage characteristic of a typical bipolar transistor.

The design goals for many transistors include a low-collector series resistance and a high-collector-emitter breakdown voltage. Unfortunately, most of the methods for reducing the collector series resistance also reduce the collector-emitter breakdown voltage. This indicates the need for a design compromise. The breakdown voltage is increased by including a thin, lightly doped epitaxial layer in the structure. The collector series resistance is reduced by inserting a heavily doped n-type "buried" layer into the substrate before the epitaxial layer is grown, as shown in Figure 9-14. Most of the collector current follows the path indicated in the figure. The major components of the collector series resistance are the vertical paths between the base-collector junction and the buried layer, and the buried layer and the collector-contact diffused region. One technique for reducing the resistance due to the latter path is to perform an additional processing step, a deep-collector diffusion. This diffusion results in heavily doped phosphorus region that is driven through the epitaxial layer to provide a low-resistance path from the collector contact to the buried layer. This process is inserted between the isolation diffusion and the base diffusion. The cost of adding a deep-collector diffusion must be justified by the improvement in the performance of the devices in the highly competitive market place. Increasing the area of the emitter will reduce the collector series resistance, but this also significantly reduces the frequency response of the device. The optimum design for the collector region consists of the highest doping level in the thinnest epitaxial layer which will support the specified breakdown voltage. The information necessary for designing the collector region is contained in Figure 9-12.

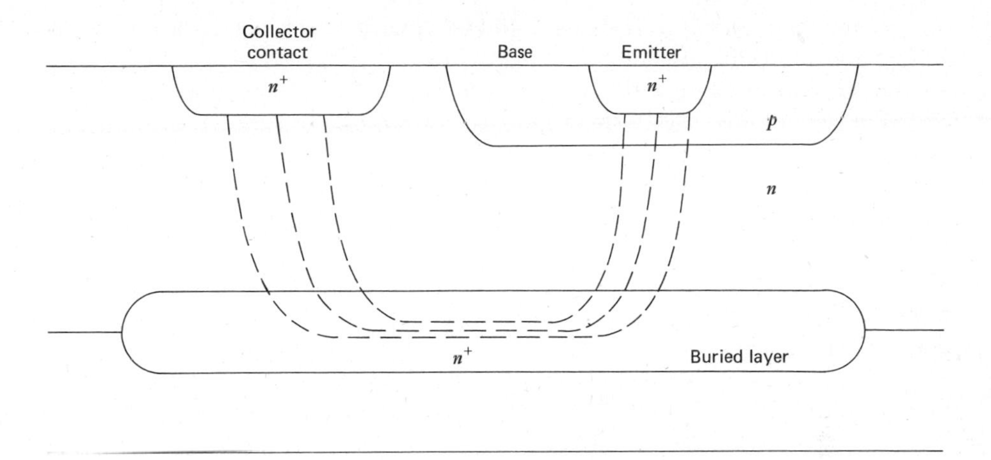

Figure 9-14. The dominant current path between the emitter and collector contact for an integrated circuit transistor with a buried layer.

The collector-emitter breakdown voltage with the base shorted to the emitter, BV_{CEs}, is, under most types of breakdown, equal to BV_{CBo}. If, however, the breakdown is due to punch-through,

$$BV_{CEs} = BV_{CBo} - BV_{EBo} \tag{9-33}$$

The punch-through effect is influenced by the thickness of the epitaxial layer, since reach-through forces an expansion of the base-collector space-charge region into the base. This effect is called "premature punch-through" (19). Under these circumstances, the breakdown voltage can be estimated by

$$BV_{CBo} = \frac{q}{\varepsilon_s}\left(\frac{Q_{Bo}x_1}{q} - \frac{N_{dC}x_1^2}{2} + \frac{Q_{Bo}x_B}{2q}\right) \tag{9-34}$$

where $x_1 = x_{epi} - x_{BC}$, x_B is the active base width, and Q_{Bo} is the built-in charge in the base. If the voltage calculated from this expression is less than that determined from Figure 9-12, premature punch-through is the breakdown mechanism.

THE STANDARD PROCESS

Based on the discussion of the previous sections, it is now possible to develop a standard process for the fabrication of bipolar transistors in a double-diffused epitaxial integrated circuit. The designer has reasonable latitude in the selection of β_F, ω_α, and BV_{CBo}. In actual practice, the final processing sequence is developed by estimating the diffusion time-temperature cycles, simulating the process in a computer program like SUPREM (1), and making a series of process runs with slight variations until the electrical characteristics are satisfactory and reproducible.

The design procedure is outlined in flowchart form in Figure 9-15. In the following paragraphs, this design sequence is used to determine a process for a transistor with $\beta_F = 45$, $\omega_\alpha = 2\pi \times 5 \times 10^9 \text{rad} \cdot \text{s}^{-1}$, a minimum BV_{CBo} of 25 V, and a base sheet resistance of 200 Ω/□. The geometry of the device is shown in Figure 9-6. In this design procedure, extensive use is made of the figures and equations of Chapter 7.

1. The base Gummel number is calculated from Equation 9-18, which assumes that emitter efficiency dominates β_F. Using the design value for G_E of $5 \times 10^{13}\ \text{cm}^{-4} \cdot \text{s}$,

$$G_B = \frac{G_E}{\beta_F} = 1.11 \times 10^{12}\ \text{cm}^{-4} \cdot \text{s}$$

The average base doping concentration in a transistor is typically $10^{17}\ \text{cm}^{-3}$ which indicates that the average minority carrier diffusion coefficient in the base, $\tilde{D}_{nB}$, is $15\ \text{cm}^2 \cdot \text{s}^{-1}$ from Figure 7-3. Then, from Equation 9-19,

$$\frac{Q_{Bo}}{q} = G_B \tilde{D}_{nB} = 1.67 \times 10^{13}\ \text{cm}^{-2}$$

The active base width can be calculated by combining Equations 9-29 and 9-30, and choosing a worst case value for ν of 4, to result in

$$x_B = \pi \left(\frac{\tilde{D}_{nB}}{2\omega_\alpha} \right)^{1/2} = 0.49 \times 10^{-4}\ \text{cm}$$

2. The collector doping density and the distance from the base-collector junction to the buried layer, $x_{epi} - x_{BC}$, can be determined from the breakdown voltage specification and Figure 9-12. Assuming that the base-collector junction will be located approximately 2 μm from the surface, the maximum allowable collector doping is $8 \times 10^{15}\ \text{cm}^{-3}$ and

$$x_{epi} - x_{BC} \simeq 1.2\ \mu\text{m}$$

3. To determine if premature punch-through would result from this structure, the values obtained thus far are inserted into Equation 9-34,

$$BV_{CBo} = \left(\frac{q}{\varepsilon_s} \right) \left(\frac{Q_{Bo} x_1}{q} - \frac{N_{dC} x_1^2}{2} + \frac{Q_{Bo} x_B}{2q} \right)$$
$$= 372\ \text{V}$$

($x_1 = x_{epi} - x_{BC}$). This indicates that, for premature punch-through to occur, the device would have to have an avalanche breakdown voltage in excess of 372 V. Therefore the values selected to this point are valid and we can progress to the next step.

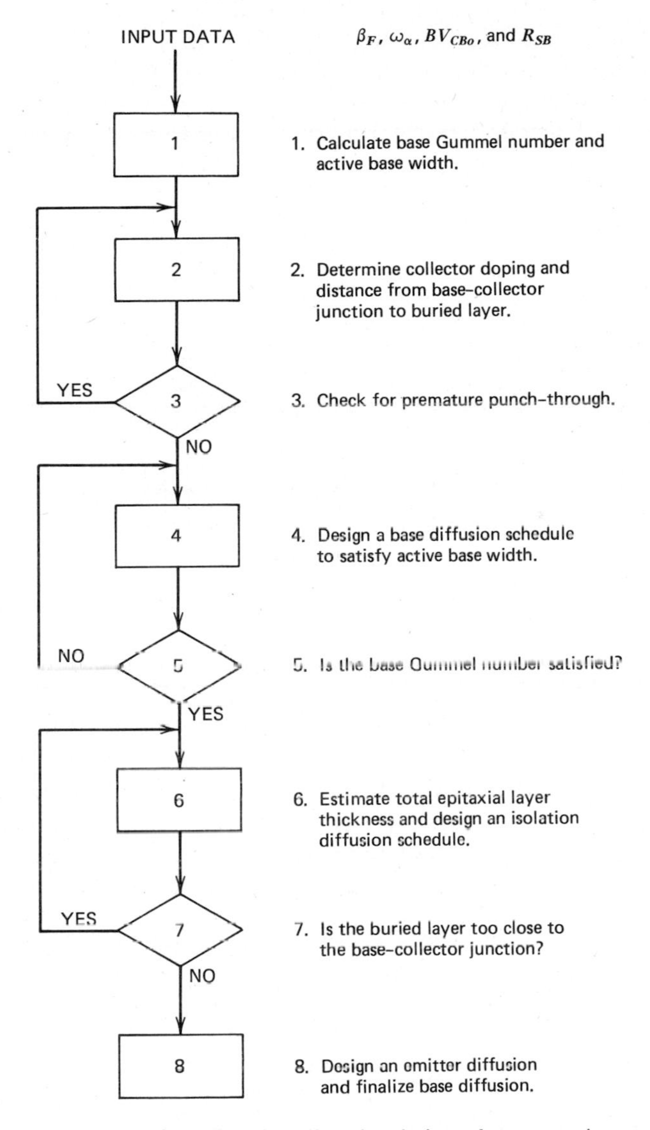
INPUT DATA
β_F, ω_α, BV_{CBo}, and R_{SB}
1
2
3
4
5
6
7
8
YES
NO
NO
YES
YES
NO
1. Calculate base Gummel number and active base width.
2. Determine collector doping and distance from base-collector junction to buried layer.
3. Check for premature punch-through.
4. Design a base diffusion schedule to satisfy active base width.
5. Is the base Gummel number satisfied?
6. Estimate total epitaxial layer thickness and design an isolation diffusion schedule.
7. Is the buried layer too close to the base-collector junction?
8. Design an emitter diffusion and finalize base diffusion.

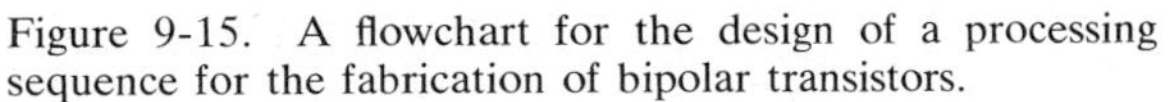

Figure 9-15. A flowchart for the design of a processing sequence for the fabrication of bipolar transistors.

4. The procedure for determining the diffusion schedule for the base-collector junction is as follows. A value is selected for the junction depth, x_{jC}. This number is multiplied by the desired base sheet resistance, R_{SB}, and the result is used to determine the final surface concentration of the base diffusion, N_{OB}, assuming a Gaussian profile, from Figure 7-11. For $x_{jC} = 2\ \mu\text{m}$, $R_{SB}x_{jC} = 400\ \Omega \cdot \mu\text{m}$, which indicates that

$$N_{OB} = 4 \times 10^{18}\ \text{cm}^{-3}$$

using the $N_B = 10^{16}\ \text{cm}^{-3}$ curve, since this is the closest value to N_{dC}. Assuming a two-step diffusion, the junction occurs when

$$N(x_j, t) = N_{OB} \exp\left(\frac{-x_{jC}^2}{4D_{2B}t_{2B}}\right) = N_{dC}$$

which leads to

$$D_{2B}t_{2B} = 1.61 \times 10^{-9}\ \text{cm}^2$$

This is all that is needed at this point, since the actual time-temperature cycle will be determined at the end of the design procedure.

5. It is now necessary to use Equation 9-22 to calculate Q_{Bo}/q for the base diffusion profile and compare the result with the previously calculated desired value. If the resulting Q_{Bo}/q is too large, β_F will be too low, and x_{jC} must be increased, and Step 4 repeated. If Q_{Bo}/q is too small, β_F will be larger than specified, and x_{jC} must be decreased to adjust β_F to the desired value. In order to use Equation 9-22, it is necessary to determine the locations of x_{EB} and x_{CB}, the boundaries of the active base. The base-collector space-charge region width can be determined from Figure 9-7. Using $V = 0.7$ V for the unbiased junction, the total space-charge region width is 0.5 μm, with the portion in the base side 0.19 μm and the portion in the collector side 0.31 μm. Then, if $x_{jC} = 2\ \mu\text{m}$, $x_{CB} = 1.81\ \mu\text{m}$ and $x_{EB} = x_{CB} - x_B = 1.32\ \mu\text{m}$. Then

$$\frac{Q_{Bo}}{q} = 2(\pi Dt)^{1/2} N_{OB}\left[\int_0^{t_2} \left(\frac{1}{2\pi}\right)^{1/2} \exp\left(\frac{-y^2}{2}\right) dy - \int_0^{t_1} \left(\frac{1}{2\pi}\right)^{1/2} \exp\left(\frac{-y^2}{2}\right) dy\right] - \int_{x_{EB}}^{x_{CB}} N_{dC}\, dx$$

with $t_1 = x_{EB}/(2Dt)^{1/2}$ and $t_2 = x_{CB}/(2Dt)^{1/2}$. The result is

$$\frac{Q_{Bo}}{q} = 5.3 \times 10^{12}\ \text{cm}^{-2}$$

which is smaller than the required value. Repeating Steps 4 and 5 several times resulted in

$$x_{jC} = 1.45\ \mu\text{m}$$
$$N_{OB} = 6.5 \times 10^{18}\ \text{cm}^{-3}$$
$$D_{2B}t_{2B} = 7.845 \times 10^{-10}\ \text{cm}^2$$

and

$$\frac{Q_{Bo}}{q} = 1.66 \times 10^{13} \text{ cm}^{-2}$$

which is in excellent agreement with the desired value.

6. The total epitaxial layer thickness, w_{epi}, is an important consideration due to the diffusion of the buried layer into the collector region during the isolation diffusion process. The isolation diffusion must drive a boron junction all the way through the epitaxial layer, and is the largest time-temperature cycle of the fabrication process. This diffusion is calculated as if the epitaxial layer is infinite in extent, and it is desirable to locate a junction w_{epi} from the surface. At this point in the design, we have established a desired value for the final position of x_{epi}. Since $x_{epi} - x_{BC}$ must be 1.2 μm, and x_{BC} is the sum of x_{jC} and the collector portion of the base-collector transition region,

$$x_{epi} = 1.2 + 1.45 + 0.31 = 2.96 \ \mu\text{m}$$

The buried layer diffuses into the epitaxial layer during both the epitaxial growth and the isolation diffusion. In order to reduce the amount of epitaxial penetration by the buried layer, it is desirable to have a relatively low surface concentration on the buried layer before the epitaxial layer is grown. The buried layer, however should have as low a sheet resistance as possible. To satisfy these goals, the buried layer is diffused into the substrate by a two-step process with a deep junction, typically 5 μm from the substrate surface. Assuming a sheet resistance of 25 $\Omega/\square$, the final surface concentration of the buried layer before epitaxial growth is $2 \times 19 \text{ cm}^{-3}$, from Figure 7-12, for a background concentration in the p-type substrate of 10^{15} cm^{-3}. If the epitaxial layer is grown at 0.1 μm/min at 1150°C, the buried layer penetrates about 0.1 of the way into the epitaxial layer during this process, as shown in Figure 7-18. As an approximation, it is assumed that the portion of the buried layer which is in the epitaxial layer at the start of the isolation diffusion acts as a predeposit for a limited source diffusion during the isolation diffusion, and that this diffusion has as its origin the point at which $x = 0.9w_{epi}$. Using Equation 7-34, the quantity of arsenic transported into the epitaxial layer during the growth of a 5-μm epitaxial layer is

$$Q = 2N_{01}\left(\frac{D_1 t_1}{\pi}\right)^{1/2} = 2.19 \times 10^{14} \text{ cm}^{-2}$$

where $D_1 = 3.15 \times 10^{-14} \text{ cm}^2 \cdot \text{s}^{-1}$ as determined from Equation 7-14, and $N_{01} = 2 \times 10^{19} \text{ cm}^{-3}$, the surface concentration of the buried layer before epitaxial growth. The total epitaxial layer thickness can now be estimated from

$$0.9w_{epi} = x_{epi} + x_2$$

where x_2 is the penetration of the buried layer into the epitaxial layer during the isolation diffusion. We now assume a value for x_2, 2 μm, and calculate the isolation diffusion for a depth of

$$w_{epi} = \frac{x_{epi} + x_2}{0.9} = 5.5\ \mu\text{m}$$

Assuming an isolation diffusion sheet resistance of $50\ \Omega/\square$, the final surface concentration, N_{0I}, is found to be $7 \times 10^{18}\ \text{cm}^{-3}$ from Figure 7-11. This corresponds to a $D_{2I}t_{2I}$ of $1.12 \times 10^{-8}\ \text{cm}^2$. If this diffusion is performed at 1200°C, where the boron diffusion coefficient is $4.4 \times 10^{-12}\ \text{cm}^2 \cdot \text{s}^{-1}$, the time is 42.3 min. The arsenic diffusion coefficient at 1200°C for a Gaussian is 9×10^{-13}. Then x_2 is found from Equation 7-6 to be 2.44 μm with a peak concentration of $2.6 \times 10^{18}\ \text{cm}^{-3}$.

7. Unfortunately, the buried layer is too close to the base-collector junction so that Step 6 must be repeated. The result is a final epitaxial layer thickness of

$$w_{epi} = 6.4\ \mu\text{m}$$

with

$$x_{epi} = 3.1\ \mu\text{m}$$

$$D_{2I}t_{2I} = 1.57 \times 10^{-8}\ \text{cm}^2$$

$$N_{0I} = 5.5 \times 10^{18}$$

$$t_{2I} = 60\ \text{min}$$

8. The final process to be defined is the emitter diffusion. The position of x_{jE} can be determined from the known final position of x_{jC} which is 1.45 μm. The base portion of the base-collector transition region was determined to be 0.19 μm. The active base width is 0.49 μm. The base portion of the base-emitter transition region under forward bias is approximately 0.1 μm. Therefore,

$$x_{jE} = 1.45 - 0.19 - 0.49 - 0.1 = 0.67\ \mu\text{m}$$

The net impurity concentration at x_{jE} must be zero. This implies that

$$N_{dE}(x_{jE}) = N_{aB}(x_{jE}) - N_{dC}(x_{jE}) \tag{9-35}$$

or

$$N_{dE} = N_{0B} \exp\left(\frac{-x_{jE}^2}{4D_{2B}t_{2B}}\right) - N_{dC}$$

$$= 1.55 \times 10^{18}\ \text{cm}^{-3}$$

Using the approximation for a constant source, solid solubility limited phosphorus diffusion given by Equations 7-32, 7-33, and 7-34, at 950°C

$$\text{erfc}\ z_1 = \frac{2N(x_{jE}, t)}{N_2(x_0)} \exp\left[\left(\frac{\alpha}{2D_2}\right)\left(\frac{3x_{jE}}{4}\right)\right]$$

$$= 0.274$$

which corresponds to, from Figure 7-4,

$$z_1 = 0.755$$

The emitter diffusion time, t_E, is given by

$$t_E = \left(\frac{3\alpha x_{jE} + 2D_2 z_1^2}{9\alpha^2}\right) + \left[\left(\frac{3\alpha x_{jE} + 2D_2 z_1^2}{9\alpha^2}\right)^2 - \frac{x_{jE}^2}{9\alpha^2}\right]^{1/2}$$

$$= 47.8 \text{ min}$$

It is also important to determine the enhancement factor, D_{TAIL}/D_i^0, for this case, since it has an effect on the base diffusion. From Equation 7-23,

$$D_{TAIL} = D_i^0 + \left(\frac{n_s^3}{n_e^2 n_{ie}}\right)\left[1 + \exp\left(\frac{0.3}{kT}\right)\right] D_i^-$$

which, using the data from Figures 7-5, 7-6, 7-7, 7-8, and 7-9 yields

$$D_{TAIL} = 1.26 \times 10^{-13} \text{ cm}^2 \cdot \text{s}^{-1}$$

and

$$D_{TAIL}/D_i^0 \simeq 37$$

Note that n_{ie} is affected by both the phosphorus and boron doping, and is approximately 10 times n_i for this case.

It is now possible to determine a diffusion schedule for the base process. For multiple time-temperature cycles, the effective Dt product is given by

$$(Dt)_{eff} = D_1 t_1 + D_2 t_2 + D_3 t_3 + \cdots \quad \text{(9-36)}$$

The $D_{2B}t_{2B}$ product may be written as the sum of two terms, one for the base drive-in, and one for the emitter diffusion. During the emitter diffusion, the base diffusion coefficient at 950°C is enhanced by the same factor as that of the "tail" portion of the phosphorus concentration. The result is that the boron diffusion coefficient at 950°C is 4.8×10^{-14} cm$^2 \cdot$ s^{-1} and a Dt product of 1.38×10^{-10} cm^2. The Dt product for the boron drive-in is then

$$Dt = D_{2B}t_{2B} - D_{950}t_E = 6.47 \times 10^{-10} \text{ cm}^2$$

If this process is performed at 1100°C where the boron diffusion coefficient is 4.6×10^{-13} cm$^2 \cdot$ s^{-1}, the resulting time is

$$t_{1100} = 23.4 \text{ min}$$

The boron predeposits for the base and isolation diffusions are considered to be solid solubility limited diffusions. The diffusion coefficient is approximated by Equation 7-14. At 900°C the effective boron diffusion coefficient is

$$D_{eff} = \left(\frac{1.225}{z}\right)^2 \left(\frac{N_{01}}{n_i}\right) D_i = 5.44 \times 10^{-15} \text{ cm}^2 \cdot \text{s}^{-1}$$

Table 9-1. **Bipolar Standard Process**

Design Values: $\beta_F = 45$ $\quad\omega_\alpha = 2\pi \times 5 \times 10^9\ \text{rad}\cdot\text{s}^{-1}$

$BV_{CB_o} = 25\ \text{V}$ $\quad R_{SB} = 200\ \Omega/\square$

Dimensions: $w_{epi} = 6.4\ \mu\text{m}$ $\quad x_{epi} = 3.1\ \mu\text{m}$

$x_{jC} = 1.45\ \mu\text{m}$ $\quad x_{jE} = 0.67\ \mu\text{m}$

$x_B = 0.49\ \mu\text{m}$ $\quad$ buried layer $R_S = 25\ \Omega/\square$

Diffusion Processes:

	T (°C)	t (min)	R_S ($\Omega/\square$)	N_0 (cm^{-3})
Isolation				
Predeposit	950	39		1.55×10^{20}
Drive-in	1200	60	50	5.5×10^{18}
Base				
Predeposit	900	17.4		1.2×10^{20}
Drive-in	1100	23.4	200	6.5×10^{18}
Emitter	950	47.8	~20	8.4×10^{20}

with $z = 2.8$ from Figure 7-4 for $N_{01} = 1.2 \times 10^{20}\ \text{cm}^{-3}$ and $N_B = N_{dC}$. The predeposit time can be found from a solution to Equation 7-36.

$$N_{0B} = \left(\frac{2N_{01}}{\pi}\right)\left(\frac{D_{1B}t_{1B}}{D_{2B}t_{2B}}\right)^{1/2}$$

which results in

$$t_{1B} = 17.4\ \text{min}$$

Similarly, for the isolation diffusion, at 950°C,

$$t_{1I} = 39\ \text{min}$$

The standard process is summarized in Table 9-1. The values determined in this example were used as input data for a process simulation using SUPREM II (1). The output plot from this simulation is shown in Figure 9-16.

Transistor Surface Geometry Design

The doping profile of a bipolar transistor is largely responsible for establishing the common-emitter forward current transfer ratio, the junction breakdown voltages, the sheet resistance of the base diffusion, and the intrinsic cutoff frequency of the device. The surface geometry is the determining factor in the current rating and the actual frequency response.

The Optimum Collector Current. The common-emitter forward current transfer ratio, β_F, is a function of the collector current, as shown in Figure

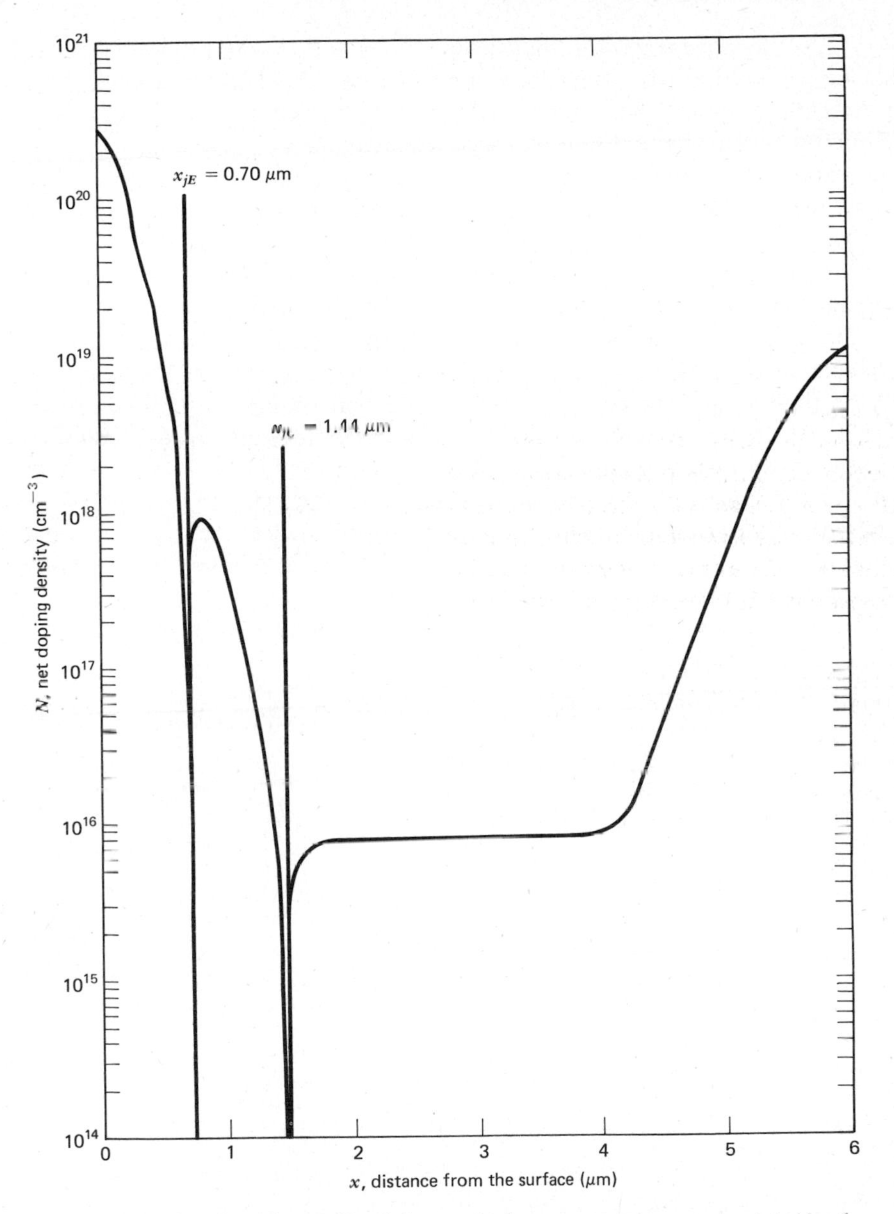

Figure 9-16. The impurity profile of the standard process transistor as simulated by the SUPREM II computer program (1).

9-17. The decrease in β_F at low current levels is due to a reduction in emitter efficiency resulting from recombination of carriers in the base-emitter junction space-charge region. The decrease in β_F at high current levels is also due to a reduction in emitter efficiency. There are two effects that contribute to this reduction, high injection and emitter crowding. The design approach is to select a geometry such that the maximum β_F occurs at the desired current level, by requiring that the onsets of high injection and emitter crowding occur at the same current level. Clearly, the transistor may be used at current levels ranging from $\frac{1}{10}$ to 10 times the current for maximum β_F.

High injection occurs when the injected electron density in the base adjacent to the emitter-base space-charge region approaches the doping density at that point. The quasineutrality condition requires a large excess hole density to effectively neutralize the large injected electron density. This has two effects on the current gain. The distribution of excess holes produces an electric field in the base in a direction which aids the transport of injected electrons across the base, thus increasing the base transport factor. As the injection level increases, the effect of the field due to high injection is much larger than that due to the graded base. This effect can be modeled as an increase in the effective diffusion coefficient, approaching a factor of two in the limit of very

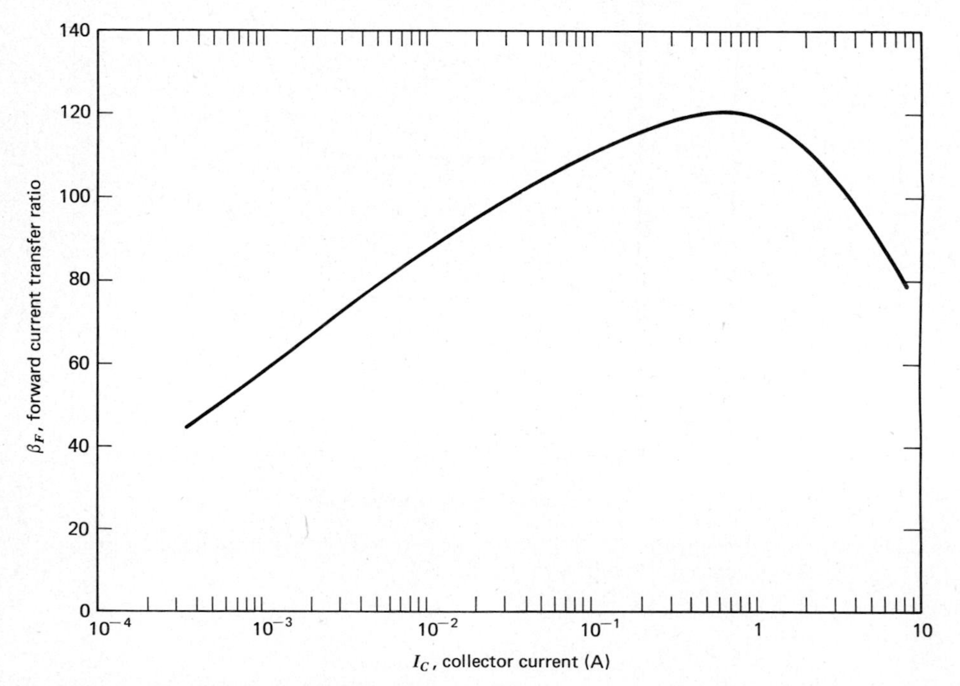

Figure 9-17. The common-emitter forward current transfer ratio as a function of collector current for a typical transistor.

high injection. The other effect of high injection is a significant reduction in emitter efficiency due to the increase in the majority charge in the base near the base-emitter junction. The decrease in emitter efficiency more than offsets the increase in the base transport factor.

Emitter crowding occurs for high emitter currents due to the lateral base current under the emitter. This effect is shown for a typical planar transistor in Figure 9-18. The voltage drop due to the lateral base current results in a position dependent bias voltage on the emitter-base junction. A small variation in forward-bias junction voltage results in a large variation of current because of the exponential dependence of the current as given by

$$I_E = I_S\left[\exp\left(\frac{qV_{EB}}{kT}\right) - 1\right] \tag{9-37}$$

where I_S is the saturation current of the junction, and kT/q is 26 mV at room temperature. To estimate the effect of emitter crowding, a simple model will be used, in which it is assumed that the base current is independent of position. The lateral voltage drop across the base, V_B, is given by

$$V_B = I_B R_{SB} \frac{l}{h} \tag{9-38}$$

where R_{SB} is the sheet resistance of that portion of the base under the emitter and is given by

$$R_{SB} = \frac{\rho_B}{x_B} = \frac{1}{Q_{Bo}\tilde{\mu}_{pB}} \tag{9-39}$$

where Q_{Bo} is the base charge and $\tilde{\mu}_{pB}$ is the average mobility of holes in the

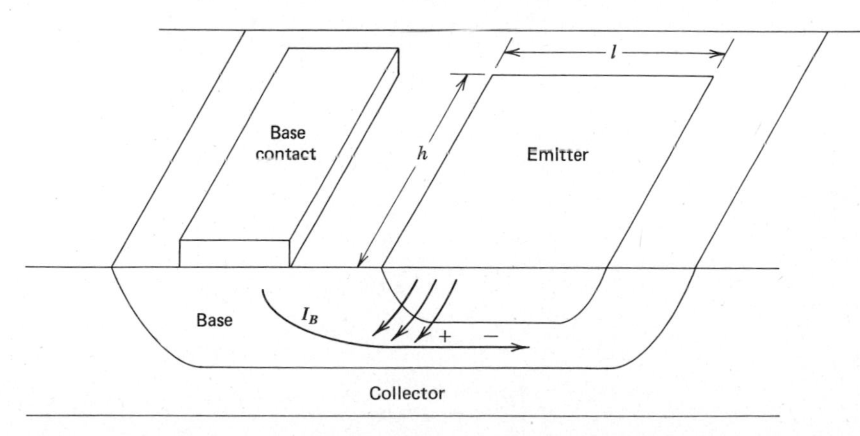

Figure 9-18. The lateral base current, which results in a nonuniform bias of the base-emitter junction, and, as a consequence, emitter crowding.

base. The base current is related to the emitter current by

$$I_B = I_E(1-\alpha_F) \tag{9-40}$$

and

$$V_B = I_E(1-\alpha_F)R_{SB}\frac{l}{h} \tag{9-41}$$

Significant emitter crowding occurs when $V_B = 2kT/q$. If I_{E1} is the current at which significant emitter crowding occurs, then

$$I_{E1} = \frac{2kT}{q}\frac{h}{lR_{SB}(1-\alpha_F)} \tag{9-42}$$

If emitter-crowding effects are ignored, the high injection condition occurs over the entire emitter area. The emitter current at the onset of high injection, I_{E2}, can be estimated in the following manner. Again, a simple model, the uniformly doped base transistor, is used, since it is only necessary to arrive at a reasonable estimate. The collector current in this type of transistor is given by

$$I_C = qA_E D_{nB}\frac{dn}{dx} = -\frac{qA_E D_{nB} n(x_{BE})}{x_B} \tag{9-43}$$

and the emitter current is

$$I_E = -\frac{I_C}{\alpha_F} = \frac{qA_E D_{nB} n(x_{BE})}{\alpha_F x_B} \tag{9-44}$$

where $A_E = lh$ is the emitter area, and $n(x_{BE})$ is the electron density at the edge of the base adjacent to the emitter-base space-charge region. At the onset of high injection, $n(x_{BE}) = p_o(x_{BE}) = N_{aB}$. Then

$$I_{E2} = \frac{qlhD_{nB}N_{aB}}{\alpha_F x_B} \tag{9-45}$$

Since the diffusion coefficients for electrons and holes are within a factor of 2 of one another, and, using the Einstein relation between mobility and diffusion coefficient,

$$qD_{nB}N_{aB} \approx \frac{kT}{q\rho_B}$$

and

$$I_{E2} \approx \frac{lhkT}{\alpha_F q x_B \rho_B} \tag{9-46}$$

The design equation is determined by setting $I_{E1} = I_{E2}$ and eliminating l

between Equations 9-42 and 9-46 to yield

$$\frac{I_{E1}}{h}=\frac{kT}{q\rho_B}\left[\frac{2}{\alpha_F(1-\alpha_F)}\right]^{1/2}$$

or

$$\frac{I_{E1}}{h}\approx\frac{kT}{q\rho_B}[2\beta_F(\text{max})]^{1/2} \tag{9-47}$$

Typical values for the resistivity of the active base region range between 0.1 and 0.2 $\Omega\cdot$cm and the range of β_F is between 25 and 200. The resulting range of the ratio of emitter current to the length of the emitter stripe is 0.09 to 0.52 mA/μm, indicating a design rule of 0.25 mA/μm. Note that the width of the emitter stripe, l, is not involved in this expression. The quantity h is usually called the effective emitter periphery. It is the length of the emitter stripe parallel to a base contact stripe. The current rating of a transistor can be doubled by adding a base contact stripe on the other side of the emitter stripe.

In many low-power integrated circuits, the transistor currents are so low that the design rule for optimum emitter current to emitter periphery can not be applied. In these cases, the minimum geometry that can be obtained is limited by the photolithographic process. The result is that these transistors are operated at currents below the optimum level with lower gains. As better techniques are developed for small pattern definition, the design rule will be applicable at lower current levels.

For high current transistors, a double-base contact geometry and an extended emitter stripe length are used. There is a limit to the length of the emitter stripe, however. This limit is imposed by the voltage drop along the length of the emitter contact metal, similar to the emitter crowding effect. For an emitter stripe of length h and width l, the resistance of the metal stripe is given by

$$R=R_S(\text{Al})\frac{h}{l} \tag{9-48}$$

where the sheet resistance of the aluminum, $R_S(\text{Al})$, is 0.05 $\Omega/\Box$. Assuming that a voltage drop of greater than $kT/q=0.026$ V is undesirable, and that the current in the stripe is independent of position, the maximum emitter stripe length is given by

$$h_{max}=\left(\frac{0.026l}{2.5\times10^{-4}\times0.05}\right)^{1/2}=45.6l^{1/2} \tag{9-49}$$

with both h_{max} and l in micrometers. For $l=12.5$ μm, $h_{max}=161$ μm. For a double-base stripe transistor, the maximum current is 80 mA. For higher current levels, an interdigitated structure of emitter and base contact stripes is used.

The Transistor Frequency Response. There are several effects that contribute to the frequency limitations of a bipolar transistor. Some of these effects are intrinsic to the transistor, like the base transit time, while others are strongly influenced by the parasitic elements introduced by the geometrical layout. The common-base cutoff frequency, ω_α, can be written

$$\omega_\alpha = (\tau_E + \tau_B + \tau_X + \tau_C)^{-1} \tag{9-50}$$

where each of the terms represents a time delay associated with the transistor. The base transit time, τ_B, has been discussed in a previous section. The other terms, the emitter capacitance charging time, τ_E, the collector space-charge layer transit time, τ_X, and the collector capacitance charging time, τ_C, are discussed in the following paragraphs.

The delay associated with charging the emitter space-charge region capacitance is represented by the time constant

$$\tau_E = r_E C_{TE} \tag{9-51}$$

where r_E is the incremental resistance of the emitter-base junction, and C_{TE} is the transition (space-charge) region capacitance of the emitter-base junction. The incremental resistance of the junction is given by

$$r_E = \frac{kT}{qI_E} \tag{9-52}$$

The transition region capacitance is that of a forward-biased junction. This capacitance is not to be confused with the diffusion capacitance whose frequency effects are included in the base transit time. The transition region capacitance under forward bias has not been accurately modeled. A reasonable approximation for C_{TE} is given by

$$C_{TE} = A_E \left[\frac{q\varepsilon_s N_{aB}(x_{EB})}{V_o}\right]^{1/2} \tag{9-53}$$

where V_o is the built-in voltage across the junction, which can be estimated from

$$V_o \approx \frac{kT}{q} \ln\left[N_{aB}(x_{EB})\frac{N_{dE}}{n_i^2}\right]$$

The value for $N_{aB}(x_{EB})$ can be obtained from Figure 9-19, and $N_{dE} \approx 10^{20}\,\text{cm}^{-3}$. The area of the base-emitter junction, A_E, includes the surface and sidewall areas.

The collector space-charge region transit time is given by

$$\tau_X = \frac{x_d}{2v_1} \tag{9-54}$$

where x_d is the collector space-charge region width, and v_1 is the carrier

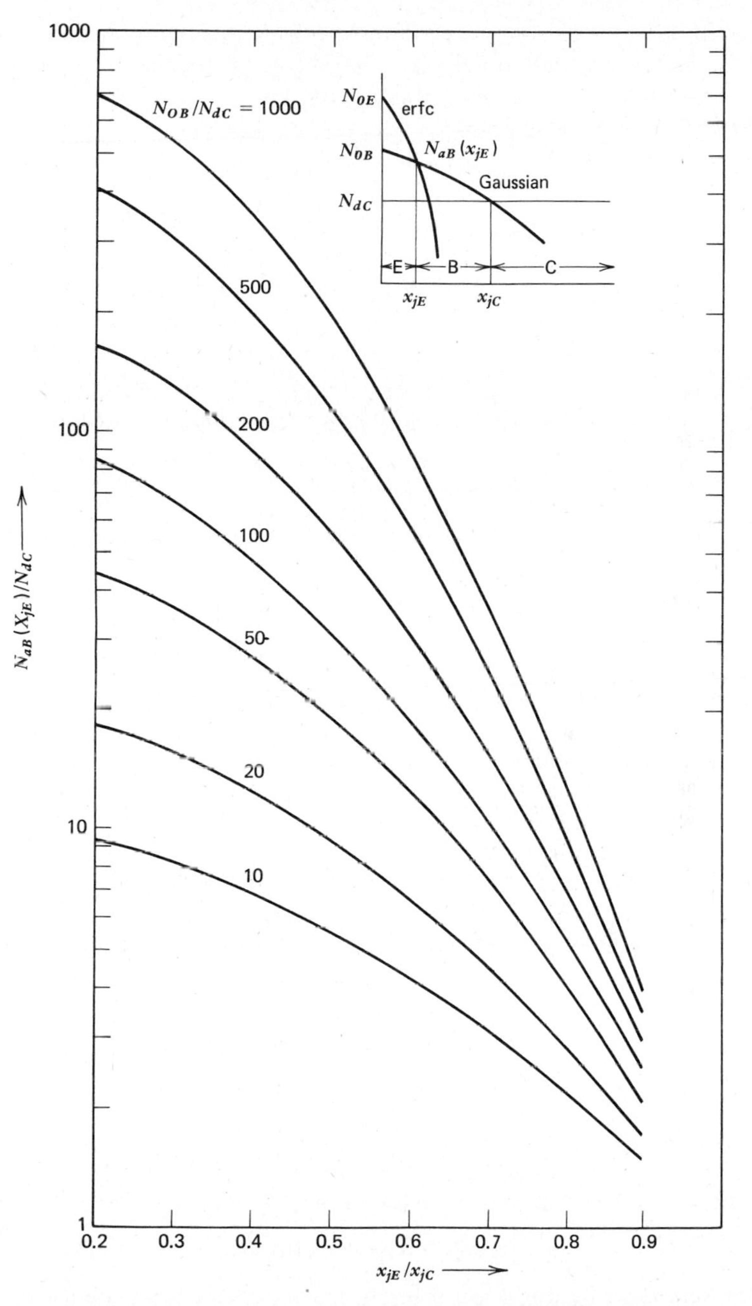

Figure 9-19. The base doping concentration at the base-emitter junction as a function of the ratio of base-emitter junction depth to base-collector junction depth in a double-diffused transistor (27).

velocity in the reverse-biased junction space-charge region. This velocity is called the scattering limited velocity and is approximately 8×10^6 cm · s^{-1}. The space-charge region width can be estimated by

$$x_d \approx \left(\frac{2\varepsilon_s}{qN_{dC}} V_{CB}\right)^{1/2} \tag{9-55}$$

where V_{CB} is the magnitude of the reverse bias voltage on the collector-base junction.

The collector capacitance charging time is given by

$$\tau_C = r_C C_{TC} \tag{9-56}$$

where r_C is the collector series resistance, and C_{TC} is the space-charge region capacitance of the collector-base junction. The collector series resistance can be approximated by

$$r_C = \frac{2\rho_C x_{\text{epi}}}{A_E} \tag{9-57}$$

where ρ_C is the resistivity of the epitaxial layer. The factor of 2 represents the paths from the base-collector junction to the buried layer and from the buried layer to the collector contact. The collector transition capacitance is given by

$$C_{TC} = \frac{\varepsilon_s A_C}{x_d} \tag{9-58}$$

where A_C is area of the collector-base junction.

The dominant terms in ω_α are τ_E and τ_B. The designer has some control over both of these quantities. The emitter time-constant is determined by the geometrical layout, and the base transit time is determined by the impurity profile. It is frequently useful to make one of these time constants very small so that the other dominates the frequency response.

A careful analysis of the frequency dependence of α_F indicates that it is a multipole function of ω. It can be modeled as a single-pole function with an excess phase shift at ω_α, as given by

$$\alpha_F(\omega) = \frac{\alpha_F(0)\exp[j((K_\theta - 1)/K_\theta)(\omega/\omega_\alpha)]}{1 + j(\omega/\omega_\alpha)} \tag{9-59}$$

where K_θ is the phase correction constant which is typically 0.7 for a graded base transistor. The common-emitter cutoff frequency, ω_β, is given by

$$\omega_\beta = K_\theta[1 - \alpha_F(0)]\omega_\alpha \tag{9-60}$$

An important figure of merit for transistors is the gain-bandwidth product, f_T, the frequency at which $|\beta_F|$ is unity. It is given by

$$f_T = \frac{\alpha_F(0)K_\theta\omega_\alpha}{2\pi} \tag{9-61}$$

This frequency is approximately equal to the maximum frequency of oscillation, f_{max}, which is the frequency at which the maximum available power gain is unity. This is given by

$$f_{max} = \left(\frac{f_T}{8\pi r'_B C_{TC}} \right) \tag{9-62}$$

where r'_B is the resistance between the base contact and edge of the emitter, and C_{TC} is the collector-base transition region capacitance. For a high-frequency transistor, the emitter stripe width, l, and the distance between the emitter stripe and the base contact are equal, and represent the smallest line widths in the layout. Under these conditions, the $r'_B C_{TC}$ product is independent

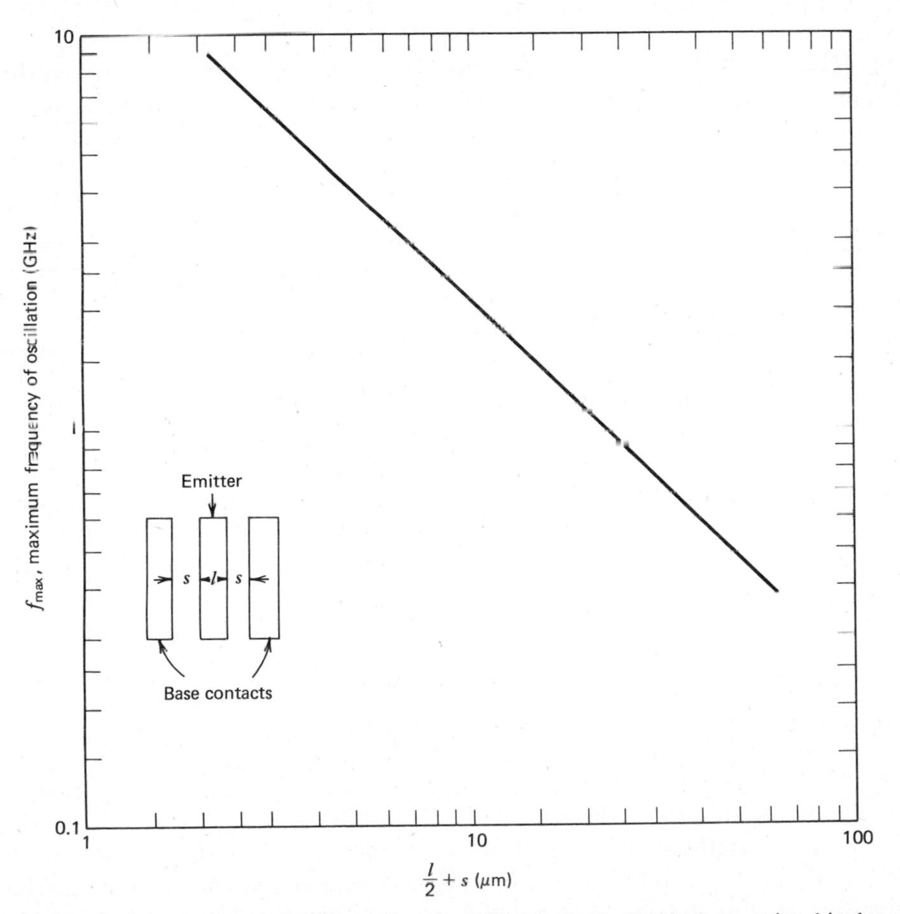

Figure 9-20. The maximum frequency of oscillation for a double base stripe bipolar transistor as a function of layout geometry (28).

of the emitter stripe length, h, and

$$f_{max} \approx \frac{2 \times 10^6}{l} \tag{9-63}$$

where f_{max} has the units of hertz and l is in centimeters. Since f_{max} is usually only slightly larger than f_T, a design rule l can be stated as

$$l = \frac{2 \times 10^6}{f_T} \text{ cm} \tag{9-64}$$

For a double-base stripe geometry, an empirical relationship for f_{max} is shown in Figure 9-20.

Transistor Layout Geometry

The layout geometry of a bipolar transistor is determined by the emitter dimensions and the photolithographic design rules, as established by the processing facility. The emitter dimensions are selected on the basis of Equation 9-64, which indicates that

$$l = \frac{2 \times 10^{10}}{f_T} \ \mu\text{m} \qquad (f_T \text{ in Hz}) \tag{9-65}$$

and Equation 9-47, which indicates that

$$h = 4I_E \ \mu\text{m} \qquad (I_E \text{ in mA}) \tag{9-66}$$

The photolithographic design rules specify a minimum line width, a, on a given mask and a minimum overlap, b, from one mask to another to allow for misalignment errors. For example, if the minimum line width on a mask is 10 μm, the minimum distance between this pattern and a pattern on another mask that must be aligned with this pattern is 5 μm on all sides. This is shown in Figure 9-21. The smallest pattern on a transistor mask is the emitter contact window. Once the dimensions of this pattern have been selected, all of the other dimensions for a minimum geometry transistor follow automatically. In Figure 9-22, a composite layout of a minimum geometry transistor is shown. The individual masks are shown in Figure 9-23. Note that the collector contact diffusion window and the collector contact window are the same size. This is possible because a slight misalignment of this pattern will result in the parallel connection of an ohmic contact and a Schottky barrier rectifier, which appears, electrically, as an ohmic contact.

If the photolithographic design rules are such that the desired value of l can not be achieved, it is possible to use a nonconventional layout to eliminate the need for changing the photolithographic process. This technique makes use of a nonreregistered emitter pattern which is called a washed emitter structure. In

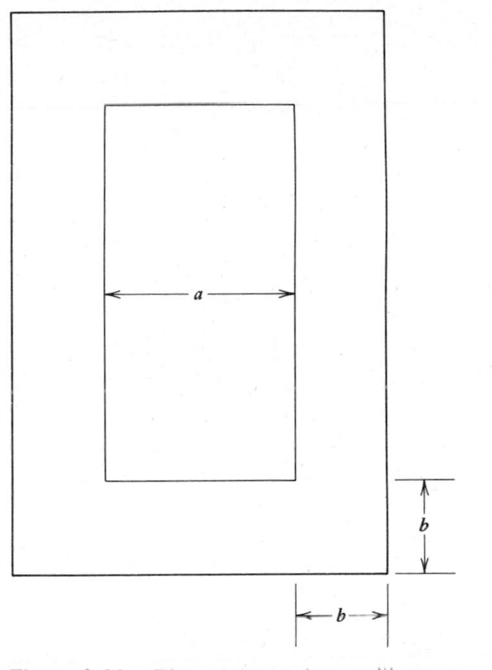

Figure 9-21 The minimum line width on a mask, *a*, and the minimum overlap dimension, *b*.

this case, the minimum line width is used for the emitter diffusion window, and there are no emitter or collector contact windows on the contact mask. After the emitter diffusion, a brief oxide etch is performed to remove the phosphorus glass from the emitter and collector contact diffusion regions. This is followed by a conventional photolithographic process to define the base contact areas.

The isolation diffusion pattern may have to be separated from the base diffusion pattern by a distance larger than the minimum design rule. The isolation diffusion must be driven entirely through the epitaxial layer, and is frequently overdriven to ensure isolation. Since the epitaxial layer is several micrometers thick in most circuits, the lateral diffusion of this process must be considered in the layout. Even if the base diffusion and isolation diffusion do not merge, a parasitic *pnp* transistor, with the *npn* base diffusion serving as the emitter, the epitaxial layer as the base, and the isolation diffusion as the collector, is formed. This transistor is biased in the forward active condition when the *npn* transistor is in saturation. For this reason, the epitaxial layer thickness should be added to the minimum line width to determine the minimum separation between the base and isolation diffusion patterns. Parasitic effects are discussed in more detail later in this chapter.

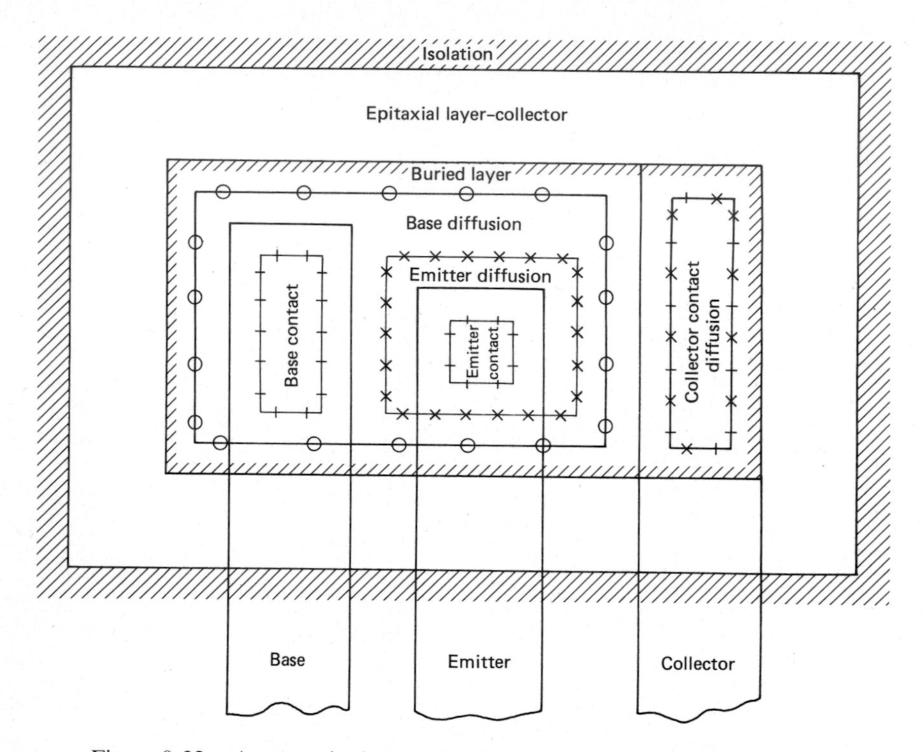

Figure 9-22. A composite layout of a minimum geometry bipolar transistor.

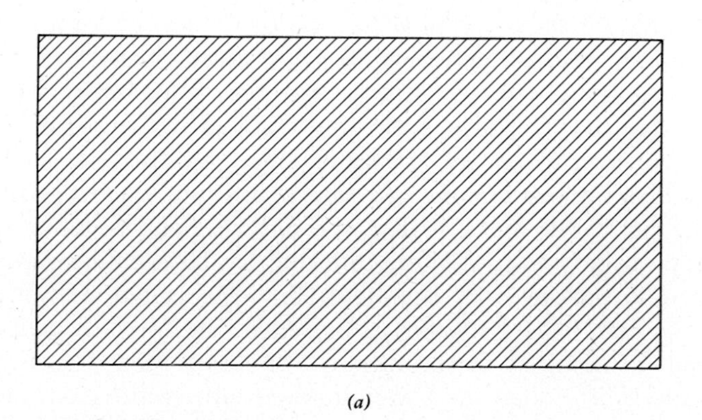

(a)

Figure 9-23. The individual masks for the transistor of Figure 9-22. (*a*) Buried layer. (*b*) Isolation. (*c*) Base. (*d*) Emitter. (*e*) Contacts. (*f*) Metal delineation.

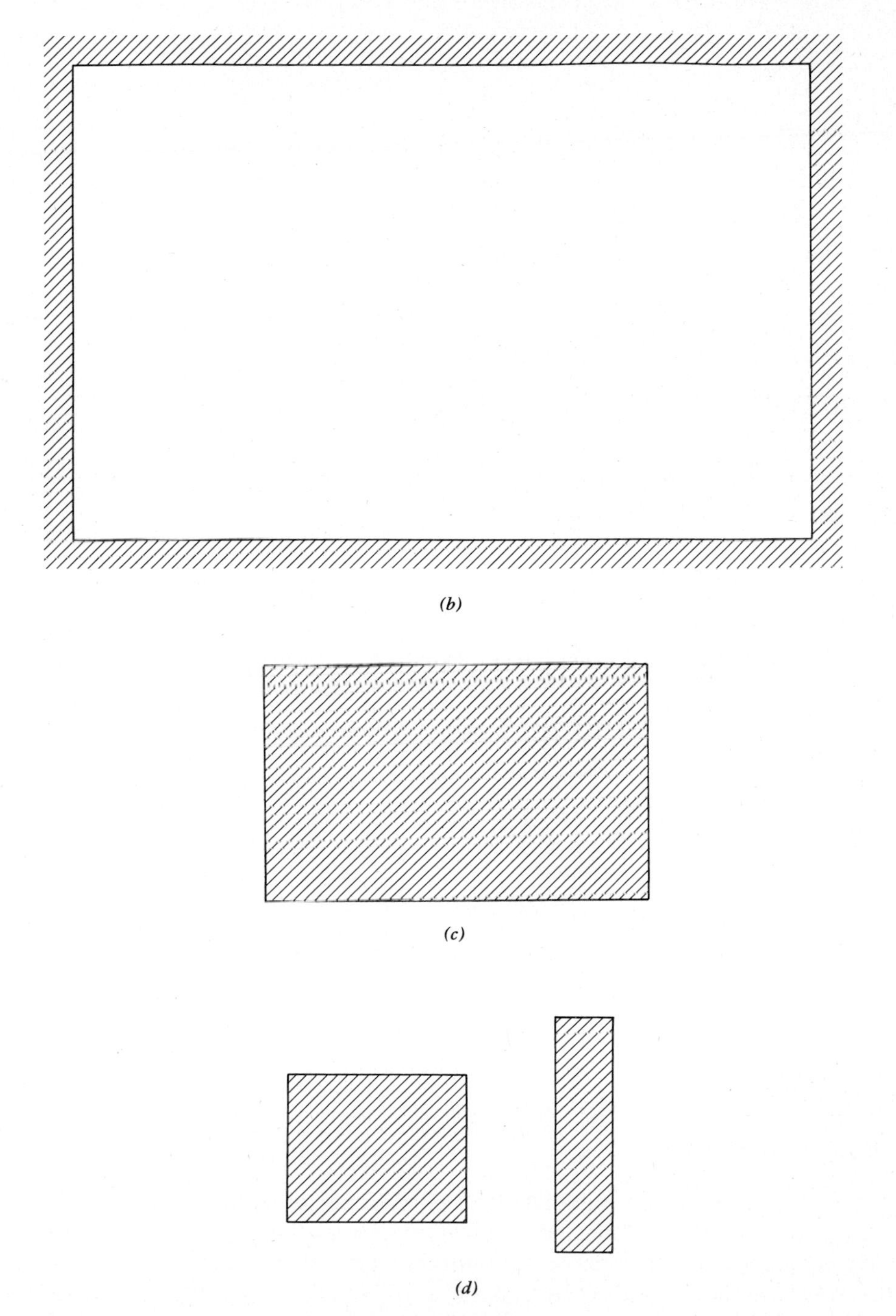

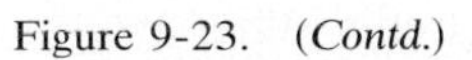
Figure 9-23. (*Contd.*)

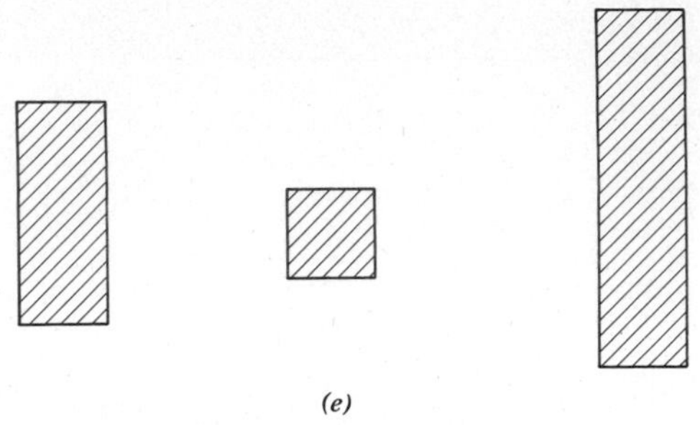

(e)

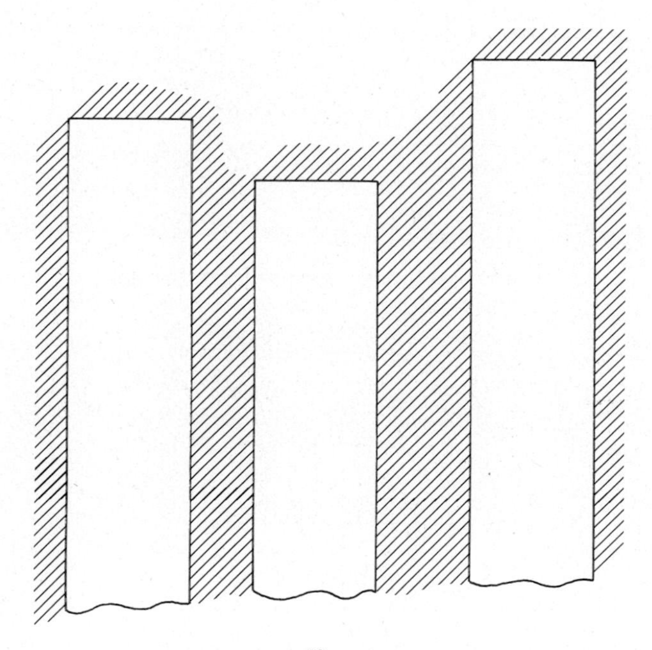

(f)

Figure 9-23. (*Contd.*)

PNP Transistors

Most of the transistors fabricated in monolithic integrated circuits are *npn* structures, like those discussed in the previous sections, but, using the same processes, it is also possible to fabricate *pnp* structures. In general, the electrical characteristics of these devices are significantly inferior to those of the *npn* devices in the same circuit. They are most frequently used as active load devices in operational amplifiers and as the injector transistors in integrated injection logic.

There are two types of *pnp* transistors used in integrated circuits, the substrate *pnp* structure and the lateral *pnp* structure. These devices are shown in Figures 9-24 and 9-25.

In the substrate *pnp* transistor, the emitter is formed by the *p*-type diffusion used to form the base of the *npn* structure, the base is formed by the epitaxial layer, and the collector is the substrate. The use of this type of transistor is severely restricted by the fact that the collector is connected to the substrate, which, to reduce parasitic effects, is usually connected to the most negative potential in the circuit. Therefore, substrate *pnp* transistors can only be used in emitter-follower circuits. These devices have a uniformly doped base, and a base width that is determined by the thickness of the epitaxial layer and the depth of the *p*-type diffusion. The frequency response is primarily determined by base transport, due to the wide base and the low mobility of the injected holes. The β_F of these devices can be as high as 100, but the f_T is typically 10 MHz for conventional processes. Optimizing the characteristics of the substrate *pnp* transistors may have adverse effects on the characteristics of the *npn* transistors in the same structure.

The lateral *pnp* transistor is much more versatile than the substrate *pnp* device, since it can be used in any circuit configuration, but it also has relatively poor electrical characteristics compared to its *npn* counterpart. It is common practice in laying out a lateral *pnp* transistor to surround the emitter with the collector to improve the current gain. A buried layer is used to reduce the resistance between the base contact and the active base, and to reduce the gain of the parasitic *pnp* substrate transistor which is biased in the active region when the emitter of the lateral transistor is forward biased. The base width of the lateral transistor is determined by the photolithographic process and the

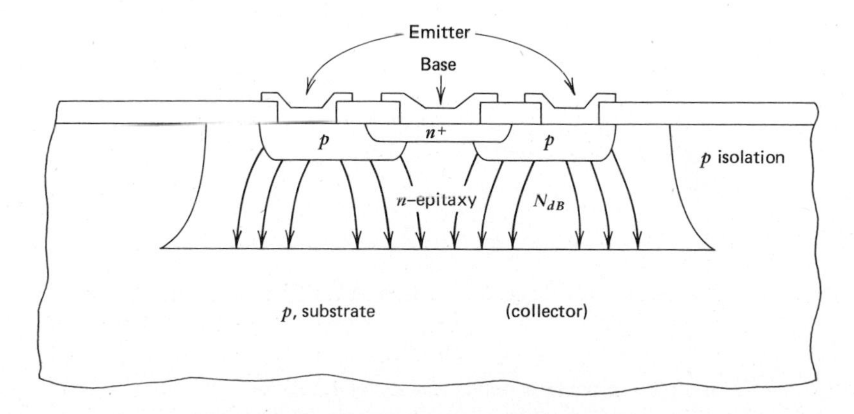

Figure 9-24. A substrate *pnp* transistor (3).

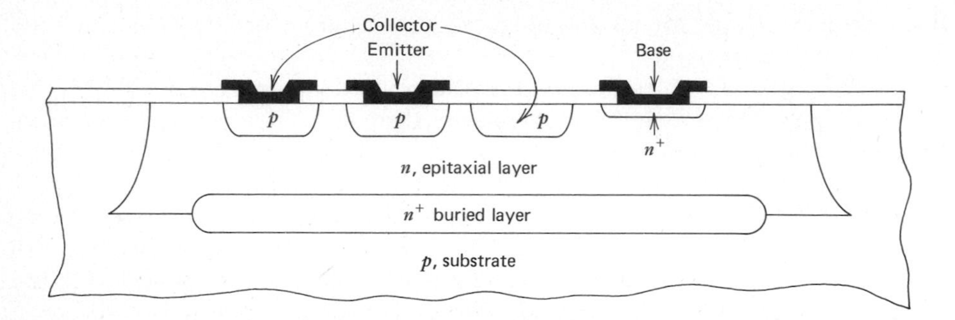

Figure 9-25. A lateral *pnp* transistor (3).

lateral diffusion of the *p*-type diffusion that is used to form the base of the *npn* transistors in the standard process. The narrowest part of the base occurs at the surface where some of the injected holes recombine due to the interface recombination velocity. This interface recombination velocity is typically 1 to 5 cm $\cdot$ s^{-1}, resulting in a negligible component of the base current. Since the lateral diffusion is a function of depth and both the emitter and collector are

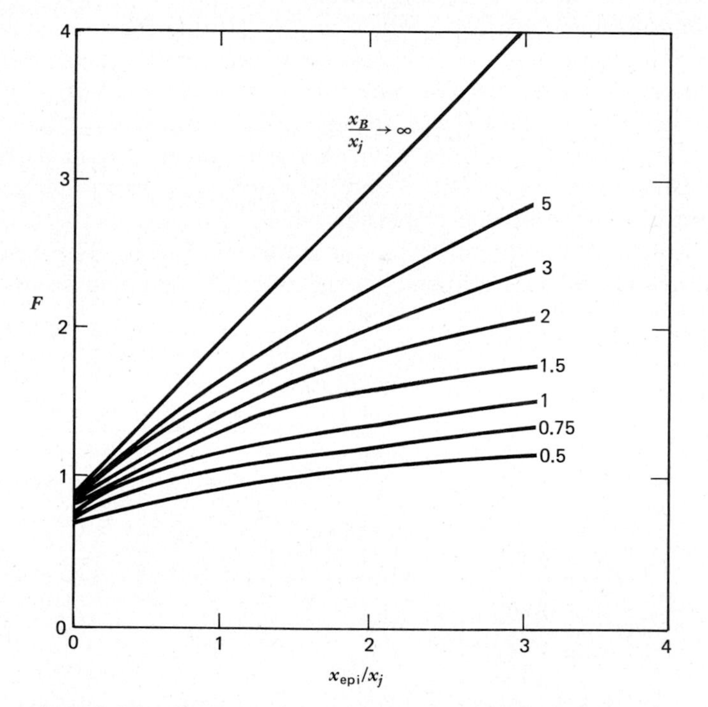

Figure 9-26. Geometry dependent factor F relating actual collector current to that predicted by a one-dimensional model for a lateral *pnp* transistor (3).

formed by this diffusion, the base width increases as a function of depth. The effective area of the emitter is the product of the perimeter of the emitter area, P_E, and the depth of the p-type diffusion, x_j, with most of the current concentrated at the surface. An empirical analysis of lateral *pnp* transistors indicates that the collector current may be approximated by

$$I_C = \frac{FqP_E x_j D_{pB} n_i^2}{N_{dB} x_B} \exp\left(\frac{qV_{EB}}{kT}\right) \tag{9-67}$$

where F is a geometry dependent factor that can be determined from Figure 9-26. Typical values for β_F for lateral *pnp* devices range from 5 to 20 with some devices having values as high as 100. It is difficult to control the base width of these transistors if attempts are made to make base widths less than 4 μm. As a result of this, the frequency response is poor, with f_T ranging from 1 to 10 MHz. The variation of β_F with collector current is more pronounced than that of an *npn* device, with high injection occurring at lower current levels due to the relatively low doping level in the epitaxial layer. Since the base is lightly doped, punch-through is the dominant breakdown mechanism, and BV_{CBo} is in the range of 6 to 10 V. Improvements in the photolithographic processes which result in the ability to accurately define smaller line widths will result in better characteristics for lateral *pnp* transistors.

DIODES

The diodes used in monolithic integrated circuits make use of the junctions formed for the transistors. There are six possible diode configurations, each exhibiting slightly different electrical characteristics. The basic diode connections are shown in Figure 9-27. The effect of the substrate connection makes

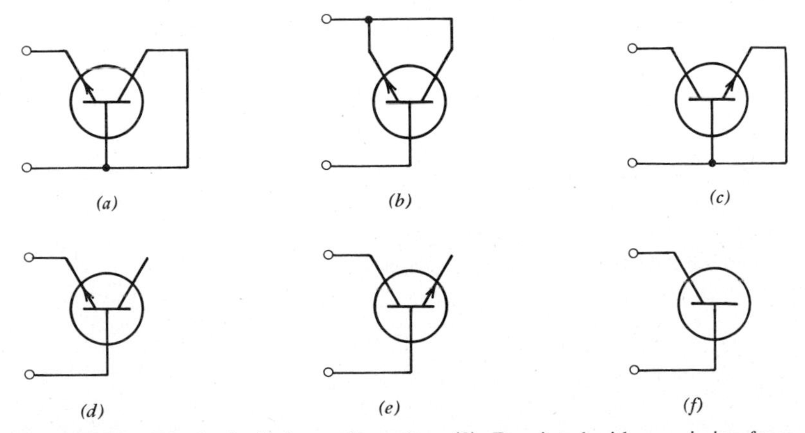

Figure 9-27. The basic diode configurations (2). Reprinted with permission from Motorola, Inc.

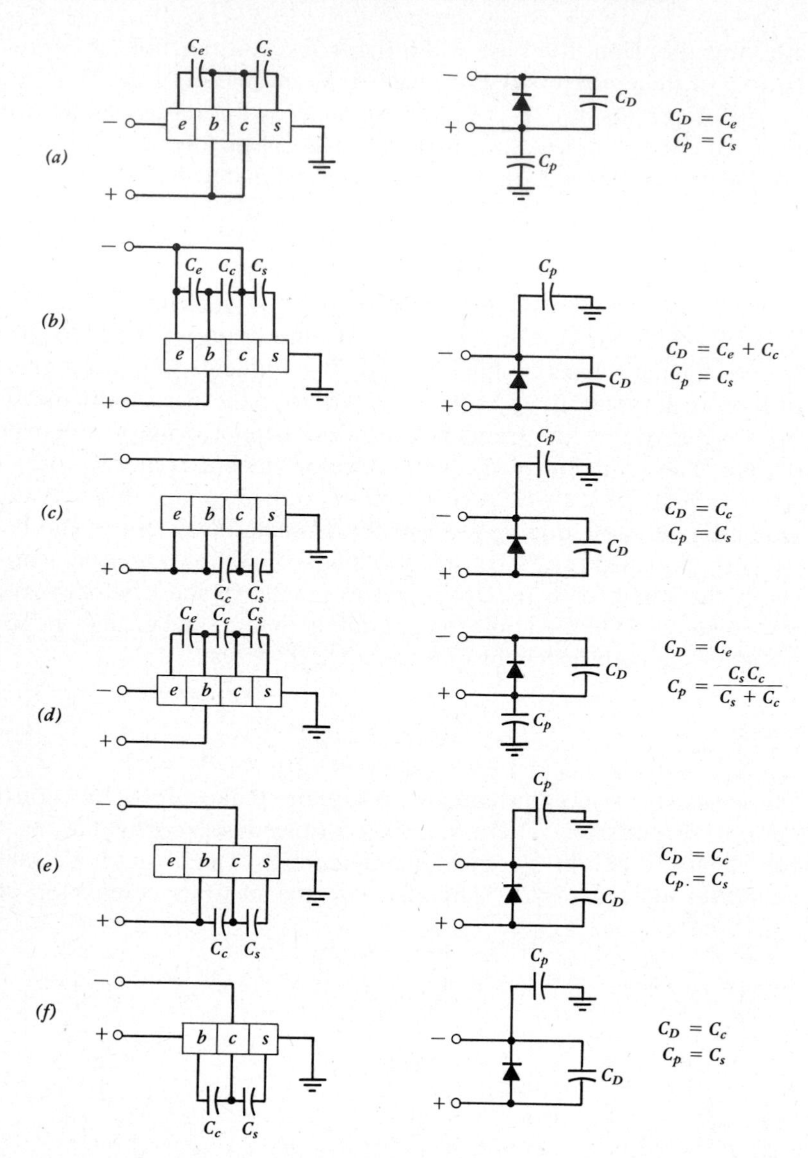

Figure 9-28. The effective capacitances of the diode configurations (2). Reprinted with permission from Motorola, Inc.

each of these configurations a three-terminal device. The capacitances associated with the basic diodes are shown in Figure 9-28. The forward current-voltage characteristics of these structures are shown in Figure 9-29. A significant consideration in some diode applications is the storage time of minority carriers associated with switching transients. Typical storage times for the basic configurations are indicated in Figure 9-30.

The selection of a diode configuration depends on the particular application. Configuration (a), with the base shorted to the collector, is a popular choice for many applications since it has the lowest forward voltage drop and the shortest storage time. Another distinct advantage of configuration (a) is that it is the only diode connection that assures that the parasitic *pnp* substrate transistor is biased at cutoff under all operating conditions. The other "diode" configurations will have slightly different currents in their two leads because of the gain of the parasitic transistor when the base-collector junction is forward biased. The major restriction on the use of configurations (a), (b), and (d) is that their reverse breakdown voltages are those of base-emitter junctions that are typically less than 8 V.

For small signal diodes, the minimum photolithographic design rules will

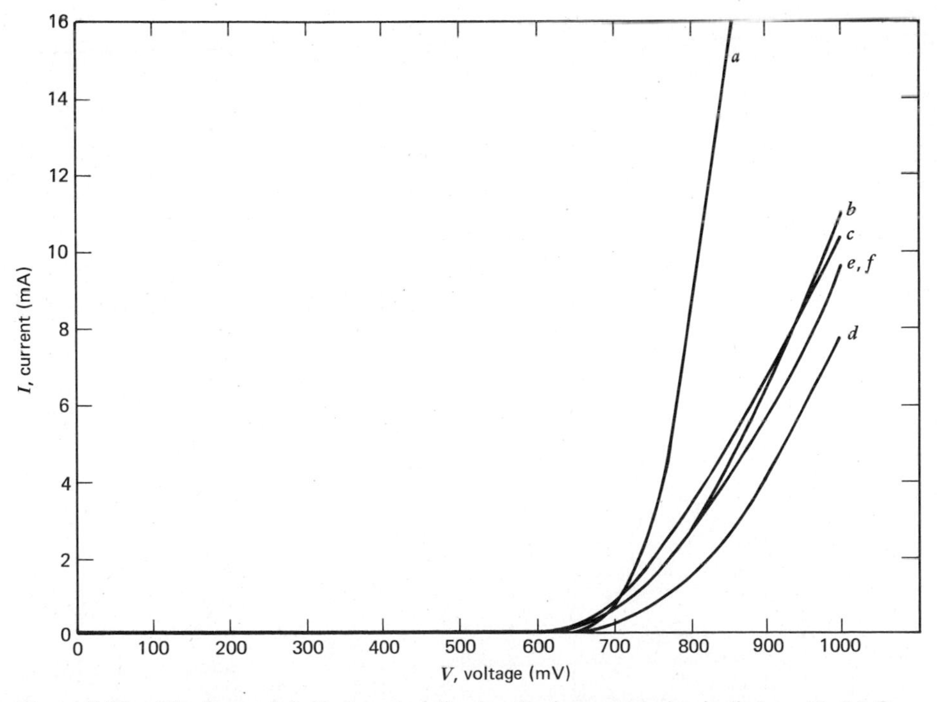

Figure 9-29. The forward I–V characteristics for the integrated circuit diodes with 0.1 Ω·cm collector and 200 Ω/□ base (2). Reprinted with permission from Motorola, Inc.

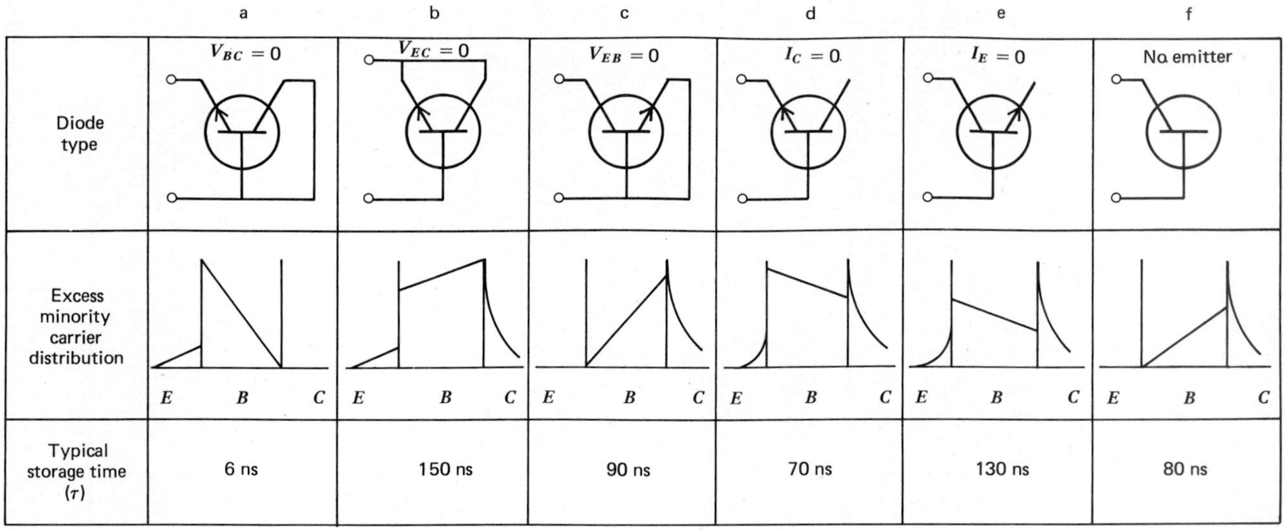

	a	b	c	d	e	f
Diode type	$V_{BC} = 0$	$V_{EC} = 0$	$V_{EB} = 0$	$I_C = 0$	$I_E = 0$	No emitter
Excess minority carrier distribution	E B C	E B C	E B C	E B C	E B C	E B C
Typical storage time (τ)	6 ns	150 ns	90 ns	70 ns	130 ns	80 ns

Figure 9-30. Typical storage times for the basic diode configurations (29). From *Basic Integrated Circuit Engineering*, by D. J. Hamilton and W. G. Howard. Copyright © 1975. Used with permission of McGraw-Hill Book Company.

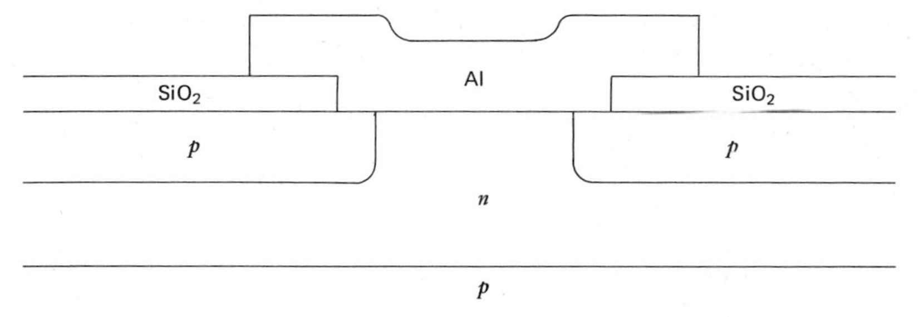

Figure 9-31. An n-type silicon-aluminum Schottky barrier diode with a p-type guard ring.

limit the size of the devices. For larger current levels, a maximum current density of $0.032\ \text{mA} \cdot \mu\text{m}^{-2}$ is recommended.

For certain applications, a different kind of diode, called a Schottky barrier, is used in monolithic integrated circuits. The Schottky barrier diode has a lower forward voltage drop than a junction diode formed in the same material, but its primary advantage is that it is a majority carrier device with essentially zero storage of minority carriers. In the standard process, aluminum forms a Schottky barrier with lightly doped n-type silicon, like the epitaxial layer used for the collector region. Improved characteristics can be obtained by adding a platinum silicide, Pt_5Si_2, layer between the silicon and the aluminum. Pt_5Si_2 is formed by sputtering platinum onto the wafer that has been patterned with oxide windows in the desired locations and heating the wafer to 600°C in air. The silicide forms in the areas not protected by the oxide. A platinum etch removes the undesired platinum without attacking the silicide. The reverse characteristic of the Schottky barrier diode is significantly improved by adding a p-type guard band around the device to reduce the effects of the electric field at the edges of the device. A typical Schottky barrier diode structure is shown in Figure 9-31. A popular application of Schottky barrier diodes is to connect them in parallel with the base-collector junctions of bipolar transistors to prevent these junctions from becoming forward biased when the transistors are driven into saturation. This produces a significant reduction in the switching times for these transistors.

CAPACITORS

There are two basic types of capacitors used in monolithic bipolar integrated circuits, the capacitance of reverse biased pn-junctions and the metal-oxide-semiconductor (MOS) capacitor. Junction capacitance depends on the magnitude of the bias voltage, restricting the use of this type of device to applications where this is an acceptable or desirable characteristic. The MOS capacitor, as constructed in bipolar integrated circuits, has characteristics that closely resemble those of discrete, nonpolar capacitors. For both types of capacitor, the capacitance density is low, severely limiting the range of capacitor values available for the circuit designer.

Junction Capacitors

The capacitance per unit area, C/A, for the transition region of a reverse biased *pn*-junction is given by

$$\frac{C}{A}(V)=\frac{C_o}{A(1+V/V_o)^r} \tag{9-68}$$

where C_o/A is the transition region capacitance per unit area at zero bias ($V=0$), V is the magnitude of the bias voltage, V_o is the built-in voltage across the junction, and r is $\frac{1}{2}$ for a step junction and $\frac{1}{3}$ for a linearly graded junction. The built-in voltage is given by

$$V_o=\left(\frac{kT}{q}\right)\ln\left(\frac{N_aN_d}{n_i^2}\right) \tag{9-69}$$

where $kT/q=0.0259\text{ V}$ at room temperature, n_i is the intrinsic carrier concentration ($1.5\times10^{10}\text{ cm}^{-3}$ at room temperature), and N_a and N_d are the net acceptor and donor concentrations on the p- and n-sides of the junction. For a step junction,

$$\left(\frac{C_o}{A}\right)_{\text{step}}=\left[\frac{\varepsilon_s qN_aN_d}{2V_o(N_a+N_d)}\right]^{1/2} \tag{9-70}$$

where ε_s is the permittivity of silicon ($1.04\times10^{-12}\text{ F}\cdot\text{cm}^{-1}$). For a linearly graded junction

$$\left(\frac{C_o}{A}\right)_{\text{linear}}=\left(\frac{\varepsilon_s^2 qa}{12V_o}\right)^{1/3} \tag{9-71}$$

where a is the grading constant. For most cases, these approximations are sufficient for estimating junction capacitance, assuming a step junction for the base-emitter and a linearly graded junction for the collector-base. If a more accurate determination of junction capacitance is required, it can be made by using the computer-generated curves of Lawrence and Warner (16).

The voltage-variable capacitance defined by Equation 9-68 is often unwieldy for computational purposes. In many cases, the capacitance can be represented by an equivalent value determined by averaging over the voltage range to which the capacitor is subjected. The average value is given by

$$C_{\text{avg}}=\frac{1}{V_2-V_1}\int_{V_1}^{V_2}C(V)\,dV \tag{9-72}$$

or

$$C_{\text{avg}} = \frac{C_o V_o}{(V_2 - V_1)(1-r)} \left[\left(1 + \frac{V_2}{V_o}\right)^{1-r} - \left(1 + \frac{V_1}{V_o}\right)^{1-r} \right] \tag{9-73}$$

where V_1 and V_2 are the minimum and maximum voltages applied to the capacitor.

There are three types of junction capacitors used in monolithic bipolar integrated circuits. These are shown in Figure 9-32. The capacitor formed by the emitter and isolation diffusions has the largest capacitance density, approximately 1.28×10^{-3} pF · μm^{-2} for zero bias, but has a breakdown voltage near 5 V and has one side connected to the substrate, so that its circuit configuration is limited. The next largest capacitance density is associated with the base-emitter junction. The breakdown voltage for this junction is approximately

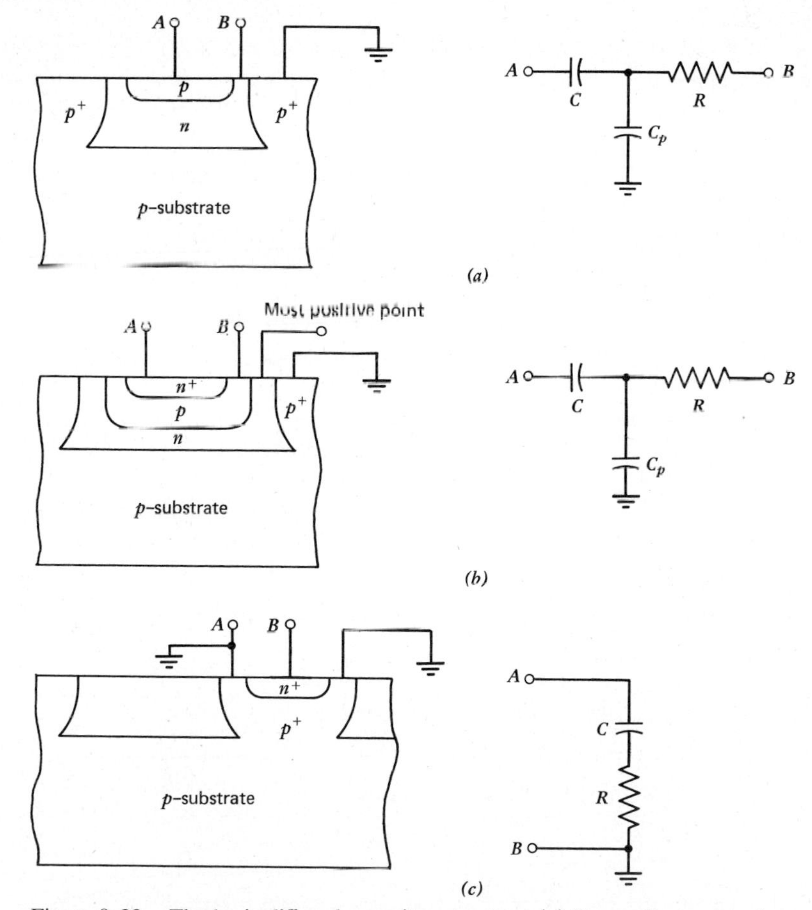

Figure 9-32. The basic diffused capacitor structure. (*a*) Base-collector junction. (*b*) Base-emitter junction. (*c*) Emitter-isolation junction (27).

6 V, and the zero bias capacitance density is 1.04×10^{-3} pF · μm^{-2}. If the circuit design calls for coupling or bypass capacitors, this is the logical choice if the bias voltage does not exceed 5 V. In general, large valued capacitors should be avoided because of the large area required for even a modest sized capacitor. The collector-base junction has the highest breakdown voltage, usually greater than 30 V, and the lowest capacitance density, 2.08×10^{-4} pF · μm^{-2} at zero bias. This structure is seldom used as a capacitor.

MOS Capacitors

In monolithic bipolar integrated circuits, the MOS capacitors make use of the n^+ emitter diffusion as one plate, a silicon dioxide layer as the dielectric, and the aluminum metallization pattern as the other plate, in a parallel plate configuration. This structure is shown in Figure 9-33.

As discussed in Chapter 10, the MOS capacitor has an interesting voltage dependence, since it is possible to accumulate, deplete, or invert the semiconductor surface depending on the polarity and magnitude of the bias voltage. Heavy doping of the semiconductor surface, such as occurs as a result of the emitter diffusion, makes it impossible to deplete the surface under normal operating conditions, resulting in a capacitor that is essentially independent of bias voltage.

The capacitance of the MOS structure is given by

$$C = \frac{\varepsilon_{ox} A}{d} \tag{9-74}$$

where ε_{ox} is the permittivity of the oxide (3.46×10^{-13} F · cm^{-1}), A is the area and d is the oxide thickness. Typical oxide thicknesses range from 700 to 3000 Å, resulting in capacitance densities from 4.8×10^{-4} to 1.12×10^{-4} pF · μm^{-2}. Some standard integrated circuits, particularly internally compensated operational amplifiers, are made with MOS capacitors with values between 30 and 50 pF.

The breakdown electric field in an MOS capacitor dielectric is on the order

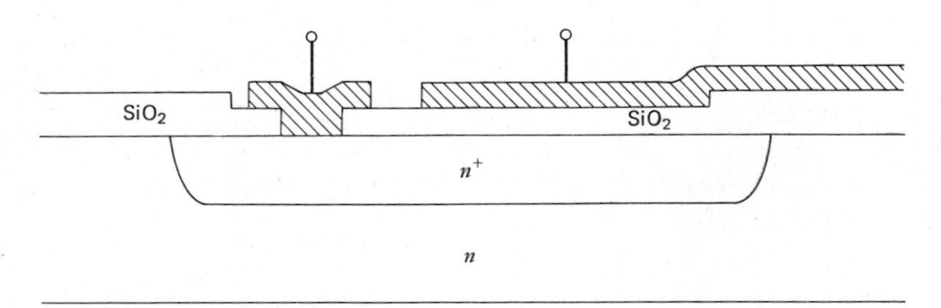

Figure 9-33. The MOS capacitor for bipolar integrated circuit applications.

of $600 \times 10^6\ \text{V} \cdot \text{m}^{-1}$, indicating that the breakdown voltage for an MOS capacitor with an oxide thickness of 700 Å is 42 V. The conductance of the oxide is extremely low, on the order of 10^{-9} mho for a typical 30-pF capacitor, and can usually be neglected.

RESISTORS

There are several types of resistors which can be fabricated in monolithic bipolar integrated circuits. By far the most popular of these makes use of the base diffusion, which has a typical sheet resistance of 200 Ω/□. The other resistor structures include the emitter diffusion on top of an isolation diffusion with a sheet resistance of 2 to 5 Ω/□, the epitaxial or collector resistor with a sheet resistance on the order of 1000 Ω/□, and thin-film resistors deposited on the oxide, the sheet resistance of which depends on the material deposited. The basic resistor structures are shown in Figure 9-34.

Diffused Resistors

The layout of diffused resistors is governed by power density considerations, or, for insignificant power levels, by limits imposed by the photolithographic process. The maximum power density for diffused resistors depends on the package used, but is generally between 3.2 and 8.0 $\mu\text{W} \cdot \mu\text{m}^{-2}$ of resistor surface area. A relatively large minimum resistor width is necessary to achieve a reasonable tolerance in resistor values, since the same sized processing flaw, like an etching undercut of 0.5 μm, results in a 20% error in a 2.5-μm-wide resistor, but only a 2% error in a 25-μm-wide resistor. It is suggested that 12.5 μm be selected as the minimum width for resistor designs.

The basic layout dimensions of a straight resistor are determined by multiplying the sheet resistance, R_S, by the ratio of the length, l, to the width, w, or

$$R = R_S \frac{l}{w} \tag{9-75}$$

If w is the minimum width, l can be determined from this equation. It is then necessary to check the power density and adjust w and l if necessary. Resistor terminations are made on the surface, and must be considered in determining the final value for the length of the resistor. Particular geometries have been examined and the equivalent number of squares have been determined. Some of these are shown in Figure 9-35. If a straight resistor is inconvenient, it is possible to put right-angle bends into the resistor pattern. Due to current crowding in the vicinity of the bend, each corner square at a bend contributes 0.65 square to the overall resistor pattern. Serpentine resistors are popular for large resistor values.

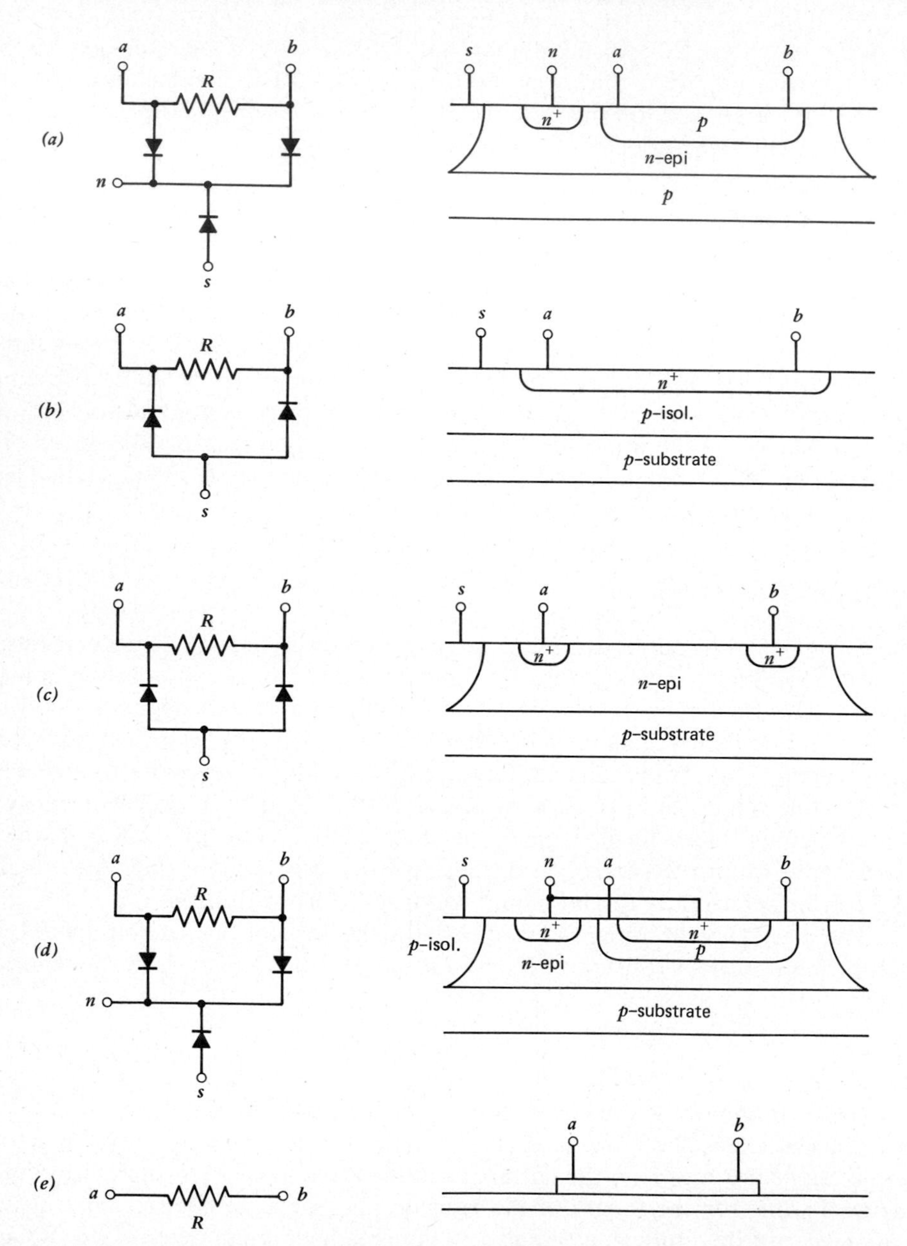

Figure 9-34. The basic resistor structure. (*a*) Base diffusion. (*b*) Emitter-isolation diffusion. (*c*) Epitaxial or collector region. (*d*) Pinch resistor. (*e*) Thin-film resistor (29). From *Basic Integrated Circuit Engineering*, by D. J. Hamilton and W. G. Howard. Copyright © 1975. Used with permission of McGraw-Hill Book Company.

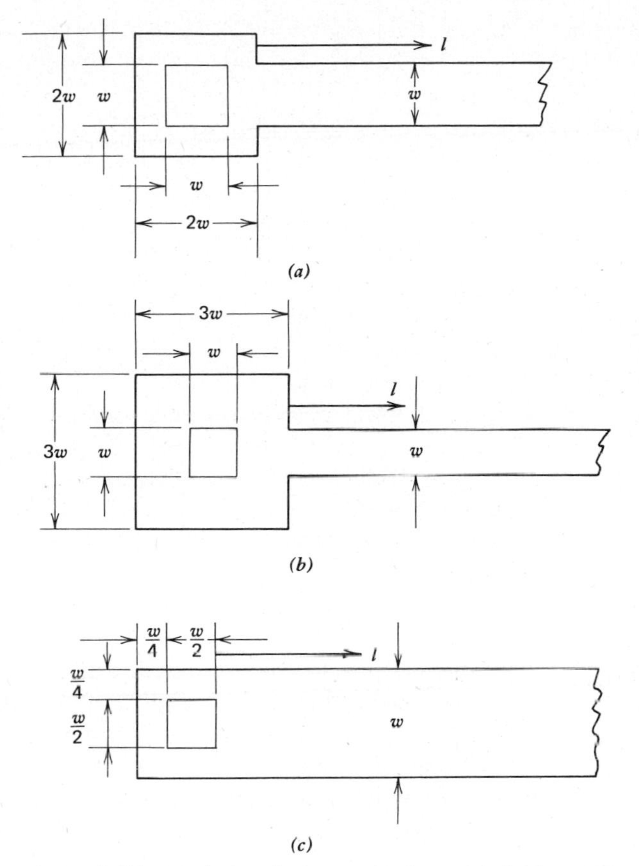

Figure 9-35. Typical resistor terminations. (*a*) 0.33 square (27). (*b*) 0.65 square (27). (*c*) 0.14 square (30).

There are a number of parasitic elements associated with diffused resistors. Since they are formed by diffusion, a *pn* junction is created, resulting in a junction capacitor. In the case of a resistor formed by the base diffusion, it is common practice to connect the epitaxial region to the most positive potential in the circuit to bias the substrate *pnp* transistor at cutoff. The result of this connection is that the resistor has the characteristics of a distributed *RC* transmission line. Since the value of the capacitor depends on the magnitude of the reverse bias, and the resistor is not an equipotential surface due to ohmic voltage drop with current, the capacitance per unit length is a function of position. If an average capacitance is assumed, using transmission line analysis, the cutoff frequency of the resistor can be estimated by

$$\omega_0 = \frac{2R_S}{C_{avg}R^2w^2} \tag{9-76}$$

The maximum value of C_{avg} is approximately 2×10^{-4} pF · μm^{-2}. It should be recognized that large valued resistors and high cutoff frequencies are not compatible.

Resistors made from the base diffusion process are inherently self-isolating components. Only one isolated region is necessary for all of the resistors formed by this process, and, as mentioned above, this region is connected to the most positive potential in the circuit through an n^+ diffused region to form an ohmic contact. The typical leakage current density for the junction between the resistor and the epitaxial layer is on the order of 160 pA · μm^{-2}. This leakage current does not present a problem unless it is comparable in magnitude to the intended resistor current. This only occurs in high-valued resistors, which will not perform as anticipated.

A consideration of the thermal effects on resistors must include the temperature effects on both mobility, which decreases with increasing temperature, and carrier concentration, which increases with increasing temperature. The net effect is a temperature coefficient of 2500 ppm/°C for base diffusion resistors, and 100 ppm/°C for emitter diffusion resistors.

The combination of the high-temperature coefficient of resistance and the long-range nonuniformity of the impurity concentrations incorporated into the wafer by the diffusion process has influenced the philosophy of the use of diffused resistors in monolithic integrated circuits. The most popular technique is to design for resistor ratios rather than values, so that circuit operation is relatively insensitive to thermal effects and impurity variations from one side of the wafer to the other. Typical diffused resistor tolerances are ±10%. The impurity concentration variation within an individual circuit is usually negligible, and resistors located in close proximity exhibit close thermal tracking.

Epitaxial Resistors

Resistors formed by using the isolation diffusion to isolate regions of the epitaxial layer must be relatively large in area, compared to diffused resistors. Allowance must be made for lateral diffusion of the deep isolation process in the layout of these components. The lateral diffusion also makes precise control of resistor values difficult. In order to form ohmic contacts, it is necessary to perform an n^+ diffusion into each contact area.

Pinch Resistors

If very large valued resistors are required in monolithic bipolar integrated circuits, the pinch resistor structure is sometimes used. The surface geometry of this structure is shown in Figure 9-36. The sheet resistance for the pinch resistor is the same as that of the active base region under the emitter, as given

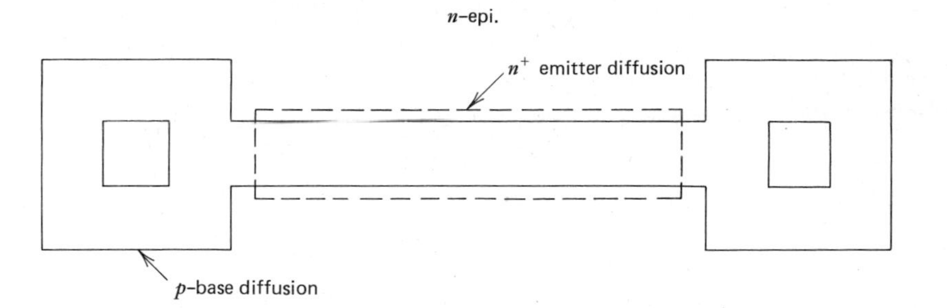

Figure 9-36. The layout geometry of a pinch resistor.

by Equation 9-39, which is repeated here for convenience

$$R_{SB} = \frac{1}{Q_{Bo}\tilde{\mu}_{pB}} \tag{9-39}$$

where Q_{Bo} is the base charge and $\tilde{\mu}_{pB}$ is the average mobility of holes in the base. Q_{Bo} is on the order of $2 \times 10^{-6}\ C \cdot cm^{-2}$ and $\tilde{\mu}_{pB}$ is typically 100 $cm^2 \cdot V^{-1} \cdot s^{-1}$, resulting in a value for R_{SB} on the order of 5000 Ω/□. Just as it is difficult to control the transistor current gain, it is difficult to control the sheet resistance of pinch resistors. Since the pinch resistor structure is essentially the same as that of a junction field-effect transistor, the sheet resistance can be increased by reverse biasing the n^+ region with respect to the body of the resistor, but the bias level must be maintained below 6 V to prevent breakdown.

Thin-Film Resistors

When precision resistors with a low-temperature coefficient of resistance are required in a monolithic integrated circuit, thin-film resistors are used. These resistors are formed by the photolithographic patterning of thin-film resistor materials that have been deposited on top of the oxide. In some cases, laser trimming can be performed on these resistors.

JUNCTION FIELD-EFFECT TRANSISTORS

In some circuits, it is desirable to include junction field-effect transistors (JFET's) in the same integrated circuits as bipolar transistors. It is very difficult, using the standard bipolar process, to produce both JFET's and bipolar transistors with acceptable characteristics.

Both n- and p-channel structures can be fabricated using processes similar to those used for bipolar transistors. There are two basic geometries for

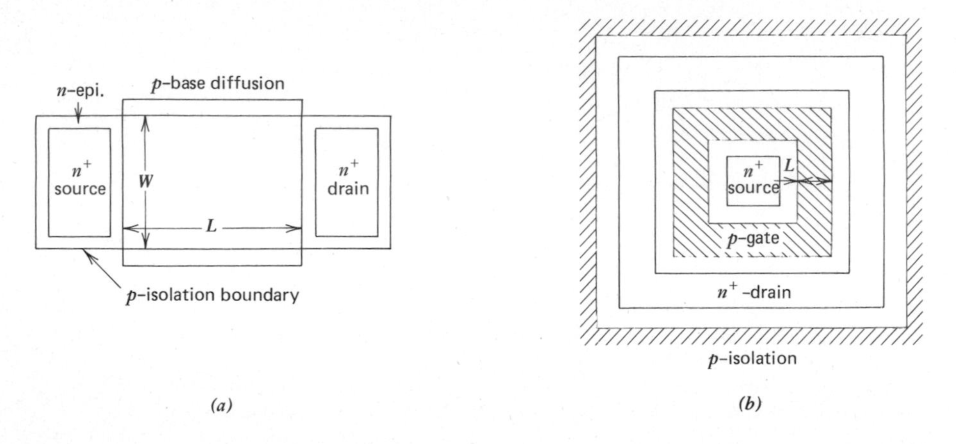

Figure 9-37. The surface layouts of (a) an open geometry and (b) a closed geometry n-channel JFET (29). From *Basic Integrated Circuit Engineering*, by D. J. Hamilton and W. G. Howard. Copyright © 1975. Used with permission of McGraw-Hill Book Company.

n-channel devices. These are shown in Figure 9-37. The open-geometry n-channel JFET is severely limited in use, because the gate is electrically connected to the substrate. The closed geometry device does not suffer from these restrictions. The basic geometry for a p-channel JFET is very similar to that of a pinch resistor, as shown in Figure 9-38. The major drawback to this device is that the gate-channel junction breaks down at a voltage below the pinchoff voltage for the device. The solution to this problem is to perform an additional n-type diffusion in the gate region, before the emitter diffusion so that the gate-channel junction is deeper than the emitter-base junction.

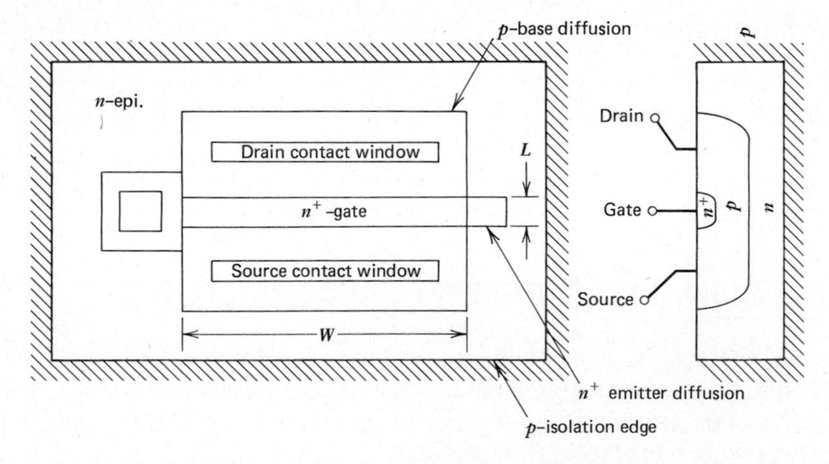

Figure 9-38. A p-channel JFET (29). From *Basic Integrated Circuit Engineering*, by D. J. Hamilton and W. G. Howard. Copyright © 1975. Used with permission of McGraw-Hill Book Company.

The basic gain parameter for a JFET is the transconductance, g_m, which, under saturation conditions, is given by

$$g_{m\,\mathrm{sat}} = \frac{W}{L} q\mu_C N_C t \left\{ 1 - \left[\frac{2\varepsilon_s}{qN_C t^2} (\phi_i - V_{GS}) \right]^{1/2} \right\} \tag{9-77}$$

where W, L, and t are the width, length, and thickness of the channel, μ_C and N_C are the majority carrier mobility and dopant density in the channel, ϕ_i is the built-in voltage of the gate-channel junction, and V_{GS} is the bias voltage from gate to source. This equation is true for conditions when the voltage from drain to source, V_{DS}, is greater than

$$V_{DS\,\mathrm{sat}} = \frac{qN_C t^2}{2\varepsilon_s} (\phi_i - V_{GS}) \tag{9-78}$$

In this model, it is assumed that the channel is uniformly doped, which is more applicable to the n-channel structure, and a more careful model may be necessary for a p-channel double diffused device. The major design choice is the ratio of the channel width to the channel length. The frequency response of the JFET is dominated by the transit time, τ_t, of carriers along the length of the channel. The cutoff frequency, ω_c, can be approximated by

$$\omega_c \approx \frac{1}{\tau_t} = \frac{\mu_C V_{DS}}{L^2} \tag{9-79}$$

for $V_{GS}=0$, and $V_{DS}=(V_{DS})_{\mathrm{sat}}$,

$$\omega_c \approx \frac{\mu_C t^2 q N_C}{2\varepsilon_s L^2} = \frac{\sigma_C t^2}{2\varepsilon_s L^2} \tag{9-80}$$

where σ_C is the conductivity of the channel.

PARASITIC EFFECTS

Junction isolated monolithic bipolar integrated circuits contain a number of parasitic elements that can have significant effects on the operation of the circuits. For this reason, it is difficult to transfer a discrete component electronic circuit to a monolithic integrated circuit design. Computer simulation and "breadboard" integrated circuits, in which individual components can be wire-bonded into the desired circuit configuration, are valuable aids in the prediction of integrated circuit performance.

Many of the parasitic elements are associated with the use of pn junction structures for practically every component and isolation function in the circuit. Since there is a capacitance and a leakage current associated with each reverse-biased junction, unless these are desired features, the circuit performance is degraded by their presence. Although the capacitance density of the junction between the epitaxial layer and the substrate is low, typically 6×10^{-5} pF $\cdot$ μm^{-2}, the areas associated with the isolation junctions are much larger than those associated with the individual components. The parasitic distributed capacitance associated with diffused resistors, which is discussed in a previous section, can impose a frequency limitation on the circuit. The semiconductor regions that serve as conductive paths between the contacts and the active parts of the devices add parasitic resistance to the circuit. These parasitic resistors are in series with the bases and collectors of transistors, and with the "plates" of the capacitors. The most interesting parasitic is the unintentional substrate *pnp* transistor. In many cases, the device is biased in the cutoff region of operation, but when the typical *npn* transistor is biased in the saturation region of operation, the parasitic substrate *pnp* transistor is biased in the forward active region of operation. Fortunately, because of a wide base and the inclusion of the buried layer, this transistor has a very low current gain, and, thus, the circuit degradation is not severe. Another parasitic effect that is possible in a monolithic bipolar integrated circuit is the turn-on of a *pnpn* structure, particularly in the presence of ionizing radiation.

There are several processing sequences that can be used to substantially

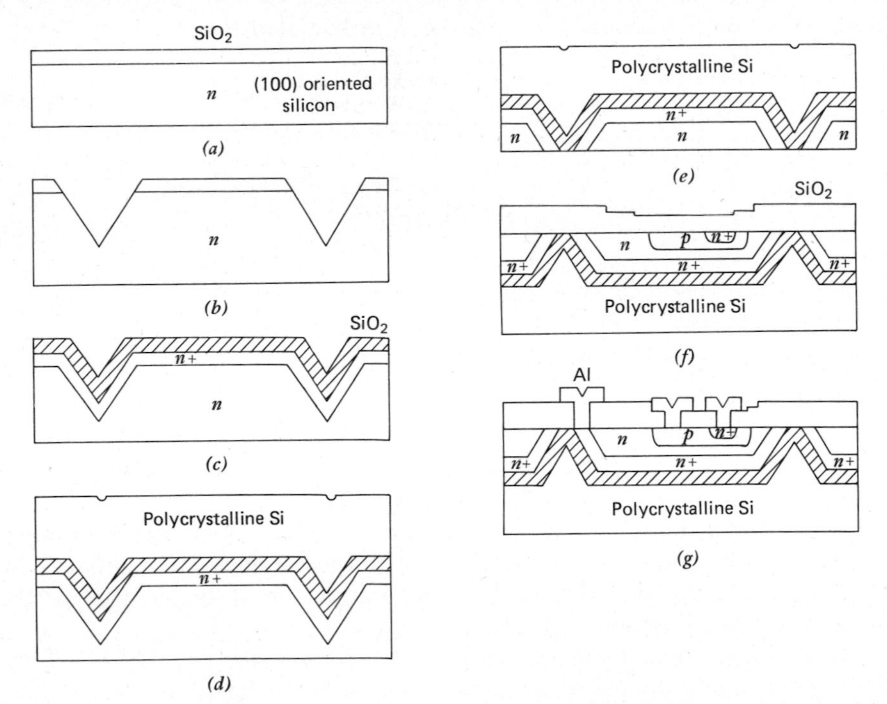

Figure 9-39. One technique for fabricating oxide isolated bipolar integrated circuits. (*a*) Oxidize. (*b*) Anisotropic etch. (*c*) n^+-diffusion and oxidation. (*d*) Polycrystalline silicon deposition. (*e*) Lap and polish. (*f*) Turn over and perform base and emitter diffusions. (*g*) Metallize and pattern.

reduce or eliminate many of the parasitic effects. One such technique, the deep collector diffusion, adds an n^+ diffusion between the isolation and base diffusions in which the buried layer is connected to the surface with a low-resistance path. A relatively expensive alternate processing sequence is called dielectric isolation. One of several possible techniques for achieving this goal is shown in Figure 9-39. Dielectric isolation essentially eliminates the parasitic effects associated with the substrate, and, in the scheme shown in the figure, even eliminates the need for a deep collector diffusion.

LIMITATIONS (20, 21)

The trend toward smaller obtainable geometries by the photolithographic processes and ion implantation brings up questions about the fundamental limitations of fabricating monolithic bipolar integrated circuits. In a bipolar transistor with a metallurgical base width of 1000 Å, the transition regions associated with the base-emitter and base-collector junctions reduce the active base width to approximately 300 Å, and punch through becomes the dominant breakdown mechanism. To increase the breakdown voltage, the base doping level can be increased, but this leads to Zener breakdown and reduces the current gain. It is more common to reduce the power supply voltage, which also reduces the power dissipation in the circuit. The voltages in the circuit must be sufficient to forward bias junctions (0.6 V) resulting in supply voltage lower limits of approximately 1.2 V. The basic design equations for bipolar transistors are based on the statistical concepts of diffusion and mean free paths. For very small geometries, the ideas of graded base dopings and collisions within the base region become questionable. New models must be developed to predict transistor operation in these very small devices.

SUMMARY

The design of devices for monolithic bipolar integrated circuits includes both the impurity profile and the geometrical surface layout. The most important device, the one upon which the process is based, is the double diffused *npn* bipolar transistor. The process developed for the transistor must be capable of producing all of the other components in the circuit, including diodes, capacitors, and resistors. The doping profile determines the breakdown voltages of the collector-base and emitter-base junctions, the current gain, and has an effect on the frequency response. The geometrical layout determines the current rating and has a significant influence on the frequency response. Diodes in monolithic integrated circuits make use of either the base-emitter or base-collector junction with the other junction usually short-circuited. Capacitors are formed by reverse biased junctions and MOS structures. Most resistors in monolithic bipolar integrated circuits are formed by the base diffusion. For higher valued resistors, the epitaxial layer or a pinch resistor structure is used. It is possible, usually with some variations in the process, to include *pnp* bipolar transistor and JFET structures in monolithic bipolar integrated circuits. Parasitic effects can severely affect the performance of an integrated circuit.

REFERENCES

1. D. A. Antoniadis et al., *SUPREM II—A Program for IC Process Modeling and Simulation*, Technical Report No. 5019-2, Stanford Electronics Lab., June 1978.
2. C. S. Meyer et al., *Analysis and Design of Integrated Circuits* (New York: McGraw-Hill, 1968).
3. R. S. Muller and T. I. Kamins, *Device Electronics for Integrated Circuits* (New York: Wiley, 1977). Figures reprinted by permission of the publisher.
4. R. M. Burger and R. P. Donovan, *Fundamentals of Silicon Integrated Device Technology, Vol. II* (Englewood Cliffs, N.J.: Prentice-Hall, 1968).
5. J. Lindmayer and C. Y. Wrigley, *Fundamentals of Semiconductor Devices* (Princeton, N.J.: Van Nostrand, 1965).
6. G. Radi, *Solid State Electronics* 15, 8 (1972).
7. H. J. J. DeMan, *IEEE Transactions on Electron Devices* ED-18, 10 (1971).
8. R. B. Fair, *IEEE Transactions on Electron Devices* ED-20, 7 (1973).
9. R. P. Mertens et al., *IEEE Transactions on Electron Devices* ED-20, 9 (1973).
10. M. S. Mock, *Solid State Electronics* 16, 11 (1973).
11. M. S. Mock, *Solid State Electronics* 17, 8 (1974).
12. B. L. H. Wilson, *Solid State Electronics* 20, 1 (1977).
13. H. C. DeGraaf et al., *Solid State Electronics* 20, 6 (1977).
14. E. J. McGrath and D. H. Navon, *IEEE Transactions on Electron Devices* ED-24, 10 (1977).
15. *Standard Math Tables*, Chemical Rubber Co., 1959.
16. H. Lawrence and R. M. Warner, Jr., *Bell System Technical Journal* 39, 3 (1960).
17. P. R. Wilson, *Solid State Electronics* 17, 5 (1974).
18. H. F. Wolf, *Silicon Semiconductor Data* (Oxford: Pergamon Press, 1969).
19. F. W. Hewlett, Jr. et al., *Solid State Electronics* 17, 5 (1973).

20. B. Hoeneisen and C. A. Mead, *Solid State Electronics* 15, 8 (1972).
21. P. Rohr et al., *Solid State Electronics* 17, 7 (1974).
22. D. A. Antoniadis and R. W. Dutton, *IEEE Transactions on Electron Devices* ED-26, 4 (1979).
23. E. L. Heasell, *IEEE Transactions on Electron Devices* ED-26, 6 (1979).
24. J. G. Fossum, *Solid State Electronics* 19, 4 (1976).
25. H. Yagi et al., *Conference Abstracts of the Electron Devices Meeting*, IEEE, 1974.
26. C. C. Allen et al., *Journal of the Electrochemical Society* 113, 5 (1966).
27. S. K. Ghandhi, *The Theory and Practice of Microelectronics* (New York: Wiley, 1968). Figures reprinted by permission of the publisher.
28. J. S. Kilby, *Solid State Design* 5, 7 (1964).
29. D. J. Hamilton and W. G. Howard, *Basic Integrated Circuit Engineering* (New York: McGraw-Hill, 1975).
30. G. R. Madland et al., *Integrated Circuit Engineering*, (Cambridge, MASS.: Boston Technical Publishers, 1966).

PROBLEMS

9-1. To obtain a better understanding of the factors involved in determining the common-base current gain of a bipolar transistor, (a) calculate the collector multiplication factor as a function of V_{CB}/BV_{CBo} for $n = 2$ and 4 and plot the results on the same graph, (b) calculate and plot the base transport factor as a function of x_B/L_{nB} over the range from 0.001 to 10, and (c) calculate and plot the emitter efficiency as a function of G_B/G_E over the range from 0.001 to 1.

9-2. The base diffusion for a transistor structure results in a Gaussian with a sheet resistance of 180 Ω/□ and a junction depth of 1.8 μm. The collector region is uniformly doped with a dopant concentration of 10^{15} cm^{-3}. Find the surface concentration and Dt product for this diffusion, and plot the profile on semilog paper. If the emitter-base junction is located 0.8 μm from the surface, find the base Gummel number for this structure (a) by assuming an average value for D_{nB} and (b) by dividing the base region into ~0.1 μm intervals and, using an appropriate value for D_{nB} and N_{aB} in each interval, performing an approximate numerical integration over the base region.

9-3. Assuming that the emitter efficiency is the dominant factor in determining transistor current gain, plot a curve of common emitter current gain as a function of base Gummel number.

9-4. For the transistor structure of Problem 9-2, the emitter doping profile is assumed to be a complementary error function with a surface concentration of $10^{21}\ cm^{-3}$. Construct a graph of the doping profile of this transistor structure. Using Equation 9-26 and Figure 9-8, estimate the effective doping profile of the emitter.

9-5. Estimate the common-emitter current gain and alpha cutoff frequency for the transistor structure of Problems 9-2 and 9-4.

9-6. It is desired to have a 5.6-V reference diode in a bipolar integrated circuit. If the base-emitter junction is to be used for this purpose, and the junction depth is to be $1\ \mu m$ with $x_{jC} - x_{jE} = 0.5\ \mu m$, what base sheet resistance would result? Repeat for $x_{jE} = 2\ \mu m$ and $x_{jE} = 3\ \mu m$.

9-7. Derive Equations 9-31 and 9-32.

9-8. For a base-collector breakdown voltage of 40 V, find an appropriate collector doping density and distance between the base-collector junction and the buried layer. What base-collector junction depth should be used?

9-9. For a collector doping level of $5 \times 10^{15}\ cm^{-3}$, an emitter area $5 \times 5\ \mu m$, a buried layer sheet resistance of $20\ \Omega/\square$, a collector contact opening $10 \times 5\ \mu m$, and a distance from the edge of the emitter closest to the collector contact of $25\ \mu m$, estimate the collector series resistance if the distance from the collector-base junction to the buried layer is 5, 7.5, and $10\ \mu m$. Repeat this calculation for buried layer sheet resistances of 10, 50, and $\infty\ \Omega/\square$ (no buried layer). Assume that the base width is $1\ \mu m$.

9-10. Design a diffusion schedule for the buried layer in the standard process.

9-11. Design a two-step diffusion schedule for the emitter of the standard process transistor, with a sheet resistance of $12\ \Omega/\square$. Use hD_i^0 as an approximate diffusion coefficient for phosphorus for the drive-in. What effect does this have on the base diffusion schedule?

9-12. Design a standard process for $\beta_F = 75$, $BV_{CBo} = 30$ V, $\omega_\alpha = 2\pi \times 10^8$ $rad \cdot s^{-1}$, and a base sheet resistance of $300\ \Omega/\square$. Include a deep collector diffusion.

9-13. Design a geometrical layout for a 750-mA transistor.

9-14. Estimate the gain-bandwidth product for transistors with emitter stripe widths of 25, 12, 5, and $1.5\ \mu m$. What base widths would you use with these transistors to ensure that the emitter stripe width would dominate the frequency response?

9-15. Many *pnp* transistors in integrated circuits have low current gains. How would you connect a high-gain *npn* and a low gain *pnp* transistor to have a resulting device that appears to the terminals as a high-gain *pnp* transistor? Which device controls the frequency response of the combination?

9-16. It is desired to fabricate a 90-pF capacitor in a bipolar integrated circuit, with neither terminal connected to the power supply. Design three different types of capacitors to satisfy this need. If the dc working voltage is to be 10 V, which configuration would you choose, and why?

9-17. Use Equation 9-70 to estimate the capacitance of the emitter sidewalls of the standard process transistor as a function of distance from the surface (use five equal intervals in which both the emitter and base doping are considered constants) and compare the sidewall capacitance to the total capacitance of a transistor fabricated by this process with an emitter window which is $2 \times 5\ \mu m$. For the same surface geometry, repeat this calculation for a base-emitter junction $2\ \mu m$ deep.

9-18. Design diffused resistors with a sheet resistance of $200\ \Omega/\square$ and a power dissipation of 50 mW for values of 100, 500, 1000, and 5000 Ω. Calculate the cutoff frequency for each of these resistors. Assume a minimum line width for photolithography of $12\ \mu m$.

9-19. Design a 50,000-Ω resistor for a bipolar integrated circuit.

9-20. Redesign the standard process using ion implantation for diffusion sources and an arsenic emitter.

Chapter 10
Devices for MOS Integrated Circuits

The invention of the surface field-effect transistor predated that of the bipolar transistor by almost two decades, but the solutions of practical problems, primarily related to the stability of the surface, were not available until the planar process for producing silicon bipolar transistors was well established. The metal-oxide-semiconductor field-effect transistor (MOSFET) was not a successful commercial product until the mid-1960s, a time at which the monolithic bipolar integrated circuit was well established in the market place. Since that time, because of relatively simple processing and high density, the MOS integrated circuit has emerged as the leader in large-scale integrated circuits (LSI) and very large-scale integrated circuits (VLSI), particularly in digital electronics. As in the case of bipolar integrated circuits, analog applications of MOS integrated circuits have lagged behind digital applications, primarily because of the lack of a well-established need for standard products in this area. A particularly versatile product of MOS technology is the charge coupled device (CCD), which makes use of the properties of MOS capacitors. Arrays of CCD's can be used for digital signal processing, analog signal processing, and analog image processing.

In this chapter, the surface field effect is described, along with its applications to MOS capacitors, and p-channel and n-channel MOSFET's. The emphasis is on applications to digital signal processing, since this is the major commercial use of MOS technology.

THE SURFACE FIELD EFFECT

The basic element of all MOS devices is the MOS capacitor. This structure is shown in Figure 10-1. A p-type substrate is used for this analysis, but an n-type substrate is an equally likely candidate. The doping density in the

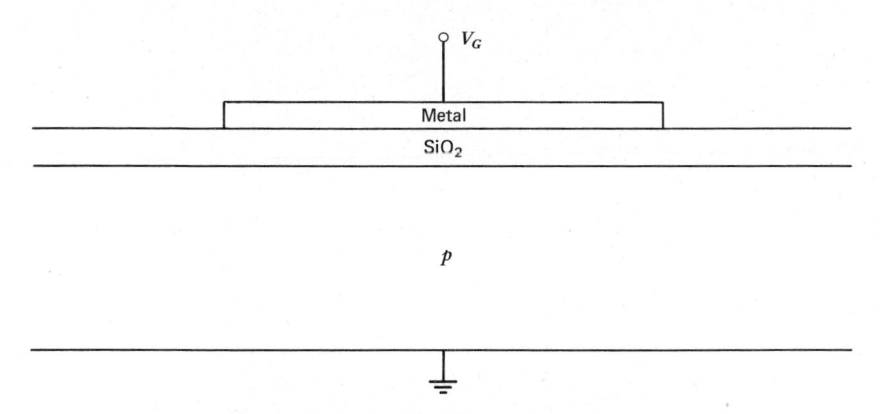

Figure 10-1. The MOS capacitor structure.

substrate, N_a, is usually on the order of 10^{15} cm^{-3}, much lower than the doping density selected for the MOS capacitor used in bipolar integrated circuits, as described in Chapter 9. Ignoring, for the moment, all of the factors that might result in bending of the energy bands of the silicon in the vicinity of the oxide-silicon interface under equilibrium conditions (zero bias voltage from the metal gate to the silicon substrate), the effects of applying a bias voltage, V_G, are as follows. If V_G is negative, holes are attracted to the surface, resulting in an increase of the lateral conductivity at the surface. This condition is called *accumulation.* If a low positive voltage is applied to the gate, holes are repelled from the surface, resulting in a negative space-charge region due to the immobile ionized acceptor atoms, in a manner similar to the formation of the space-charge region of a Schottky barrier or junction diode. This condition is called *depletion.* As V_G is made more positive, the space-charge region becomes wider. In the space-charge region, since the mobile carrier concentration is less than that in equilibrium, the rate of generation of electron-hole pairs is greater than the rate of carrier recombination. The electric field in the space-charge region is in such a direction that the generated electrons are swept to the oxide-silicon interface and the holes are swept into the bulk of the silicon. The result of slowly increasing the gate voltage, allowing time for the generation of electron-hole pairs, is that a layer of mobile electrons occurs at the oxide-silicon interface. When the density of electrons at the interface is on the same order as the equilibrium density of holes in the bulk, the surface is said to be in *strong inversion,* and the space-charge region width has reached its maximum thickness, $x_{d\max}$. A further slow increase in V_G results in an increase in the charge in the inversion layer, rather than an increase in x_d. The band structure of the system under these four conditions is shown in Figure 10-2. The time required to form an inversion layer in an MOS capacitor structure is on the order of 0.2 s. If there is a source for electrons, such as illumination or a diffused n-type region adjacent to the gate electrode, the inversion layer can form in less than 1 μs.

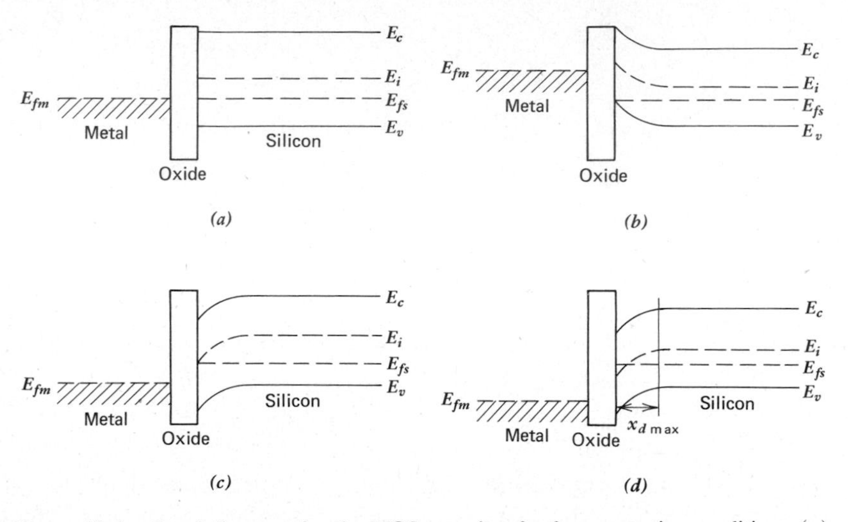

Figure 10-2. Band diagrams for the MOS capacitor for four operating conditions. (*a*) Equilibrium. (*b*) Accumulation. (*c*) Depletion. (*d*) Inversion.

The small signal capacitance is a function of the magnitude and polarity of the bias voltage between gate and substrate. The condition imposed in the previous paragraph of zero band bending under conditions of zero bias voltage occurs in a real structure at a voltage called the *flat-band voltage*, V_{FB}, which is discussed in the next section. The capacitance per unit area, $C'_{FB'}$ for a bias voltage of V_{FB} is given by

$$C'_{FB} = \frac{1}{1/C'_{ox} + L_D/\varepsilon_s} \tag{10-1}$$

where C'_{ox} is the capacitance per unit area of the oxide structure,

$$C'_{ox} = \frac{\varepsilon_{ox}}{x_{ox}} \tag{10-2}$$

where ε_{ox} is the permittivity of the oxide ($3.453 \times 10^{-13}\ \text{F} \cdot \text{cm}^{-1}$) and x_{ox} is the thickness of the oxide layer. The quantity, L_D is the extrinsic Debye length, which is given by

$$L_D = \left(\frac{\varepsilon_s kT}{q^2 N_a}\right)^{1/2} \tag{10-3}$$

and ε_s is the permittivity of the silicon ($1.036 \times 10^{-12}\ \text{F} \cdot \text{cm}^{-1}$). Under conditions of accumulation, when V_G is several kT/q less than V_{FB}, $C' = C'_{ox}$. When V_G is more positive than V_{FB}, and the surface is depleted,

$$C' = \frac{1}{1/C'_{ox} + x_d/\varepsilon_s} \tag{10-4}$$

where x_d is the width of the surface space-charge region, which is given by

$$x_d = \left[\frac{2\varepsilon_s(\phi_s - \phi_p)}{qN_a}\right]^{1/2} \tag{10-5}$$

where ϕ_s is the potential at the surface. It is given by

$$\phi_s = \frac{E_f - E_i(0)}{q} \tag{10-6}$$

where E_f is Fermi energy and $E_i(0)$ is the intrinsic Fermi energy at $x=0$, the oxide-silicon interface. The potential ϕ_p is associated with the doping in the substrate. It is given by

$$\phi_p = \frac{E_f - E_i}{q} \tag{10-7}$$

where E_i is the intrinsic Fermi energy in the substrate. This potential is negative for p-type material. At room temperature, essentially all of the acceptor atoms are ionized, and this potential is given by

$$\phi_p = \frac{kT}{q} \ln \frac{n_i}{N_a} \tag{10-8}$$

A further increase in V_G results in an inversion of the semiconductor surface. The condition of strong inversion is defined by

$$\phi_s = -\phi_p \tag{10-9}$$

Under these conditions, the surface potential remains relatively constant, and the space-charge region width is given by

$$x_{d\max} = \left(\frac{4\varepsilon_s |\phi_p|}{qN_a}\right)^{1/2} \tag{10-10}$$

The capacitance under strong inversion is given by

$$C'_{\text{inv}} = \frac{1}{1/C'_{\text{ox}} + x_{d\max}/\varepsilon_s} \tag{10-11}$$

The small-signal capacitance of the MOS structure depends on the bias voltage and the measuring frequency. The results of this type of measurement for a typical MOS device are shown in Figure 10-3. The high-frequency curve represents Equation 10-11. If the measurement is made at a low enough frequency to allow the generated carriers in the space-charge region to move to the inverted region, the measured capacitance follows the low-frequency curve. If both the gate bias and small-signal measuring voltage vary at a faster rate than can be accommodated by the generation of carriers in the space-charge region, the deep-depletion condition is observed. Since carrier generation

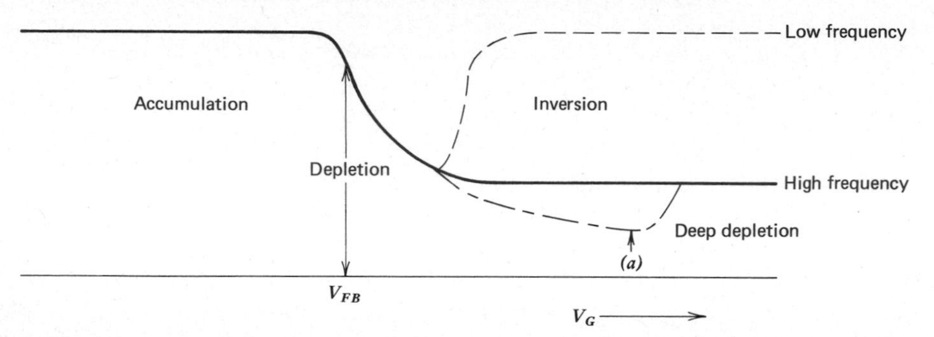

Figure 10-3. The small signal capacitance of an MOS capacitor as a function of gate bias voltage.

increases as the space-charge layer widens, the deep-depletion curve usually relaxes to the high-frequency curve at higher bias levels, as indicated by point (a) in the figure.

If there is a diffused *pn* junction adjacent to an MOS structure with an inverted surface, reverse biasing the junction increases the width of the space-charge region. This structure is shown in Figure 10-4. Under these

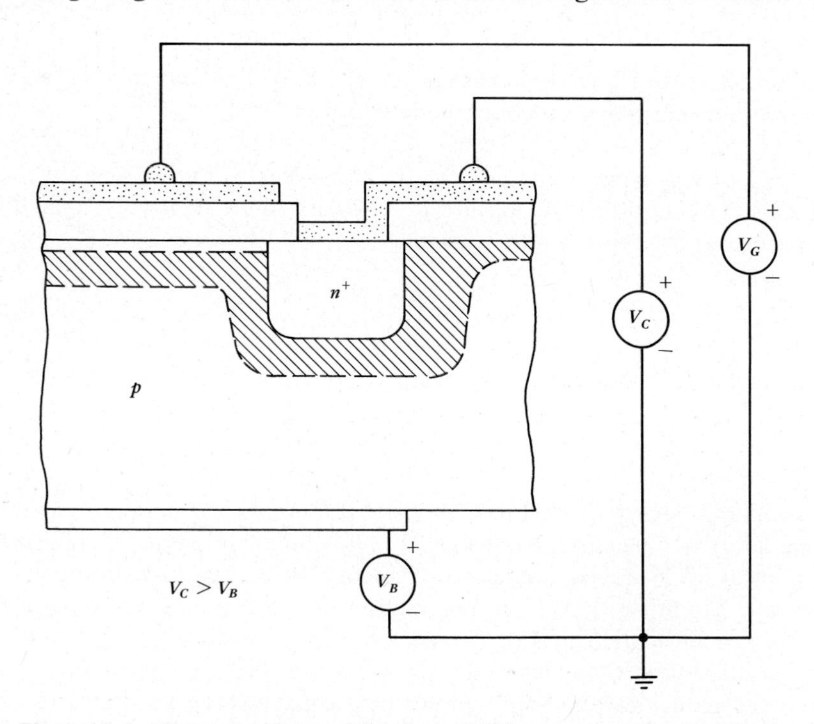

Figure 10-4. The space-charge region for an MOS capacitor with a junction in the vicinity of the boundary (18).

conditions the maximum space-charge region width is given by

$$x_{d\max} = \left[\frac{2\varepsilon_s(2|\phi_p| + V_C - V_B)}{qN_a}\right]^{1/2} \qquad (10\text{-}12)$$

where V_C is the voltage between the n-side of the diffused junction and ground, and V_B is the voltage between the substrate and ground.

The flat-band voltage, V_{FB}, depends on several quantities. One of these is the difference between the work function of the metal, Φ_M, and that of the semiconductor, Φ_S, resulting in

$$\Phi_{MS} = \Phi_M - \Phi_S = \Delta V_{FB1} \qquad (10\text{-}13)$$

Another significant quantity is due to the effect of charge at the interface between the oxide and the semiconductor. This charge density, Q'_{SS}, is a charge per unit area, and is related to the surface-state density of the semiconductor. Modern processing techniques have reduced the surface state density to the order of $10^{10}\ \text{cm}^{-2}$, but it should be recognized that the surface-state density of (111) oriented silicon is approximately three times that of (100) oriented silicon. The charge density due to interface states is positive and is assumed to be located in the oxide very near the interface. The presence of the interface charge results in an accumulation layer in n-type silicon, and a depletion or inversion layer in p-type silicon. The flat-band voltage shift due to Q'_{SS} is given by

$$\Delta V_{FB2} = -\frac{Q'_{SS}}{C'_{ox}} \qquad (10\text{-}14)$$

The other significant quantity that causes a flat-band voltage shift is charge in the oxide. In this case

$$\Delta V_{FB3} = -\frac{1}{C'_{ox}} \int_0^{x_{ox}} \frac{x}{x_{ox}} \rho(x)\, dx \qquad (10\text{-}15)$$

where $\rho(x)$ is the charge density distributed through the oxide. In the pioneer days of MOS processing, sodium ion contamination was a major problem. These ions are mobile in the oxide, resulting in a redistribution of the charge under various combinations of bias and temperature. Oxidations using HCl or phosphorus in the ambient result in stable oxides. The chemicals used in processing MOS devices are specially prepared to be low in sodium content. The total flat-band voltage is given by the sum of the three quantities, resulting in

$$V_{FB} = \Phi_{MS} - \frac{Q'_{SS}}{C'_{ox}} - \frac{1}{C'_{ox}} \int_0^{x_{ox}} \frac{x}{x_{ox}} \rho(x)\, dx \qquad (10\text{-}16)$$

In order to determine the minimum gate voltage that produces an inverted channel at the interface, it is necessary to consider the charge per unit area induced in the semiconductor, Q'_S. This is the sum of the mobile charge density, Q'_n, and the space-charge region charge density, Q'_d. For the case when the capacitor is in strong inversion and the channel is reverse biased with respect to the substrate, the mobile charge density is given by

$$Q'_n = -C'_{ox}(V_G - V_{FB} - V_C - 2\,|\phi_p|) + [2\varepsilon_s q N_a(2\,|\phi_p| + V_C - V_B)]^{1/2} \tag{10-17}$$

The voltage between the gate and ground just necessary to create a channel is called the *threshold* voltage, V_T, and is found by setting $Q'_n = 0$, resulting in

$$V_T = V_{FB} + V_C + 2\,|\phi_p| + \frac{1}{C'_{ox}}[2\varepsilon_s q N_a(2\,|\phi_p| + V_C - V_B)]^{1/2} \tag{10-18}$$

The surface field effect results in a complex capacitance versus voltage characteristic. Depending on the polarity and magnitude of the bias voltage and the frequency of the measuring signal, the capacitance of the device exhibits characteristics due to accumulation, depletion, inversion, or deep depletion of the semiconductor surface. Of particular importance in the surface field effect, from the viewpoint of MOS integrated circuit design, is the threshold voltage for the formation of an inverted channel at the oxide-silicon interface.

THE MOSFET

By far the most important type of component used in MOS integrated circuits is the MOSFET. In the most conductive state, the drain-source conductance of a MOSFET is much lower than that of the collector-emitter of a bipolar transistor in saturation. For this reason, if resistors were used for load components, they would be extremely large in value. In some cases, ion-implanted resistors are used, but, in the vast majority of circuits, MOSFET's are used for load devices. As a consequence, virtually all of the devices used in monolithic MOS integrated circuits are MOSFET's.

The first MOS integrated circuits to emerge in the market place were p-channel logic circuits. Since holes have a lower mobility than electrons, these circuits operated at relatively low frequencies, and were used with negative power supplies, usually with higher voltages than their bipolar counterparts. Special techniques were necessary to provide for interfacing with the highly popular TTL bipolar circuits which were often used in conjunction with the p-channel MOS circuits. The major reason for the early domination of p-channel MOS circuits over their n-channel counterparts was the difficulty in obtaining positive threshold voltages in n-channel MOSFET's. The application of ion implantation to threshold voltage tailoring permitted the reproducible fabrication of n-channel MOS integrated circuits. The third type of MOS

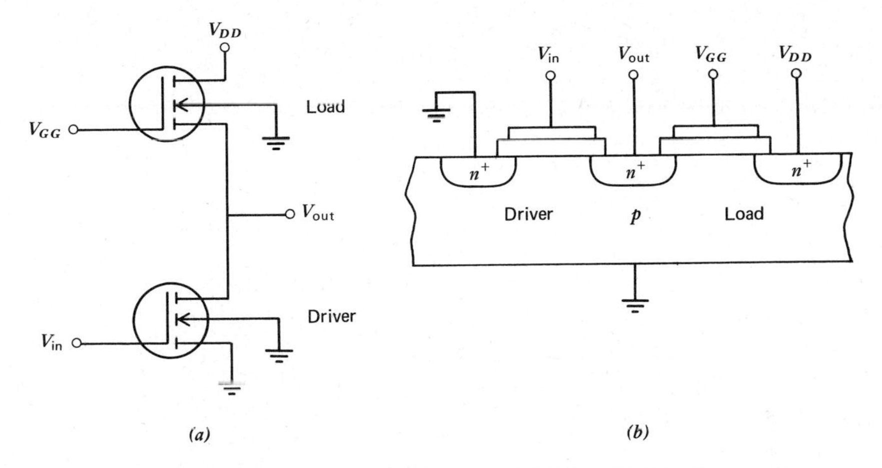

Figure 10-5. An n-channel MOS inverter. (a) Schematic. (b) Cross section.

integrated circuits makes use of pairs of n-channel and p-channel devices, and is called *complementary* MOS or CMOS.

In most cases, the MOSFET is designed so that no channel exists for zero bias between the gate and source. These devices are called *enhancement* mode MOSFET's. It is also possible to fabricate *depletion* mode MOSFET's, usually with ion-implanted channels.

The basic logic element in n- or p-channel MOS integrated circuits is the inverter, consisting of a driver MOSFET and a load MOSFET. This circuit is shown in Figure 10-5. The source of the driver transistor and the substrate are tied to ground, but the output terminal, which serves as the drain of the driver and the source of the load, is not tied to ground. This means that the channel of the driver transistor is controlled by the gate-source voltage of that device, V_{in}, but the channel of the load device is controlled by the gate-substrate voltage of that device, V_{GG}. For this reason, the model developed in the next section is based on general bias voltages to ground, rather than the specific case of source and substrate tied to ground.

A Model for an *n*-Channel MOSFET

In this section, a physical model for the operation of an n-channel MOSFET is developed. A similar model exists for p-channel devices. The model is based on the *gradual-channel* assumption, which implies that the electric fields in the direction parallel to the current are much smaller than those perpendicular to the direction of the current. This model includes the effects of the variation of the charge with position in the space-charge region along the length of the channel. The basic geometry is shown in Figure 10-6. The width of the surface of the channel is designated W.

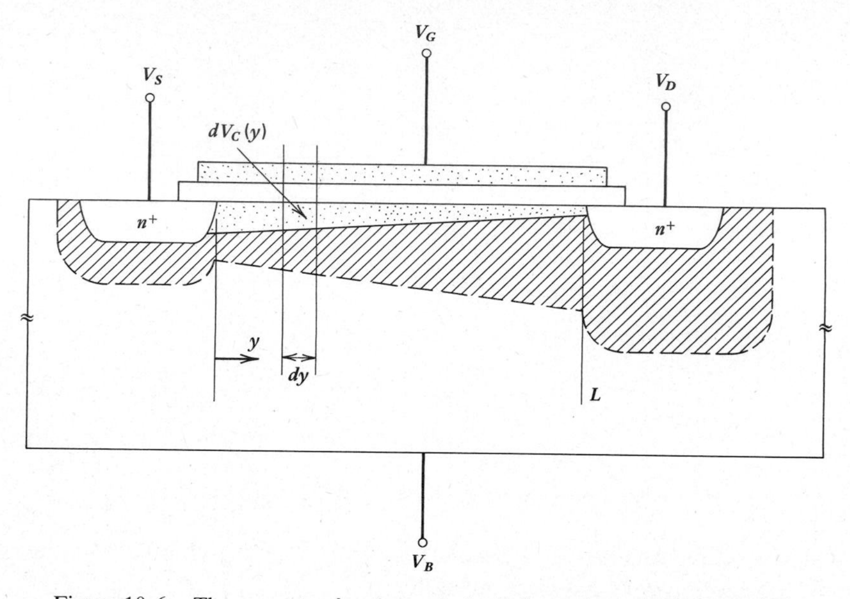

Figure 10-6. The geometry for the analysis of the n-channel MOSFET (18).

The voltage drop across an incremental length of the channel, dy, is given by

$$dV_C = \frac{-I_D dy}{W\mu_n Q'_n(y)} \tag{10-19}$$

where I_D is the drain current, and μ_n is the electron mobility in the surface region. The surface mobile charge density, Q'_n, is given by Equation 10-17, repeated here for convenience,

$$Q'_n = -C'_{ox}(V_G - V_{FB} - V_C - 2|\phi_p|) + [2\varepsilon_s qN_a(2|\phi_p| + V_C - V_B)]^{1/2} \tag{10-17}$$

In these equations, V_C is the voltage between the channel and ground, which is a function of y. Integrating Equation 10-19 along the length of the channel results in

$$I_D = \frac{\mu_n W}{L}\left\{C'_{ox}\left(V_G - V_{FB} - 2|\phi_p| - \frac{V_D}{2} - \frac{V_S}{2}\right)(V_D - V_S)\right.$$
$$\left. - \frac{2}{3}(2\varepsilon_s qN_a)^{1/2}[(2|\phi_p| + V_D - V_B)^{3/2} - (2|\phi_p| + V_S - V_B)^{3/2}]\right\} \tag{10-20}$$

This equation is valid for the voltage range from $V_D = 0$ to $V_D = V_{Dsat}$, the drain voltage at which the mobile charge at the drain end of the channel is reduced to zero. This drain voltage is given by

$$V_{D\text{sat}} = V_G - V_{FB} - 2|\phi_p| - \frac{\varepsilon_s q N_a}{(C'_{ox})^2}\left\{\left[1 + \frac{2(C'_{ox})^2}{\varepsilon_s q N_a}(V_G - V_{FB} - V_B)\right]^{1/2} - 1\right\} \tag{10-21}$$

Note that V_{Dsat} is independent of V_S. The drain current saturates at a value $I_{D\text{sat}}$, which is found by substituting Equation 10-21 into Equation 10-20. In a real MOSFET, the drain current increases slightly with increasing drain voltage greater than $V_{D\text{sat}}$ due to a decrease in the effective length of the channel.

The gain parameter for the MOSFET is the transconductance, g_m, defined by

$$g_m \equiv \frac{\partial I_D}{\partial V_G} \tag{10-22}$$

which has the value

$$g_m = \mu_n C'_{ox} \frac{W}{L}(V_D - V_S) \tag{10-23}$$

for $V_D < V_{D\text{sat}}$ and

$$g_{m\text{sat}} = \mu_n C'_{ox} \frac{W}{L}\left(V_G - V_{FB} - 2|\phi_p| - V_S - \frac{\varepsilon_s q N_a}{(C'_{ox})^2}\left\{\left[1 + \frac{2(C'_{ox})^2}{\varepsilon_s q N_a}(V_G - V_{FB} - V_B)\right]^{1/2} - 1\right\}\right) \tag{10-24}$$

for $V_D > V_{D\text{sat}}$.

The breakdown voltage of a MOSFET is essentially that of the drain-substrate junction. The breakdown voltage of this junction is primarily controlled by the substrate doping level.

The frequency response of a typical MOSFET is dominated by parasitic effects. The intrinsic frequency response is determined by the transit time along the channel, T_{tr}, which can be approximated by

$$T_{tr} \approx \frac{4}{3}\frac{L^2}{\mu_n(V_G - V_T)} \tag{10-25}$$

The frequency response predicted from the transit time is at least an order of magnitude higher than that observed for actual devices. Of particular importance to the actual frequency response are the capacitances due to the overlap of the gate with the source and drain diffused regions. These overlap capacitances are significantly reduced by special processing techniques in which the source and drain are positioned by the gate pattern. Some of these self-aligned gate techniques are described later in this chapter. Other factors contributing to the frequency limitations of MOS integrated circuits include the capacitances associated with the source-substrate junction (if the source is not tied to the substrate) and the drain-substrate junction, the capacitance between the intraconnection pattern and the substrate, and the resistance of the intraconnection pattern. In some processes, the intraconnection pattern consists, in part, of diffused regions to eliminate the need for a second level of metalization. In other processes, heavily doped polycrystalline silicon serves as the gate material and is also used as one level of intraconnection pattern. Some processes use refractory metals for the same purpose. As line widths in production processes are made smaller, the resistance of the intraconnection becomes larger, and, thus, more significant in limiting the operating frequency. Among the highest frequency MOS integrated circuits are those made by the CMOS silicon on sapphire process. These circuits are slightly slower than n-channel circuits, but operate at much lower power levels. In this fabrication technique, a thin film (usually less than 1 μm) of heteroepitaxial silicon is deposited on a sapphire substrate. The individual transistors are isolated from one another by etching away the silicon between them, leaving islands of silicon on the substrate. This process eliminates many of the parasitic capacitances associated with the more conventional structures.

The transconductance of a MOSFET, as expressed in Equations 10-23 and 10-24, can be increased by decreasing the channel length, L. These equations are good approximations as long as L is appreciably greater than the sum of the widths of the space-charge regions associated with the drain-substrate diffused regions. When this condition is not satisfied, the resulting device is called a *short-channel* MOSFET. In this type of device, under saturation conditions, the pinched-off portion of the channel is a significant part of the channel length. For this reason, the slope of the I_D versus V_{DS} curve is relatively large under saturation conditions. The threshold voltage for short-channel MOSFET's is somewhat lower than that for conventional devices with the same substrate doping. One difficulty encountered with short-channel devices is a low breakdown voltage due to punch-through when the drain-substrate space-charge region contacts the source-substrate space-charge region, with the gate-substrate voltage below the threshold voltage. This effect is reduced by increasing the doping level of the substrate and reducing the operating voltage. The maximum doping concentration in the substrate is limited to approximately $10^{19}\,\text{cm}^{-3}$ because the breakdown field in the oxide, $6\times10^{6}\,V\cdot\text{cm}^{-1}$, is insufficient to invert the surface if the doping level is greater than that level. The short channel effects become important when L is less than 3 μm.

MOS PROCESSES

In this section, several of the fabrication technologies employed in the manufacture of MOS integrated circuits are described. The emphasis is on the influence of the fabrication processes on the electrical characteristics of the

Table 10.1. **Design equations for MOSFET's**

Quantity	n-channel	p-channel
Flat-band voltage	$V_{FB} = \Phi_{MS} - \frac{Q'_{SS}}{C'_{ox}} - \frac{1}{C'_{ox}} \int_0^{x_{ox}} \frac{x\rho(x)}{x_{ox}}\, dx$	$V_{FB} = \Phi_{MS} - \frac{Q'_{ss}}{C'_{ox}} - \frac{1}{C'_{ox}} \int_0^{x_{ox}} \frac{x\rho(x)}{x_{ox}}\, dx$
Threshold voltage	$V_T = V_{FB} + \lvert V_S \rvert + 2\lvert\phi_p\rvert + \frac{1}{C'_{ox}} [2\varepsilon_s q N_a (2\lvert\phi_p\rvert + \lvert V_S - V_B \rvert)]^{1/2}$	$V_T = V_{FB} - \lvert V_S \rvert - 2\lvert\phi_n\rvert - \frac{1}{C'_{ox}} [2\varepsilon_s q N_d (2\lvert\phi_n\rvert + \lvert V_S - V_B \rvert)]^{1/2}$
Drain current	$I_D = \mu_n \left(\frac{W}{L}\right)\Big\{C'_{ox}\Big(V_G - V_{FB} - 2\lvert\phi_p\rvert - \frac{V_D}{2} - \frac{V_S}{2}\Big)(V_D - V_S) - \frac{2}{3}(2\varepsilon_s q N_a)^{1/2}[(2\lvert\phi_p\rvert + \lvert V_D - V_B\rvert)^{3/2} - (2\lvert\phi_p\rvert + \lvert V_S - V_B\rvert)^{3/2}]\Big\}$	$I_D = -\mu_p \left(\frac{W}{L}\right)\Big\{C'_{ox}\Big(V_G - V_{FB} + 2\lvert\phi_n\rvert - \frac{V_D}{2} - \frac{V_S}{2}\Big)(V_D - V_S) - \frac{2}{3}(2\varepsilon_s q N_d)^{1/2}[(2\lvert\phi_n\rvert + \lvert V_D - V_B\rvert)^{3/2} - (2\lvert\phi_p\rvert + \lvert V_S - V_B\rvert)^{3/2}]\Big\}$
Saturation voltage	$V_{D\text{sat}} = V_G - V_{FB} - 2\lvert\phi_p\rvert - \frac{\varepsilon_S q N_a}{(C'_{ox})^2}\left\{\left[1 + \frac{2(C'_{ox})^2}{\varepsilon_S q N_a}(V_G - V_{FB} - V_B)\right]^{1/2} - 1\right\}$	$V_{D\text{sat}} = V_G - V_{FB} + 2\lvert\phi_n\rvert + \frac{\varepsilon_S q N_d}{(C'_{ox})^2}\left\{\left[1 + \frac{2(C'_{ox})^2}{\epsilon_S q N_d} \lvert(V_G - V_{FB} - V_B)\rvert\right]^{1/2} - 1\right\}$
Trans-conductance $\lvert V_D \rvert \ll \lvert V_{D\text{sat}} \rvert$	$g_m = \mu_n C'_{ox} \frac{W}{L}(V_D - V_S)$	$g_m = -\mu_p C'_{ox} \frac{W}{L}(V_D - V_S)$
Bulk potential	$\phi_p = \frac{kT}{q} \ln \frac{n_i}{N_a}$	$\phi_n = \frac{kT}{q} \ln \frac{N_d}{n_i}$
Oxide capacitance density	$C'_{ox} = \frac{\varepsilon_{ox}}{x_{ox}}$	$C'_{ox} = \frac{\varepsilon_{ox}}{x_{ox}}$

MOSFET's, and the options available to the designer to achieve the desired objective for the circuit.

The important design equations for both n-channel and p-channel MOSFET's are listed in Table 10-1. The parameters that are controlled by the fabrication process are the oxide thickness and permittivity, the surface-state charge density, the substrate doping concentration, the metal-semiconductor work function difference, and the conductivity type, which determines the surface mobility.

The carrier mobility in the channel is, in general, approximately half that in the bulk for similar carrier concentrations. This is due to scattering at the surface. There is a further reduction in mobility as the gate voltage increases above the threshold voltage, because the carriers are attracted more strongly to the surface. Since electron mobility is two to three times that of hole mobility, n-channel devices are preferred.

The substrate doping influences the surface mobility, the threshold voltage, and the breakdown voltage. It is common practice to select a doping level on the order of 10^{15} to 10^{16} cm^{-3} to obtain a breakdown voltage of 20 to 30 V.

The Threshold Voltage

The threshold voltage depends on the work function difference between the gate and the silicon, the oxide thickness and permittivity, the surface-state charge density, the doping density of the substrate, and the distributed charge in the oxide. If a clean process is used to grow the gate oxide, including HCl, the distributed oxide charge is reduced to negligible levels. The surface state density, Q'_{SS}/q, in a clean process can be as low as 10^{11} cm^{-2} in (111) oriented silicon and is reduced by a factor of three by using (100) oriented silicon.

Several materials are used for the gates in MOS integrated circuits. Aluminum, the most popular metalization system for bipolar integrated circuits, was the selection for the "standard" p-channel MOS process. The work function for silicon depends on the doping level and impurity type. The work function, $q\Phi$, is defined as the energy required to remove from the material an electron with an energy at the Fermi level. The electron affinity, $q\chi$, is the energy necessary to remove from the material an electron with an energy at the conduction band edge. The electron affinity is a constant of the material, and is 4.10 eV for silicon. The voltage that is equivalent to work function for a semiconductor is given by

$$\Phi_S = \chi_S + \frac{E_g}{2q} - \phi \tag{10-26}$$

where E_g is the gap energy (1.11 eV for silicon at 300°K) and ϕ is given by

$$\phi = \frac{E_f - E_i}{q} \tag{10-27}$$

which is negative for p-type material and positive for n-type material. Thus, for silicon

$$\Phi_S = 4.65 - \phi \tag{10-28}$$

The work function of aluminum is 4.1 eV, resulting in a work function difference of

$$\Phi_{MS} = -0.55 + \phi \tag{10-29}$$

For a p-channel MOSFET, assuming a doping density of $N_d = 10^{15}\ \text{cm}^{-3}$ (the substrate is n-type), ϕ is $+0.29$ V and $\Phi_{MS} = -0.26$ V. Note that the contributions to the threshold voltage for aluminum gate MOSFET's are negative in both cases. A negative threshold voltage is desirable for p-channel devices, but a positive threshold voltage is desirable for n-channel devices.

One of the most successful gate materials is polycrystalline silicon which forms when silicon is deposited on the thermal oxide of the gate insulator. This film is typically 4000 Å thick and undoped. The electrical properties of the polycrystalline silicon are not well defined. Impurity doping of the gate material is usually performed at the same time as that of the source and drain regions. This is done primarily by diffusion, but ion implantation is sometimes used. The sheet resistance of the diffusion doped polycrystalline gate material is typically 20 to 30 Ω/□, compared to the source and drain diffusions formed by the same process which may be as low as 8 to 10 Ω/□. Ion-implanted polycrystalline silicon is high in resistivity and is sometimes used to form high valued resistors. In polycrystalline material, diffusion is dominated by rapid diffusion along grain boundaries, and ion-implanted impurities are difficult to electrically activate by the thermal processes which are effective in single-crystal materials, making it difficult to predict the work function for the gate material. However, the doping type of the gate is opposite to that of the substrate so that the work function difference is given by

$$\Phi_{GS} = -\phi_G + \phi_S \tag{10-30}$$

for an n-channel device, ϕ_G is positive and ϕ_S is negative, resulting in a negative Φ_{GS} (typically -0.6 V). For a p-channel device, the opposite signs are obtained, resulting in a positive value for Φ_{GS} (typically $+0.6$ V). The result is that a polysilicon gate can be used to reduce the negative threshold voltage in a p-channel device, but the threshold voltage shift in an n-channel is only slightly less negative than the case for an aluminum gate.

Molybdenum is also used as a gate material. The work function for molybdenum is approximately 4.3 eV, which is similar to that for aluminum. It has a sheet resistance which is much higher than that for aluminum, but lower than that which can be obtained with a polycrystalline silicon gate.

The surface state charge density, Q'_{SS}, is a function of the cleanliness of the process and has been improved by an order of magnitude over that obtained in the early days of commercial MOS processing. A typical value of Q'_{SS} for (111) oriented silicon is $8 \times 10^{-9}\,\mathrm{C \cdot cm^{-2}}$, and, for (100) oriented silicon, it is $1.5 \times 10^{-9}\,\mathrm{C \cdot cm^{-2}}$. The flat-band voltage shift due to Q'_{SS} is negative for both p- and n-channel devices. It is directly proportional to the oxide thickness. In typical processes, the gate oxide thickness is between 800 and 1200 Å, resulting in ΔV_{FB2} for (111) oriented silicon of -0.23 to -0.34 V, and, for (100) silicon, -0.042 to -0.064 V.

The other major contributions to the threshold voltage are related to the substrate doping level. For a doping concentration of $10^{15}\,\mathrm{cm^{-3}}$, and with the source tied to the substrate and ground, the threshold voltage shift due to the doping level is approximately $+1.0$ V for an n-channel device, and -1.0 V for a p-channel device.

The total threshold voltage for a p-channel silicon gate device with an 800-Å gate oxide on (100) oriented silicon with a doping concentration of $10^{15}\,\mathrm{cm^{-3}}$ is -0.44 V. For an n-channel silicon gate device with the same oxide thickness and substrate doping, $V_T = +0.36$ V. These values are strongly dependent on the cleanliness of the process, since an increase in the surface-state charge density of an order of magnitude results in a threshold voltage shift of -0.38 V which is sufficient to make the threshold voltage for an n-channel device negative.

In actual practice, the threshold voltage is tailored by ion implantation. For an n-channel device, boron ions are implanted through the gate oxide to produce a heavily doped region in the shape of a Gaussian near the oxide-silicon interface. For the purposes of analysis, the implant profile can be approximated by a rectangular distribution as indicated in Figure 10-7. The approximation is characterized by an average concentration, N_{ai}, and a characteristic depth, x_i, such that the product of N_{ai} and x_i is N_s, the total dose. Using this approximation, the threshold voltage can be written

$$V_T = V_{FB} + V_S + |\phi_p| + |\phi_{ps}| + \frac{qN_s}{C'_{ox}} + \frac{1}{C'_{ox}}[2qN_a\varepsilon_s(|\phi_p| + |\phi_{ps}| + V_S - V_B) - q^2x_iN_sN_a]^{1/2} \tag{10-31}$$

where $\phi_{ps} = (kT/q)\ln[(N_{ai} + N_a)/n_i]$. A good choice for x_i is twice the projected range. Using this technique, the threshold voltage can be reproducibly set at 1.0 V. The ion-implantation technique, using phosphorus or arsenic ions, is also used to set a negative threshold voltage for depletion-mode devices which are used as load transistors in many n-channel MOS integrated circuits.

The threshold voltage has another important effect in MOS integrated circuits. Unlike the bipolar transistor, the MOSFET is essentially self-isolating, but only if the surface of the silicon is only inverted under the gate and not in the regions between devices. To prevent this undesirable surface inversion, a

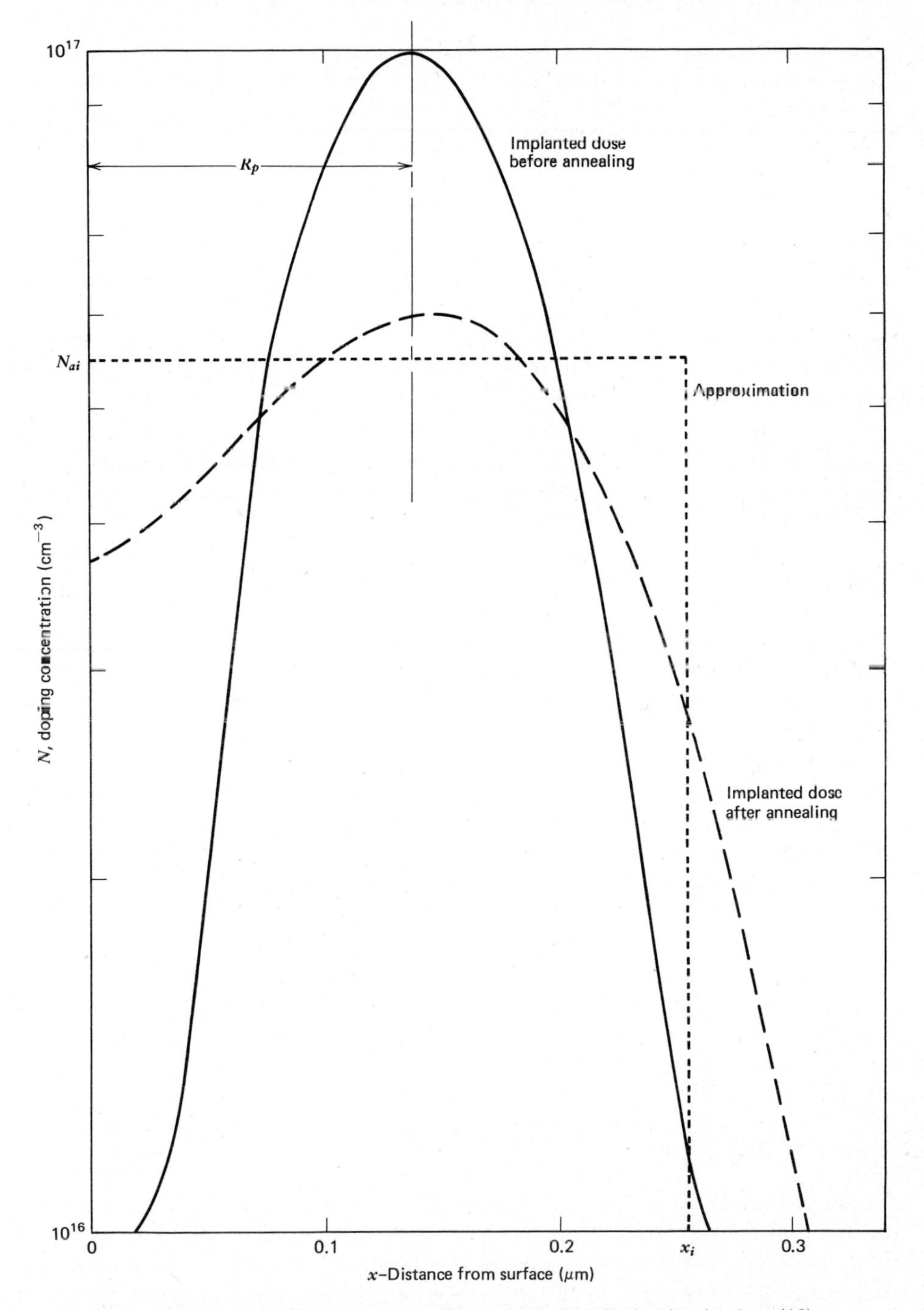

Figure 10-7. An approximate model for a threshold tailoring ion implant (18).

thick oxide, called the *field oxide*, is grown or deposited in these regions. The field-oxide is typically 10,000 to 15,000 Å thick. To further increase the threshold voltage in the field regions, channel stopping heavily doped diffusions or implants are used to make it very difficult to invert the surface.

Standard Processes

There are three processes that have achieved a certain degree of standardization throughout the industry. They are the thick-oxide metal-gate p-channel process, the self-aligned silicon-gate n-channel process, and the metal-gate CMOS process. These are the basic processes by which most MOS integrated are fabricated.

The standard p-channel process is illustrated in Figure 10-8. Note that the contact windows are included on the gate-oxide mask so that it is not necessary to etch the thick oxide more than once. The contacts are opened a second time by etching through the thin oxide grown during the gate oxidation. The diffused regions can be used for cross-unders in the intraconnection pattern.

The standard n-channel process is shown in Figure 10-9. In this process, ion implantation is used for the channel-stops and for threshold tailoring. The mask used to define the polycrystalline silicon is also used to define the source and drain diffusion windows. Arsenic is used to dope the source, drain, and polycrystalline silicon gate. Note that the electrical contacts to the silicon can be made by either the polycrystalline silicon or the aluminum metalization. The combination of a polycrystalline silicon layer and the aluminum layer allows for two layers of intraconnections.

The standard CMOS process is illustrated in Figure 10-10. The substrate is n-type because it is relatively easy to obtain a negative threshold voltage for the p-channel device. The p-well is implanted with boron and diffused to obtain a suitable positive threshold for the n-channel device. The p^+ and n^+ guard-band diffusions provide isolation by destroying the gain of the lateral *npn* and *pnp* bipolar transistors. The substrate *npn* parasitic transistor is biased in the cutoff region of operation, since the substrate is tied to the positive supply and the source of the n-channel device is either grounded or positive with respect to ground.

Special Structures

There are many processes used for the fabrication of MOS integrated circuits in addition to the standard processes. In this section, some of these processes are described. Since new processes are being developed continuously, this section is not complete, but it provides the reader with a reasonable cross section of the wide variety of processes available to the MOS designer.

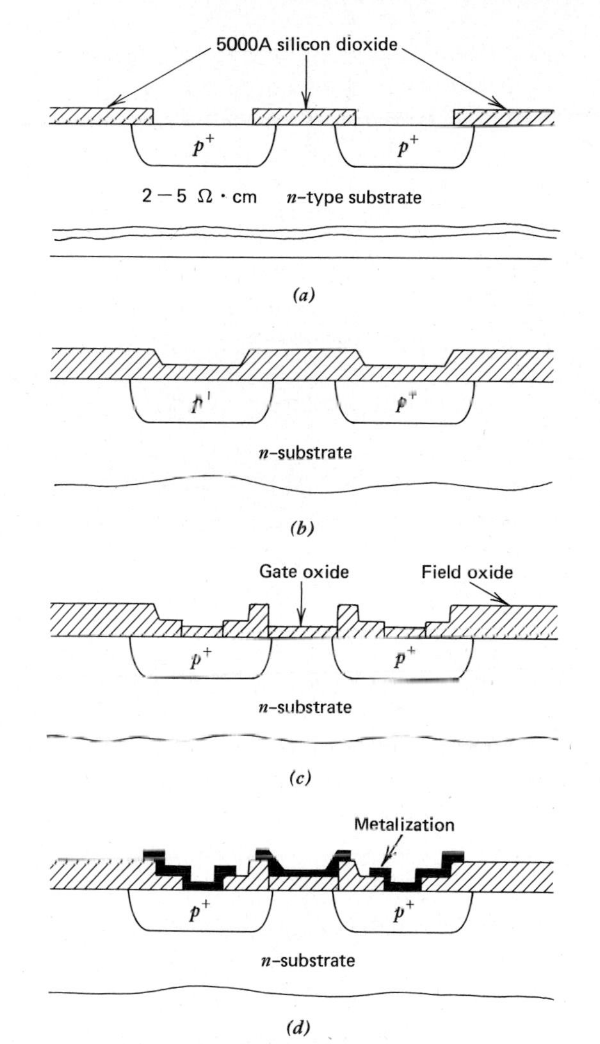

Figure 10-8. The processing sequence for a *p*-channel aluminum gate MOS integrated circuit. (*a*) Oxidation and source and drain diffusion. (*b*) Oxidation. (*c*) Gate oxide growth. (*d*) Metalization and definition of intraconnection pattern.

The Dual Dielectric Process. One technique for altering the threshold voltage of MOS devices is to increase the permittivity of the gate insulator. Since the thermal oxidation of silicon produces the lowest surface-state charge density, and it is difficult to prevent the formation of less than 30 Å of oxide on a silicon surface, most attempts to fabricate devices with a higher permittivity use a combination of a thin oxide layer and a layer of the higher permittivity material

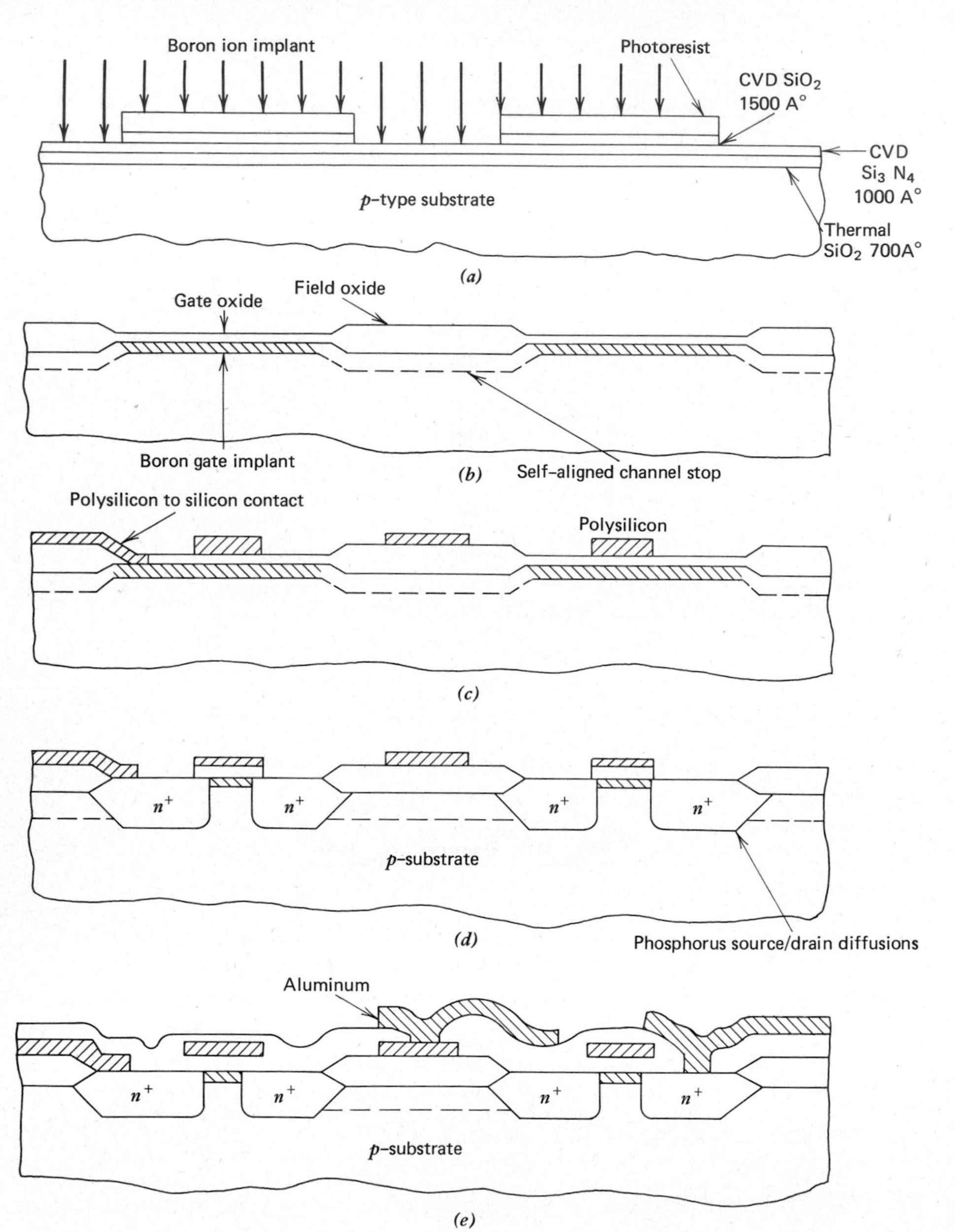

Figure 10-9. The processing sequence for an n-channel silicon gate MOS integrated circuit. (a) Channel stop implant. (b) Field oxide growth and threshold tailoring implant. (c) Polycrystalline deposition and definition. (d) Source and drain diffusion. (e) Metalization and pattern definition.

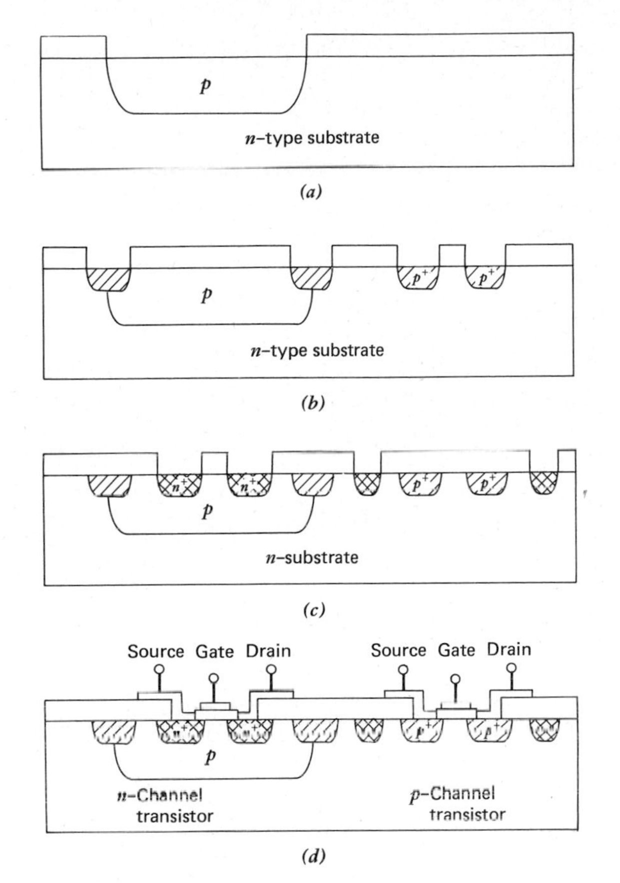

Figure 10-10. The processing sequence for a CMOS integrated circuit. (*a*) *p*-type well. (*b*) p^+ diffusion. (*c*) n^+ diffusion. (*d*) Contacts.

as the gate insulator. This dual dielectric structure has an overall permittivity less than that which would result from a structure consisting only of the higher permittivity material with the same total thickness. A popular combination for a dual dielectric structure is silicon nitride (Si_3N_4) and silicon dioxide (S_iO_2). Commercial *p*-channel devices, with 200 Å of SiO_2 (relative permittivity 3.2) and 800 Å of Si_3N_4 (relative permittivity 7.5) resulting in an effective relative permittivity of 5.9, are available. A particularly interesting dual dielectric structure is the metal-nitride-oxide-silicon (MNOS) electrically alterable read-only-memory (EAROM). In this structure, the oxide layer is typically 30 Å thick. Under the proper bias conditions, electrons can be induced to tunnel from the silicon through the oxide to the oxide-nitride interface, where they are essentially permanently trapped, resulting in a shift in the threshold voltage. The memory cells containing the trapped charge thus represent a different state than those not containing the trapped charge. The application of bias of the opposite polarity results in the removal of the trapped charge, leaving the device in the other state. The resulting memory is nonvolatile and electrically alterable.

Self-Aligned Gate Structures. To reduce the capacitance due to the overlap of the gate with the source and drain, self-aligned gate processes are used. The n-channel silicon-gate standard process is an example of such a process. For metal gate devices, self-alignment is accomplished by ion-implanting regions from the source and drain to the edges of the gate, using the gate as a mask. This structure is shown in Figure 10–11. Self-alignment of the gate is an important consideration in the development of new processes for MOS fabrication.

The Floating Gate. A very popular integrated circuit is the erasable programmable read-only memory (EPROM). These circuits make use of the floating gate avalanche injection MOS structure (FAMOS). The basic structure is shown in Figure 10-12. The polycrystalline silicon gate is completely surrounded by silicon dioxide, isolating it from the rest of the device and the remainder of the circuit. Electrons can be injected into the oxide when either the source-substrate or drain-substrate junction is biased into avalanche breakdown. Some of these electrons are trapped by the isolated gate, leaving it charged. The device then acts as if a voltage were applied to the gate, producing a channel. Thus, each bit of a memory can be individually programmed to an "off" or "on" nonvolatile state. The entire memory can be erased by exposure to ultraviolet light through a window in the package. The ultraviolet

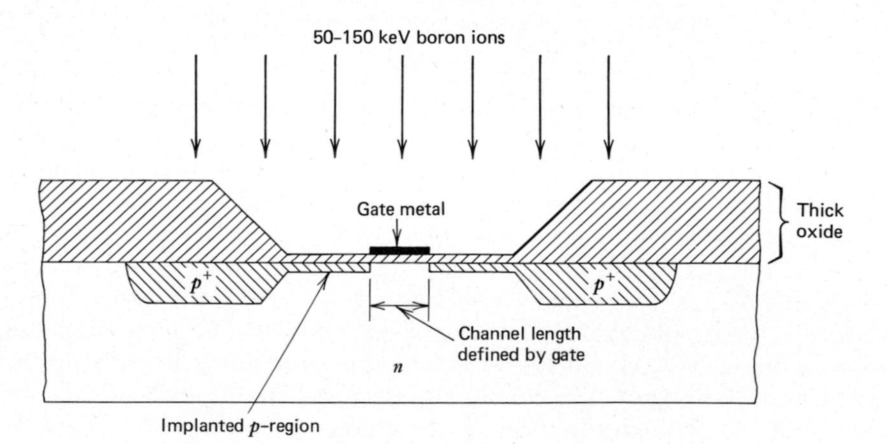

Figure 10-11. An ion-implanted self-aligned gate MOS device (2) (Courtesy of Texas Instruments, Incorporated).

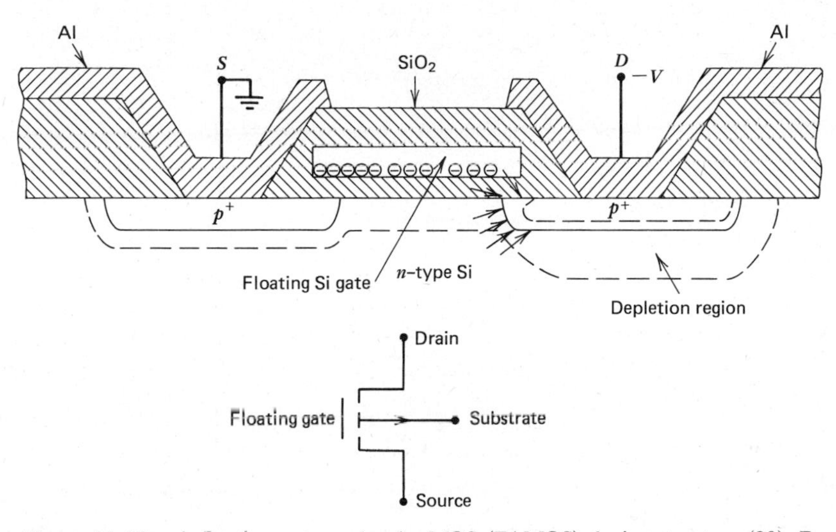

Figure 10-12. A floating-gate avalanche MOS (FAMOS) device structure (20). Reprinted from *Electronics*, May 10, 1971;

light produces electron-hole pairs in the gate oxide, and, thus, provides a conductive path from the isolated gate to the substrate.

Double-Diffused MOS. The double-diffused MOS process (DMOS) was developed as a technique for producing narrow channel lengths without relying on the photolithographic process to define these small dimensions. The basic structure is shown in Figure 10-13. The channel is formed by diffusing both p- and n-type dopants through the mask opening for the source diffusion. A modification of this process uses a silicon gate self-alignment technique to define the openings for the source and drain and uses the double-diffusion through both of the openings to produce very short channels.

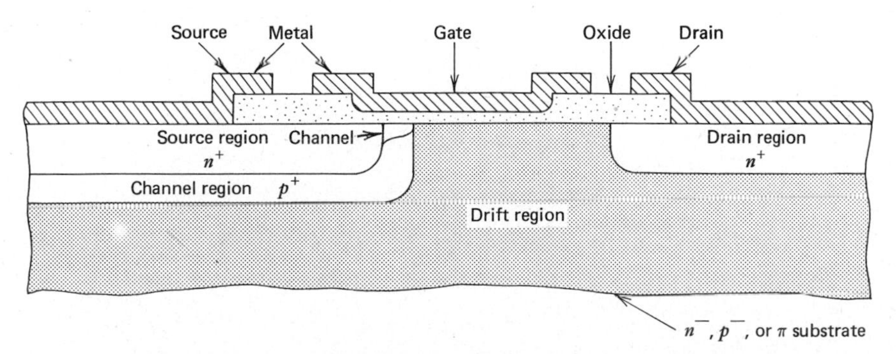

Figure 10-13. A double-diffused MOS (DMOS) device structure (8).

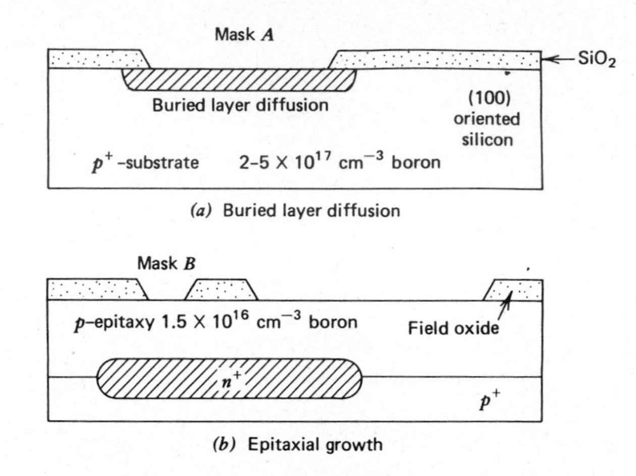

(a) Buried layer diffusion

(b) Epitaxial growth

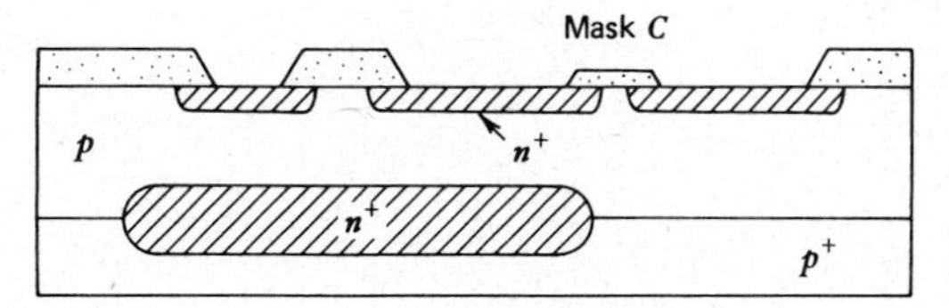

(c) Source–drain diffusion

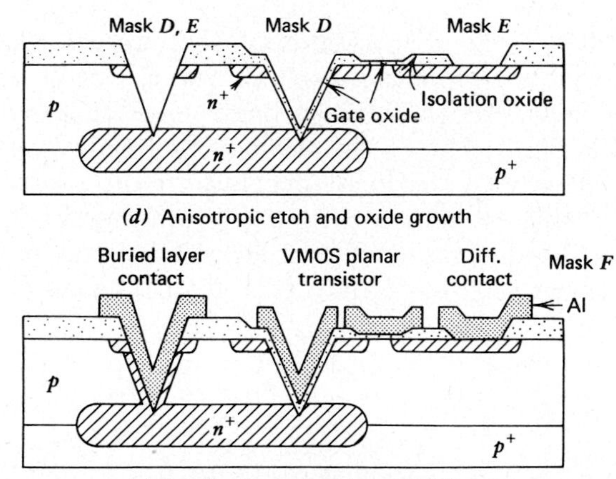

(d) Anisotropic etoh and oxide growth

(e) Metalization and pattern definition

Figure 10-14. The VMOS fabrication sequence (9). Copyright © 1978 IEEE. Reprinted from *IEEE Journal of Solid-State Circuits,* Oct. 1978, p. 617.

V-Groove MOS Structures. Another high-density MOS technology makes use of an anisotropic etch on (100) oriented silicon to produce "V" shaped grooves in the silicon, resulting in a different type of MOS transistor. This process is called VMOS. The basic process is illustrated in Figure 10-14. The channel is along the sloping surface of the groove, resulting in a channel length that is determined primarily by the distance between the drain diffusion and the buried layer.

Silicon on Sapphire Structures. The capability of depositing (100) oriented thin films of silicon on sapphire by the process of heteroepitaxy makes possible yet another type of MOS integrated circuit technology. In most cases, complementary MOS circuits are constructed by this technique. The basic structure is indicated in Figure 10-15. The structure shown is a self aligned silicon gate process, but many of the other technologies are possible. The silicon layer is usually less than 1.0 μm thick, and the devices are separated by anisotropic etching. The diffusions penetrate to the substrate. Silicon on sapphire MOS integrated logic circuits exhibit a very low speed-power product.

MOS-Bipolar Technology. In some cases, it is desirable to use both MOS and bipolar transistors on the same substrate. This type of technology is usually used in operational amplifier circuits, either for high-input impedance or to use p channel MOSFET's as active load devices for improved performance. The basic process is the same as that for bipolar integrated circuits, with the gate-oxide being grown after the emitter diffusion.

Charge-Coupled Devices. An interesting extension of MOS technology is the charge-coupled device (CCD). A unity cell of a CCD circuit consists of two or three closely-spaced MOS capacitors operated by clock pulses as indicated in Figure 10-16. The two-phase device does not require the small geometries required for the three-phase device. The clock frequency must be high enough

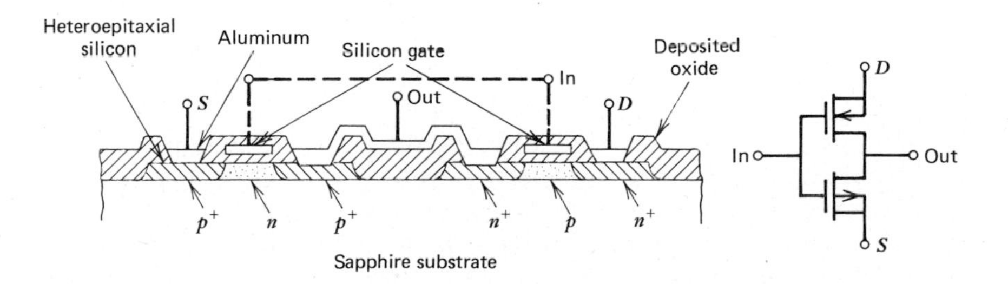

Figure 10-15. Complementary MOS on sapphire (CMOSOS) (10). Reprinted from *Solid State Technology*, "Recent SOS Technology, Advances and Applications," by R. S. Ronen and F. B. Micheletti, Vol. 18, August 1975. Copyright © Cowan Publishing Corp.

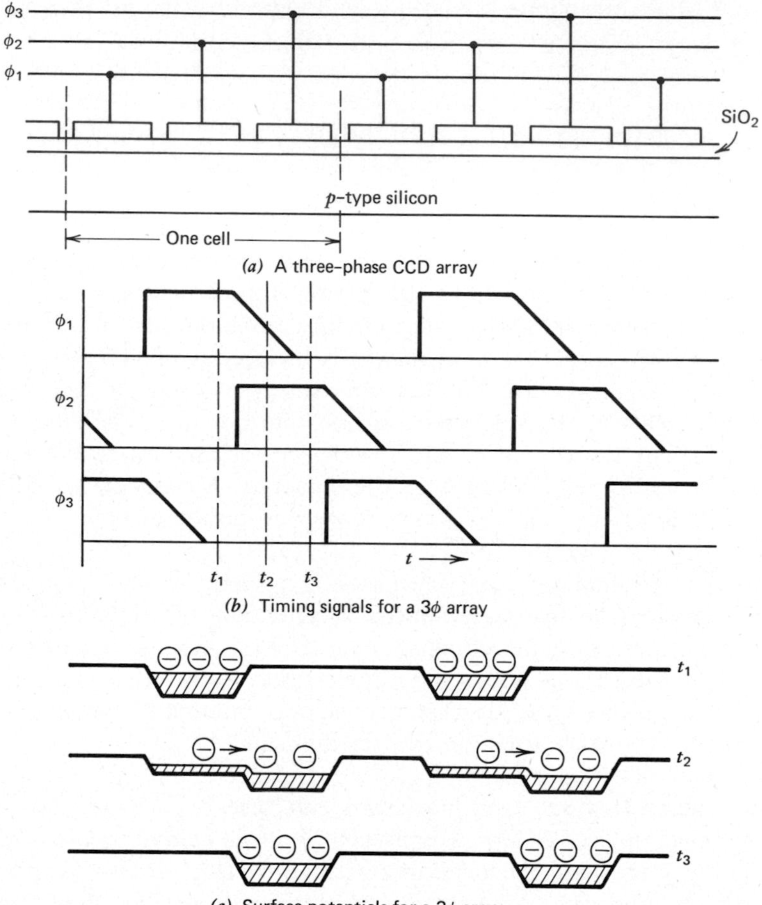

Figure 10-16. Three phase and two phase charge-coupled device (CCD) arrays (19). Reprinted with permission from *Electronic Design*, Vol. 23, No. 8, April 12, 1975, Copyrignt © Heyden Publishing Co., Inc., 1975.

to prevent the thermally generated carriers from filling the potential wells. The unusual characteristics of CCD's permit their use as digital shift-registers, analog shift-registers, and imaging devices. Since the processing for CCD's is essentially identical with that of MOSFET's, it is common practice to include both kinds of devices on the same substrate. While most CCD's make use of the oxide-silicon interface for charge storage, it is also possible to fabricate devices with buried channels, usually produced by ion implantation. This improves the charge transfer efficiency.

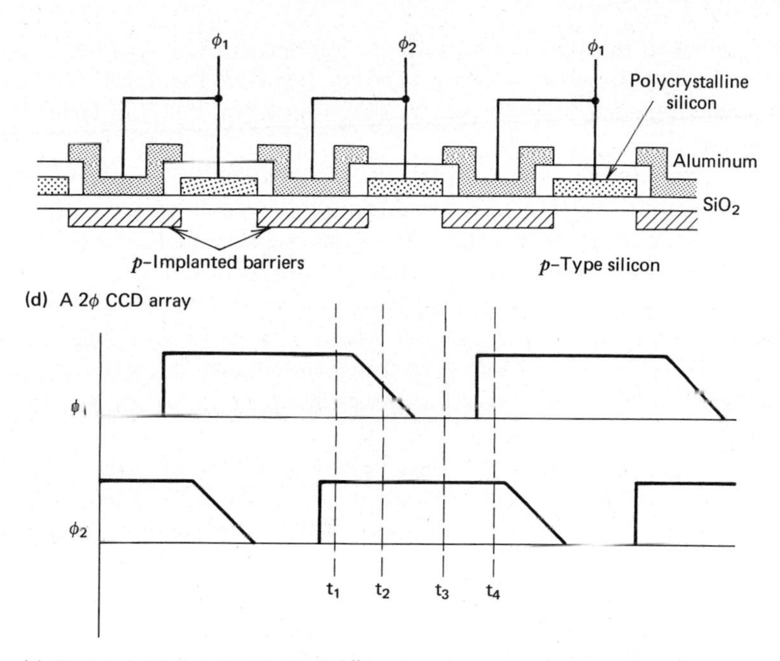

(d) A 2ϕ CCD array

(c) Timing signals for a 2ϕ "push-clock" array

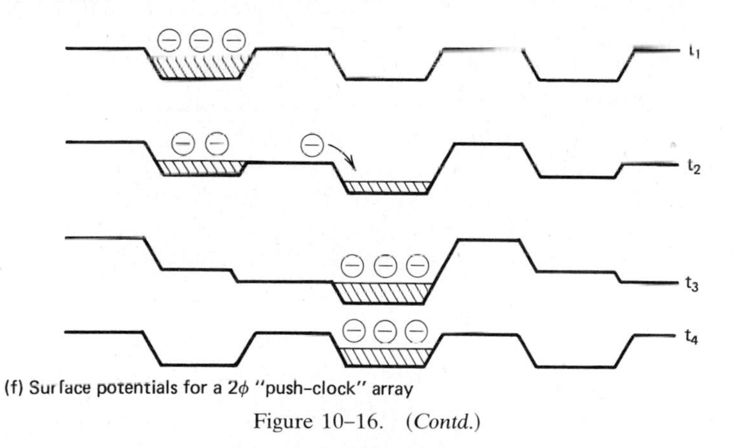

(f) Surface potentials for a 2ϕ "push-clock" array

Figure 10–16. (*Contd.*)

MOSFET SURFACE GEOMETRY

The majority of MOS integrated circuits consist almost entirely of MOSFET's intraconnected in large-scale logic or memory arrays. Each cell contains both driver and load MOSFET's. The designer at an MOS facility has available a library of compatible MOS cells stored on a computer. Each cell has been characterized electrically, so that, when a logic diagram has been entered into the computer, the entire circuit can be simulated and a testing program can be created which can be used to evaluate the finished circuits. Computer software has been developed which can be used to position the cells and draw the intraconnection pattern, resulting in a custom integrated circuit design in a very short time.

After a process has been selected for a particular MOS integrated circuit, the only important decision remaining for the device designer is the width to length ratio (W/L) for the channel. As can be seen from Equation 10-13, the transconductance of the device is directly proportional to this ratio. The channel length, L, should be as small as possible, constrained by the photolithographic process and short-channel effects. To avoid short-channel effects, L is usually 3 to 5 μm in length. The transconductance is then determined by the selection of W. The desired transconductance depends on the circuit configuration.

In MOS logic circuits, the basic element is an inverter. This usually consists of a driver device and a load device, as shown in Figure 10-17. The driver transistor is an enhancement mode MOSFET which is off for $|v_{in}|<|V_T|$, saturated for $|v_{in}-V_T|\leqq|v_{out}|$, and nonsaturated for $|v_{in}-V_T|>|v_{out}|$. The load device can have several configurations. It can be an enhancement mode MOSFET operated in a saturated or nonsaturated region of operation, a depletion mode MOSFET, or a high-valued resistor. For logic operation using an enhancement load device, if $V_{GG}=V_{DD}$, the load device is in saturation for all input conditions. If $|V_{GG}|>|V_{DD}+V_T|$, the load device is nonsaturated for all input conditions. In this configuration, V_{GG} is typically $6V_T$ and $V_{DD}=3V_T$. For a depletion load device, with an ion-implanted channel, $V_{GG}=v_{out}$, and the device is operated in the nonsaturated region of operation. Ion-implanted resistors are sometimes used as load devices. The principal advantage of nonsaturated or resistor loads is that the output resistance of these circuits is lower than that of a saturated load circuit, resulting in a smaller time constant for charging the input capacitances of logic elements connected to the

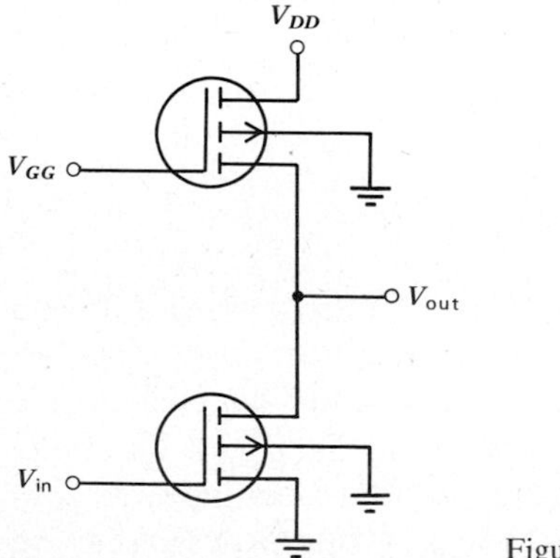

Figure 10-17. A p-channel inverter.

output terminal. This increase in speed is offset by a more complicated power supply configuration or an increase in fabrication cost.

For proper logic operation, $|v_{out}|$ must be less than $|V_T|$ when $|v_{in}|$ is greater than $|V_T|$, and $|v_{out}|$ must be greater than $|V_T|$ when $|v_{in}|$ is less than $|V_T|$. In static logic elements, this is accomplished by making the transconductance of the driver device large compared to that of the load device. The ratio

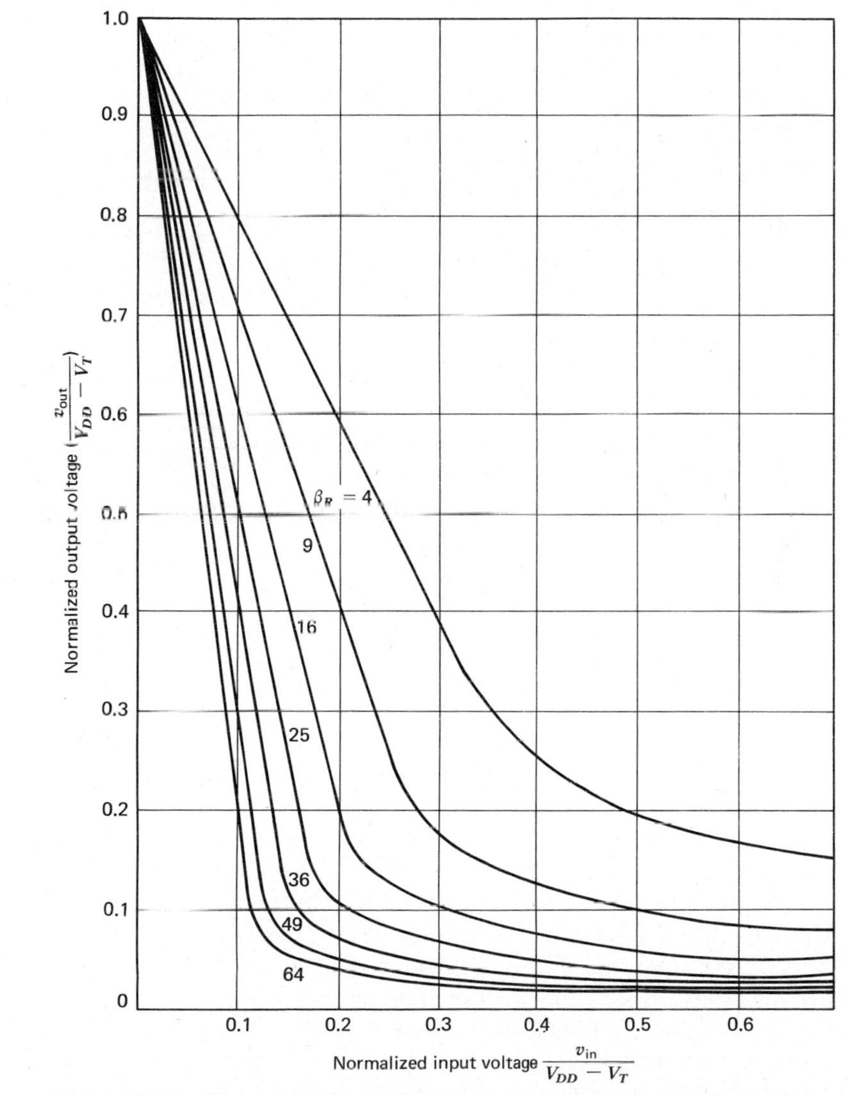

Figure 10-18. The transfer characteristics of a p-channel inverter circuit with β_R as a parameter (1). Reprinted by permission of Litton Educational Publishing, Inc.

of the transconductances is called β_R, and is given approximately by

$$\beta_R \simeq \frac{W_D/L_D}{W_L/L_L} \tag{10-32}$$

where the subscripts D and L refer to the driver and load devices, respectively. The voltage transfer characteristics for a saturated load enhancement mode inverter, with β_R as a parameter, are shown in Figure 10-18. A typical β_R is 25. Once β_R has been established, the channel dimensions can be calculated for both transistors, choosing L_D and W_L as minimum dimensions.

In complementary MOS logic circuits, the β_R is usually set at unity, since each device alternates between being the driver and being the load. In this case, the ratio of transconductances yields

$$\beta_R \simeq \frac{\mu_n W_n/L_n}{\mu_p W_p/L_p} \tag{10-33}$$

Therefore,

$$\mu_n \frac{W_n}{L_n} = \mu_p \frac{W_p}{L_p} \tag{10-34}$$

for symmetry. Since the surface mobility for electrons is approximately twice that for holes, the n-channel device should be twice as long as the p-channel device.

In dynamic logic circuits, the gate of the load device is driven by clock pulses instead of a steady potential, as in the case of static logic. In certain circuit configurations, the β_R for the driver and load devices is unimportant, and minimum geometry devices are used.

LIMITATIONS (17)

The ultimate density of circuit functions that can be realized in MOS integrated circuits is limited by the minimum area occupied by a transistor structure, since the power dissipation is decreased by most of the steps necessary to decrease the area. It should also be recognized that the area required for interconnections between devices can become significant. IBM has developed a technique for flip-chip bonding of large-scale integrated circuits in which the bonding points are distributed over the wafer area rather than concentrated at the periphery as in conventional bonding. This eliminates the need to bring all of the connecting points to the outside of the circuit, and saves considerable area.

Since dynamic logic circuits use minimum geometry transistors, they are the most likely candidates for high-density circuitry. The minimum channel length is limited by the punch-through situation brought about by the overlap of the transition regions of the drain-substrate and source-substrate junctions.

The transition region widths can be reduced by decreasing the supply voltage and increasing the doping concentration of the substrate. The lower limit of the supply voltage is determined by the reproducibility of the gate threshold voltage, and the minimum gate-oxide thickness that can be reliably manufactured. Increasing the substrate doping reduces the extension of the transition regions associated with the drain-substrate and source-substrate junctions into the substrate. This permits minimum spacing between the source and the drain. Unfortunately, for doping levels greater than $2 \times 10^{19}\,cm^{-3}$, field emission occurs at the junctions and isolation between the source and drain no longer exists. Even before this doping level is reached, at approximately $1.3 \times 10^{19}\,cm^{-3}$, oxide breakdown occurs at fields insufficient to invert the surface. For gate oxide thicknesses of 50 Å, the minimum thickness without a high probability of tunneling from gate to substrate, drain-to-source voltages as low as 0.7 V can be used. For lower voltages, the substrate doping would have to be decreased so that an inversion layer can be formed, resulting in a need for larger spacing between the source and drain.

SUMMARY

In this chapter, the basic design equations for MOS transistors used in integrated circuits have been developed. There are a wide variety of MOS processes in common use throughout the integrated circuits industry, which permit the designer of MOS circuits to have considerably more flexibility than the designer of bipolar integrated circuits. The use of ion implantation in the fabrication process allows the designer to tailor the device characteristics to suit the circuit requirements. Most of the MOS integrated circuits are digital logic circuits and memories.

REFERENCES

1. W. M. Penney and L. Lau, eds., *MOS Integrated Circuits* (New York: Van Nostrand-Reinhold, 1972).
2. W. N. Carr and J. P. Mize, *MOS/LSI Design and Applications* (New York: McGraw-Hill, 1972).
3. A. S. Grove, *Physics and Technology of Semiconductor Devices* (New York: Wiley, 1967).
4. D. J. Hamilton and W. G. Howard, *Basic Integrated Circuit Engineering* (New York: McGraw-Hill, 1975).
5. B. Hoeneisen and C. A. Mead, *Solid State Electronics* 15, (1972).
6. J. Compeers et al., *IEEE Transactions on Electron Devices* ED-4, 6 (1977).

7. R. R. Troutman, *IEEE Transactions on Electron Devices* ED-24, 3 (1977).
8. E. R. Hnatek, *A User's Handbook of Integrated Circuits* (New York: Wiley, 1973). Figure reprinted by permission of the publisher.
9. K. Hoffmann and R. Losehand, *IEEE Journal of Solid-State Circuits* SC-13, 5 (1978).
10. R. S. Ronen and F. B. Micheletti, *Solid State Technology* 18, 8 (1975).
11. M. Polinsky and S. Graf, *IEEE Transactions on Electron Devices* ED-20, 3 (1973).
12. A. B. Glaser and G. E. Subak-Sharpe, *Integrated Circuit Engineering* (Reading, Mass.: Addison-Wesley, 1977).
13. R. Melen and D. Buss, eds., *Charge-Coupled Devices: Technology and Applications* (New York: IEEE Press, 1977).
14. E. Wolfendale, ed., *MOS Integrated Circuit Design* (New York: Halsted Press, 1973).
15. J. T. Wallmark and L. G. Carlstedt, *Field-Effect Transistors in Integrated Circuits* (New York: Halsted Press, 1974).
16. R. H. Crawford, *MOSFET in Circuit Design* (New York: McGraw-Hill, 1967).
17. B. Hoeneisen and C. A. Mead, *Solid State Electronics* 15, 7 (1972).
18. R. S. Muller and T. I. Kamins, *Device Electronics for Integrated Circuits* (New York: Wiley, 1977). Figures reprinted by permission of the publisher.
19. W. F. Kosonocky and D. J. Sauer, *Electronic Design* 23, 8 (1975).
20. D. Froman-Bentchkowsky, *Electronics*, 44, 10 (1971).

PROBLEMS

10-1. Calculate the maximum width of the surface controlled space-change region for substrate doping concentrations of 10^{14}, 10^{15}, and $10^{16}\ \text{cm}^3$. Assume that all of the impurity atoms are ionized.

10-2. In an MOS capacitor, using the fact that $V_G = V_{ox} + \phi_s - \phi_p$, where V_G is the gate voltage, V_{ox} is the voltage across the oxide, and $\phi_s - \phi_p$ is the potential of the surface with respect to the p-type substrate, show that

$$\frac{C'}{C'_{ox}} = \left(1 + \frac{2\varepsilon_{ox}{}^2 V_G}{qN_a \varepsilon_s x_{ox}{}^2}\right)^{-1/2}$$

for the semiconductor surface in the depleted state. Make a plot of

C'/C'_{ox} versus V_G for $V_{FB}=0$, $N_a=10^{15}\ \text{cm}^{-3}$, and $x_{ox}=900$ Å. Include all three semiconductor surface conditions, accumulation, depletion, and inversion. Assume no interface charge.

10-3. Calculate the flat-band voltage shift associated with a negative charge density distributed (a) uniformly through the oxide with a density of 10^{15} electrons cm^{-3}, (b) as a step which is zero for the first two-thirds of the oxide (from the gate) and 3×10^{15} electrons cm^{-3} for the final one-third (nearest the oxide-silicon interface), and (c) linearly, starting at zero at the gate and reaching a value of 2×10^{15} electrons cm^{-3} at the interface. The oxide is 900 Å thick.

10-4. Derive Equation 10-21.

10-5. Find the doping density required for the substrate if the surface state density is $5\times10^{10}\ \text{cm}^{-2}$ for the threshold voltage to be -1 V for an aluminum gate p-channel MOSFET with a gate oxide thickness of 800 Å for the source grounded to the substrate.

10-6. An implant dose at 30 keV of 5×10^{10} boron atoms cm^{-2} is used to alter the threshold voltage of an n-channel MOSFET with a substrate doping density of $10^{15}\ \text{cm}^{-3}$ boron atoms, an oxide thickness of 900 Å, and a flat-band voltage of -0.84 V. If the source is grounded to the substrate, what is the new threshold voltage?

10-7. Estimate the saturation transconductance of a minimum geometry n-channel MOSFET assuming 5-μm design rules, a gate oxide thickness of 750 Å, a substrate doping density of $10^{15}\ \text{cm}^{-3}$, and a flat-band voltage of -0.44 V. The source is grounded to the substrate and the gate and drain are tied to $+5$ V.

10-8. Design a processing schedule for the fabrication of n-MOS integrated circuits.

10-9. Design a processing schedule for the fabrication of CMOS integrated circuits.

Chapter 11
Hybrid Microcircuit Component Design

Hybrid microcircuits are an important segment of the total microelectronics scene. The term "hybrid" implies that more than one basic technology is involved in the fabrication sequence. Although there are many possible combinations that could be considered under this definition, the most common hybrid technologies make use of passive components and conductor patterns formed by thick-film or thin-film deposition on ceramic substrates, with the other components necessary to form the electronic circuits attached to the conductor patterns by some type of bonding. The add-on components may be any or all of the following: transistors, diodes, integrated circuits, transformers, inductors, capacitors, and resistors. In most cases, the deposited and patterned films are restricted to the formation of resistors, conductors, and crossovers. Capacitors can be fabricated by either thick- or thin-film techniques, but the high-capacitance density of chip capacitors makes them a likely choice for most applications. Inductors can be fabricated with spiral conductor patterns, but both the inductance and Q of these structures are very low, essentially limiting their application to very high frequencies.

Thick-film hybrids are used in high-power circuits, microwave structures, multilayer interconnections for monolithic integrated circuits, and in the low-volume production of custom circuits. One of the most important applications of thick-film hybrids has been the voltage regulator for automobiles, where millions are produced annually. The major advantage of the thick-film hybrid processs is the capability of fabricating a wide range of resistor values, typically from less than 1 Ω to several megohms, on the same substrate. The use of the thick-film multilayer process for the high-density interconnection of integrated circuits, either in chip form or special packages designed for hybrid applications, has become an increasingly important technique for assembling microcomputers and memories.

Thin-film hybrids are used in high-reliability applications like communications circuits and in high-frequency and microwave structures. The thin-film process can be used to produce high quality, precise resistors, which can be used in analog-to-digital and digital-to-analog converters, as well as other applications. Thin-film hybrids are also used for interconnecting large-scale monolithic integrated circuits to form compact, reliable systems. The frequency range above 1 GHz has been dominated by thin-film hybrids, but thick-film hybrids have made some advances in this high-frequency arena.

One particularly interesting application of hybrids is in the area of biomedical engineering. Custom microcircuits for implantation into humans are usually made by hybrid techniques, since the demand for this type of circuit is limited.

In this chapter, the design and characteristics of film resistors are emphasized, since these are the primary components formed on hybrid substrates. Capacitor design, as well as crossover and multilayer techniques, are also described.

THICK-FILM DESIGN GUIDELINES

Thick-film pastes provide the designer of passive components with a flexibility unequaled in microcircuit technology. The designer has the option of using up to three resistor pastes, and may also use more than one type of conductor paste. This makes it possible to use different bonding techniques for attaching add-on components. Dielectric pastes are used for crossovers, multilayer insulators, resistor encapsulation, and, infrequently, for capacitor dielectrics.

Thick-film Resistor Design

Thick-film resistor pastes are specified in sheet resistance per 25 μm of dried thickness. This is the thickness of the print after the drying stage of the process before the substrate enters the firing furnace. This thickness can be measured with a light-section microscope, and provides a means for predicting the sheet resistance that will be obtained after firing. The value specified by the ink manufacturer is based on several specific parameters, including the furnace profile, the aspect ratio (L/W), the print thickness, and the conductor terminations for the resistors. Since any of these parameters can have a significant effect on the sheet resistance, test firings using the hybrid circuit manufacturer's process are recommended to characterize the pastes.

Thick-film resistor pastes contain functional materials, glass frit, and an organic vehicle. The first successful thick-film resistor materials used a mixture of palladium, palladium oxide, and silver as the functional materials. This system is economical, but the resistance is sensitive to firing conditions because it relies on an incomplete, high-temperature chemical reaction to determine the resistor characteristics. This resistor system has been replaced, to a great extent, by a number of systems based on ruthenium oxide or ruthenates. These resistor materials are much less sensitive to firing conditions, and have better stability and electrical characteristics than those based on the former system. Each paste manufacturer specifies a variety of conductor pastes that are compatible with their resistor systems for the formation of resistor terminations. These conductor pastes contain noble metals like silver, palladium, platinum, and gold, which are suitable for air firing. Resistor systems for base-metal conductors that require nonoxidizing firing atmospheres are under development. The firing range for typical resistor and conductor pastes is between 800 and 900°C.

Resistor pastes are supplied in compatible "families," which are designed for simultaneous firing. For increased resistor stability, it is common practice to print and dry all of the resistor patterns before the final high-temperature firing of the substrate. The standard resistor pastes are supplied in decade values from 1 Ω/□ to 10 MΩ/□ (per 25 μm of dried thickness), but most paste families do not cover this entire range. Resistor pastes can be blended to produce intermediate values.

The sensitivity of resistance to temperature variations is indicated by the temperature coefficient of resistance (TCR), which is given by

$$\mathrm{TCR}=\frac{R(T_2)-R(T_1)}{R(T_1)(T_2-T_1)}\times 10^6\ \mathrm{ppm/^\circ C} \tag{11-1}$$

where T_2 is the temperature at which the resistance, R, is to be determined for a known value of R at the reference temperature, T_1. Thick-film resistors can have TCR's which are positive, negative, or zero. Unfortunately, resistors made from pastes within a compatible family will not all have the same TCR. In general, low sheet resistance resistors have a relatively large positive TCR, and those made from high sheet resistance materials have a relatively large negative TCR. If a low TCR is desirable, intermediate sheet resistance pastes should be used. Typical values for TCR for a thick-film resistor system range from ±250 to ±50 ppm/°C. For comparison, the TCR of carbon composition resistors is typically +50,000 ppm/°C, the TCR of metal film resistors is ±25 ppm/°C, TCR's for thin-film resistors are usually less than 100 ppm/°C, and the TCR's for monolithic integrated-circuit diffused resistors range between 2000 and 5000 ppm/°C.

Resistor values are also sensitive to voltage. The voltage coefficient of resistance (VCR) is given by

$$\mathrm{VCR}=\frac{R(V_2)-R(V_1)}{R(V_1)(V_2-V_1)}\times 10^6\ \mathrm{ppm/V} \tag{11-2}$$

As long as the electric field is restricted to values below 40,000 $\mathrm{V\cdot m^{-1}}$, the

VCR for thick-film resistors is less than 50 ppm/V. For fields in excess of this value, permanent changes can occur in thick-film resistors, usually resulting in a significant decrease in the resistance. Static electricity can produce these excess fields on high-valued resistors.

Noise in resistors is usually measured by comparing the noise generated in the components with that generated in a standard resistor. For thick-film resistors, high sheet resistance components exhibit a noise value of +10 dB/decade, while low sheet resistance components have noise values of −25 dB/decade. The resistor area usually specified for noise measurements is $1250 \times 1250\ \mu m^2$.

The high-frequency characteristic of the components called "resistors" are significantly different from the low-frequency characteristics. Wire-wound resistors, for example, exhibit inductance and capacitance, due to the geometry associated with the construction of these components. Thick-film resistors retain their low-frequency characteristics up to approximately 1 MHz. Above this frequency, the effective resistance of the component changes. Components fabricated from high sheet resistance pastes tend to decrease in resistance with increasing frequency, and those made from low sheet resistance materials tend to increase in resistance with increasing frequency. The frequency dependence of resistance also is influenced by layout considerations. Thick-film resistors have been used as transmission line terminations in microwave circuits up to 10 GHz.

In general, thick-film resistors can dissipate more power than those fabricated by any other microelectronic technology. This is partly due to the ability of the thick film hybrid structure to transfer heat, and partly due to the relatively large size of the resistors. The maximum power density for thick-film resistors is $7.75\ W \cdot cm^{-2}$, compared to $3.88\ W \cdot cm^{-2}$ for thin-film resistors, and $775\ W \cdot cm^{-2}$ for monolithic diffused resistors. These are meant as guidelines rather than absolute maxima, since a lot depends on the package structure and the methods employed for heat transfer. Even though the resistors in monolithic integrated circuits have a higher allowed power density, this is predicated on the resistors occupying a relatively small portion of the circuit, and the total power dissipation is severely limited by the ability of the package to transfer heat.

Thick-film resistors can be economically trimmed to ±1%, or trimmed while the circuit is operating to assure proper functioning of the circuit. Where cost is an important consideration, and the tolerance for the resistor values is wide enough, some thick-film resistors are not trimmed at all, to save time. Trimming is either an air abrasive or laser process. In either case, up to 50% of the effective resistor area can be removed by the trimming process. For this reason, the resistor is designed so that

$$\frac{2P_R}{LW} \leq P_M \tag{11-3}$$

where P_R is the power dissipated in the resistor, P_M is the allowable power density, W is the width of the resistor, and L is the length of the resistor between terminations.

The resistor value is given by

$$R = R_S \times A.R. \tag{11-4}$$

where R_S is the sheet resistance, and A.R. is the aspect ratio, L/W. The minimum width of the resistor, $W_{\min}$, is then

$$W_{\min} = \left(\frac{2P_R}{P_M \times A.R.}\right)^{1/2} \tag{11-5}$$

In most hybrid processing facilities, a minimum dimension for resistors of 1250 μm has been established, to assure reproducibility. This design rule applies for both L and W. If the aspect ratio is greater than unity, and $W_{\min}$ is less than 1250 μm, W is set at 1250 μm and L is calculated from the aspect ratio. If the aspect ratio is less than unity, L is set at 1250 μm and the resulting W is compared to $W_{\min}$. If W is less than $W_{\min}$, both L and W are adjusted accordingly.

The spread in the values of as-fired resistors is typically between 10 and 20%. Since resistor trimming can only increase resistor values, it is common practice to use a design value of 75% of the desired value. The preferred aspect ratio range, from the standpoint of ease of printing, is from $\frac{1}{3}$ to 3. It is desirable to limit the number of different resistor pastes for a particular circuit to three, from economic considerations and printing registration difficulties. In order to minimize the number of pastes, the aspect ratio range can be extended to $\frac{1}{10}$ to 10, or series or parallel combinations of resistors can be used. In some cases, thick-film chip resistors are used in place of additional pastes. The total length of the resistor pattern should be 500 μm longer than L to provide overlapping regions with the conductor pattern used for terminations. A typical rectangular resistor layout is shown in Figure 11–1. For best results, the long direction of all of the resistor patterns should be parallel to the direction of squeegee motion.

The majority of thick-film resistors are made in rectangular patterns. The serpentine geometries popular in thin-film and monolithic integrated circuits for realizing large resistor values are not necessary in thick-film hybrids, due to the wide range of sheet resistance pastes available. They are also difficult to print uniformly. One other popular thick-film resistor geometry is called the "top hat." This geometry is recommended when the aspect ratio is greater than 6, and the power dissipation is near the maximum rating. The basic layout of a "top hat" resistor is shown in Figure 11-2. Because of current crowding, a corner square contributes only 0.5 square to the resistor aspect ratio. The width of material removed by the trimming process is typically 750 μm for air-abrasive trimming and 50 μm for laser trimming.

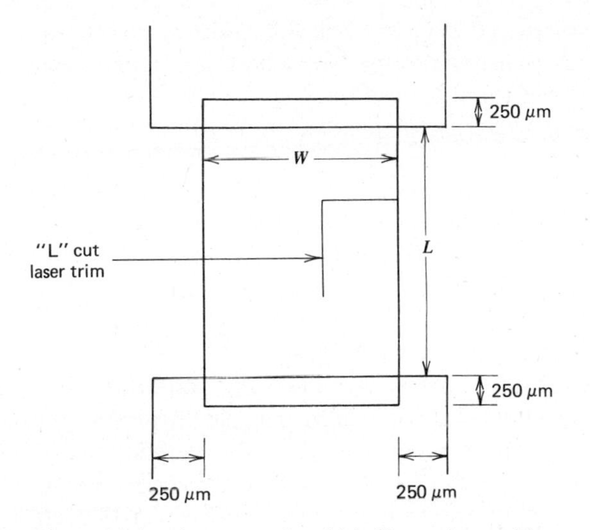

Figure 11-1. A rectangular thick-film resistor with an "L" cut laser trim.

The two popular trimming techniques make use of the same principle, removing resistor material to increase resistance. The material removed by the laser trimmer is much less, due to the narrower cut, and, since the laser trim causes localized heating, the glass in the resistor material flows to seal the cut. Air-abrasive trimming leaves an exposed cut. A popular technique with laser trimming is the "L-cut," as illustrated in Figure 11-1. The perpendicular cut proceeds until the resistance is within 2% of the final value, and the longitudinal cut is used for final trimming. The laser beam is positioned by computer controlled mirrors.

Thick-Film Dielectric Applications

Dielectric materials serve several purposes in thick-film hybrids. They are used as encapsulants to protect resistors from the environment and to halt the flow of solder into undesired regions of the substrate during reflow soldering. As

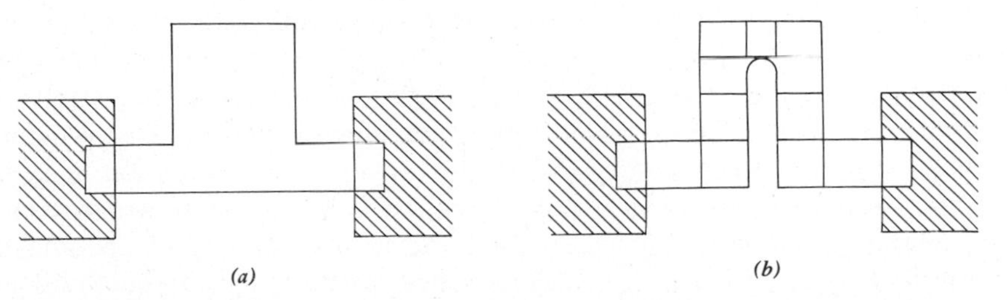

Figure 11-2. The "top-hat" geometry for thick-film power resistors.

insulators, they are used in crossovers and multilayer structures. In a few cases, it is economical to screen print capacitors, and high-permittivity dielectric pastes are available for this purpose.

Encapsulant dielectrics are glasses with a relatively low flow-point temperature, usually between 400 and 500°C. Since this firing temperature is several hundred degrees below the resistor and conductor firing temperatures, there is little effect on the resistor characteristics. These glasses are based on lead oxide-zinc-boron oxide ($PbO–Zn–B_2O_3$) systems. It is common practice to print and fire these dielectrics before resistor trimming. The entire substrate, except for soldering and bonding areas, is covered by this print, if it is determined that encapsulation is desirable.

For crossover applications, the dielectric material must be able to withstand several high-temperature firings. It is also desirable that the permittivity of the material be low, to minimize capacitive coupling between the upper and lower conductors. Glasses of various constitutions based on lead borosilicate formulations have been used successfully as crossover dielectrics. These glasses require careful processing. The lower conductor is printed, dried, and fired at the highest temperature allowed for that paste. The crossover pattern is printed, dried, and fired, and then reprinted, dried, and fired at a temperature approximately 50°C below that of the lower conductor firing. The double printing provides a thicker dielectric and reduces the chances for pin-holes, which could result in the failure of the crossover. The upper conductor pattern is printed, dried, and fired at a temperature approximately 50°C below that of the dielectric past firing. Resistor pastes are then printed, dried, and fired at a temperature approximately 50°C below that of the upper conductor paste firing. The sequence of firing temperatures reduces the tendency for the upper conductor to "swim" in the softened dielectric glass during the last two firing cycles. This type of dielectric material has been replaced in most cases by glass-ceramic crossover materials. These dielectrics are also called recrystallizable glasses. They are based on formulations containing BaO, TiO_2, B_2O_3, P_2O_5, and SiO_2, or BaO, PbO, Al_2O_3, TiO_2, and SiO_2, or similar combinations. At high temperatures, these glasses form a crystalline phase that is unaffected by subsequent high-temperature firings. Double printing is also used with glass-ceramic crossover dielectrics.

There are two basic approaches to multilayer thick-film structures. One technique makes use of glass-ceramic dielectric pastes. In this process, the lower conductor is printed, dried, and fired on the substrate. This is followed by a dielectric print using the reverse of the conductor pattern, except in areas where resistors are to be printed. After this dielectric paste is fired, a second dielectric print is made using a pattern with vias in the areas where contact between the second and first conductor patterns is desired, and openings for the resistors. After firing the second dielectric pattern, the second conductor pattern is printed, dried, and fired. This process is repeated until all of the conductor layers have been fired. Then the resistors are printed directly on the substrate, dried, and fired. This process has been used to fabricate circuits with up to ten conductor layers, but fabrication becomes difficult beyond four layers, particularly if resistors are necessary. The basic process is illustrated in Figure 11-3. One paste manufacturer has developed a resistor system that can be printed on top of the dielectric, eliminating the problem or printing resistors in the "caverns" left in the multilayer structure. Gold pastes are popular for multilayer conductors, since silver has a tendency to diffuse through the dielectric material. Some problems may be encountered due to the different thermal coefficients of expansion of the glass-ceramic and the substrate. The second multilayer technique makes use of layers of "green" ceramic containing vias and thick-film refractory metal pastes like molymanganese or platinum

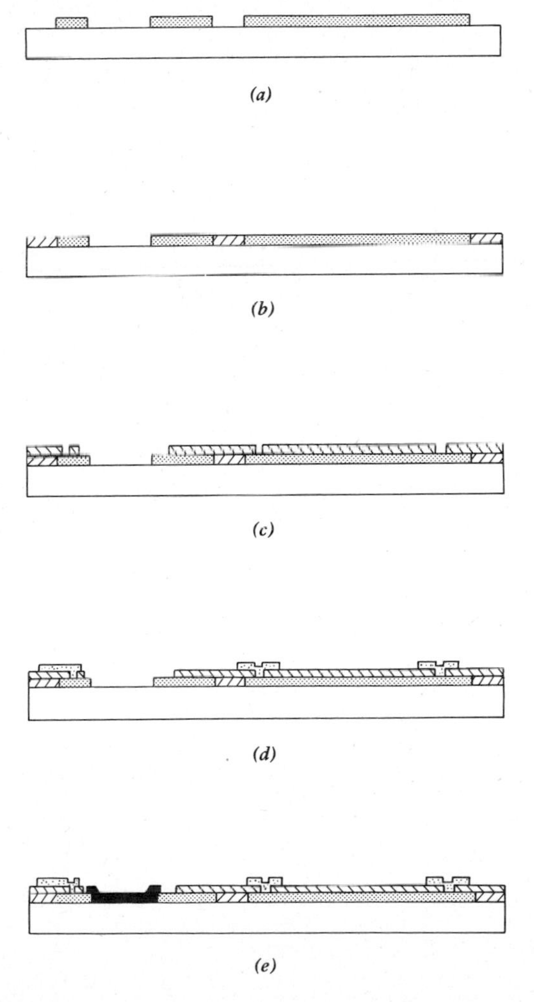

Figure 11-3. The fabrication sequence for a screen printed multilayer thick-film process. (*a*) Print first conductor layer pattern and fire. (*b*) Print first dielectric pattern and fire. (*c*) Print second dielectric pattern with vias and fire. (*d*) Print second conductor pattern and fire. (*e*) Print resistor pattern and fire.

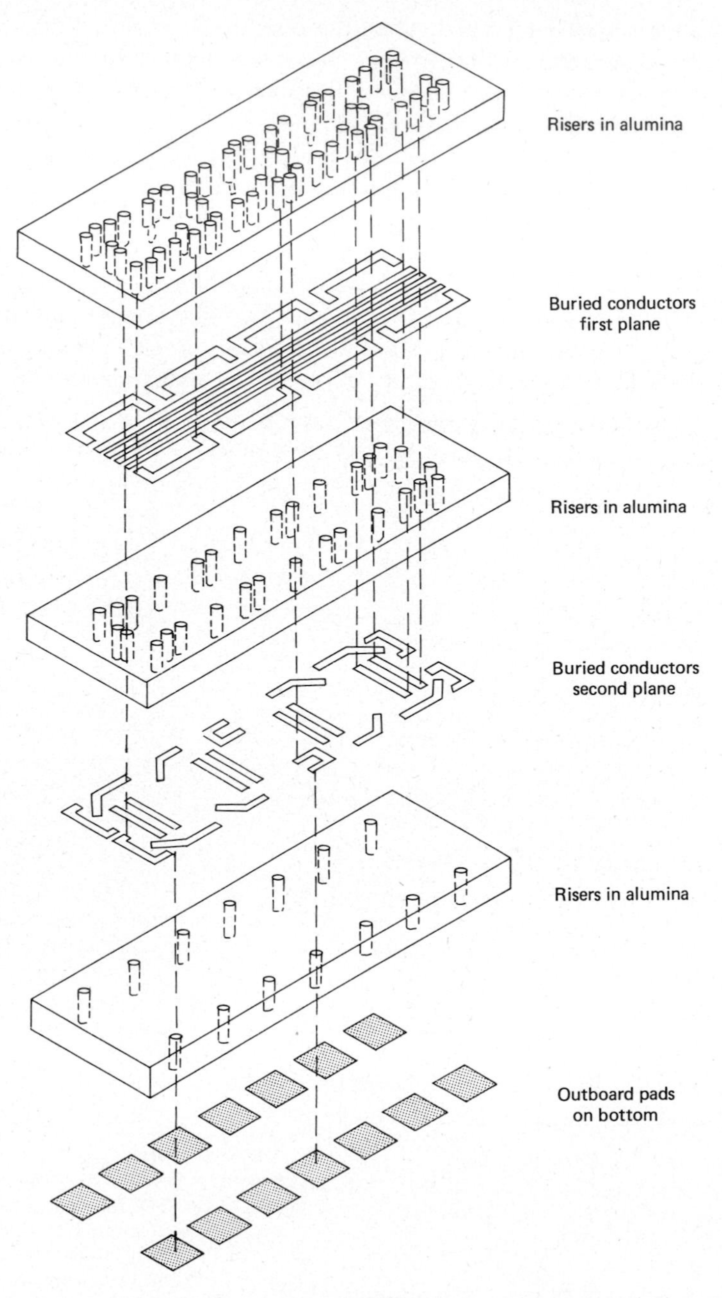

Figure 11-4. A multilayer "green" ceramic process (13).

based systems. This results in a "buried conductor" structure as shown in Figure 11-4. In this process, the entire assembly is fired in the same manner as the firing of "green" substrates in the substrate manufacturing process.

Thick-film capacitors can be fabricated in the parallel plate configuration. The capacitance, C, for this type of structure is given by

$$C = \frac{\varepsilon A}{d} \tag{11-6}$$

where ε is the permittivity of the dielectric, A is the area of overlap between the bottom and top conductors, and d is the thickness of the dielectric. The basic structure is shown in Figure 11-5. Since the thickness of the dielectric is typically 25 μm, the capacitance density is very low unless a high-permittivity material is used for the dielectric. High-permittivity dielectrics are made by mixing a high concentration of barium titanate particles with a barium titanate recrystallizable glass. Capacitance densities as high as 150,000 pF·cm^{-2} have been obtained, but these capacitors contain a ferroelectric dielectric, which implies that the capacitance exhibits an hysteresis effect and the capacitance is a function of bias voltage and temperature. Thick-film capacitors can be trimmed

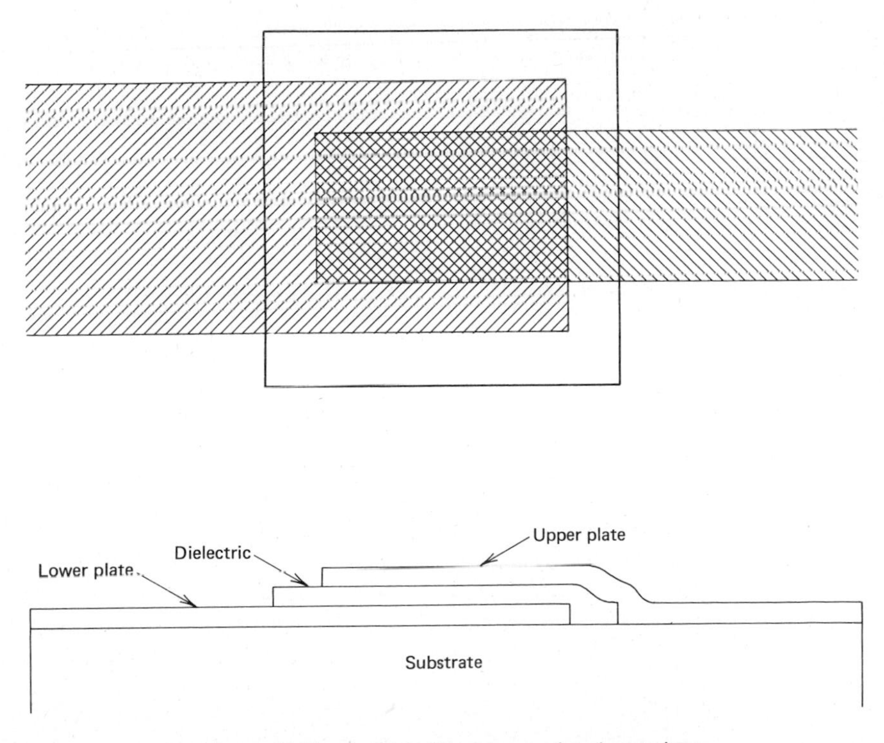

Figure 11-5. A thick-film screen printed capacitor.

by removing part of the upper electrode. In a vast majority of cases, it is more economical to use add-on chip capacitors in thick-film hybrid microcircuits, rather than printing these components.

Thick-Film conductors

Thick-film conductor inks are selected to be compatible with the resistor and dielectric inks used on the hybrid. For ease of processing and inspection, conductor patterns are usually at least 500 μm wide with a similar spacing between lines. To allow for alignment errors, the conductor pattern is usually 500 μm wider than the resistor pattern at a resistor termination. In special cases, particularly when it is neccessary to attach beam-lead or leadless inverted devices or integrated circuits to the hybrid, it is possible to print conductor patterns which are 250 μm wide by 250-μm spacings. This adds considerable cost to the processing, since special screens or etched metal patterns are required for fine-line printing.

THIN-FILM HYBRID DESIGN GUIDELINES

Thin-film hybrid microcircuits are, in general, smaller and have less power dissipation capabilities than thick-film hybrids designed for the same electronic function. They are also more expensive. As a result, thin-film hybrids have been relegated to specific applications, primarily in cases where long-term stability and precision resistors are necessary. The smaller geometries possible with thin-film technology also give this fabrication technique an advantage in high-frequency applications. As in the case of thick-film hybrids, the emphasis in thin-film hybrid design is on resistors and conductors, with the other components being added by bonding techniques. In order to take advantage of the small geometries available in thin-film hybrids, bare silicon chips or special structures like beam-lead devices and circuits are frequently used.

Thin-Film Resistor Design

The major difference between thin-film and thick-film resistor design is that one sheet resistance is used for thin-film resistors, and up to three-sheet resistances may be used for thick-film resistors. This means that the aspect ratio for thin-film resistors must cover a much wider range than that for thick-film resistors. As a result, the serpentine geometry is popular for large-valued thin-film resistors.

Common substrates for thin-film hybrids include glass, 99% alumina, glazed alumina, oxidized silicon, and sapphire. The choice of the substrate depends on the application. The major features to consider are thermal conductivity, dielectric constant, electrical conductivity, surface smoothness, and mechanical strength.

There are a number of materials which can be used for thin-film resistors. The most popular choices are tantalum nitride, nickel-chromium alloys (typically 80 parts nickel and 20 parts chromium), and the class of materials called "cermets" (ceramic metals). Tantalum nitride was developed by the Bell System as a highly reliable resistor material with excellent stability for applications expected to perform well over a 20-year period. A particular advantage of this material is that it can be anodized to form a capacitor dielectric or to trim the resistors. The conductor system used with tantalum nitride is usually a combination of a thin layer of titanium or chromium to provide for good adhesion to the tantalum nitride, followed by a layer of gold for high conductivity. The resistivity of tantalum nitride films is approximately 250 $\mu\Omega$-cm. The film thickness can be varied from 10 Å to several thousand angstroms, resulting in a sheet resistance range from 2500 to 1 $\Omega/\square$. In actual practice, the thickness is usually in the range between 100 and 500 Å, resulting in a sheet resistance range between 250 and 50 $\Omega/\square$. It is difficult to control processes for depositions less than 100 Å thick. The TCR of tantalum nitride is typically -75 ppm/°C. Nickel-chromium films have a slightly lower resistivity than tantalum nitride films, with a 100-Å layer having a sheet resistance of 150 $\Omega/\square$. The major disadvantage of nickel-chromium alloys is their susceptability to attack by moisture. To avoid this problem, they are usually protected by a layer of glass, or packaged by vacuum baking prior to an hermetic seal. The conductor materials compatible with nickel-chromium alloys include gold and aluminum. The TCR's of these films are in the range between 0 and $+50$ ppm°C, with a typical value of 30 ppm/°C. The cermet films are mixtures of metals and oxides that are frequently deposited by flash evaporation. The sheet resistance of these materials depends on both the thickness of the film and the ratio of metal to oxide. A popular choice for cermet films is chromium and silicon monoxide. The sheet resistance of a composition containing 65% Cr and 35% SiO is on the order of 600 $\Omega/\square$, while a combination of 70% Cr and 30% SiO has a sheet resistance of 300 $\Omega/\square$. These materials have a TCR that is positive and less than 100 ppm/°C. Unfortunately, increasing the ratio of SiO beyond the 60% value, which is necessary to produce high sheet resistance values, also results in negative TCR's greater than 750 ppm/°C. The stability of cermet resistors is increased by coating them with silicon monoxide and annealing at temperatures on the order of 300°C for long time periods. Cermet resistors are particularly sensitive to high voltages, and a significant permanent decrease in sheet resistance has been observed in some cases.

The design equations for thin-film resistors can be developed in the following manner. The resistance, R, is given by

$$R = \frac{R_S L}{W} \tag{11-7}$$

and the power density, P/A, is given by

$$P/A = \frac{V^2}{RLW} \tag{11-8}$$

where V is the rms voltage across the resistor. Eliminating R between these two equations yields

$$L = \left(\frac{V^2}{R_S P/A}\right)^{1/2} \tag{11-9}$$

and eliminating L results in

$$W = \left(\frac{V}{R}\right)\left(\frac{R_S}{P/A}\right)^{1/2} \tag{11-10}$$

A typical value for the power density, assuming an alumina substrate, is $3.88\ \mathrm{W \cdot cm^{-2}}$ for long term stability. Equations 11-9 and 11-10 apply for a rectangular resistor geometry.

In many cases it is impractical to layout a large valued resistor in a rectangular pattern. The preferred geometry in these cases is a serpentine or meandering pattern. In this type of geometry, heat transfer is assumed to result from the rectangular block containing the resistor, since heat transfer between parallel segments of the meander pattern is impossible. If the spacing between meanders is the same as the width of the resistor pattern, the area of the block containing the resistor is given by

$$A = 2LW \tag{11-11}$$

Each corner square is assumed to contribute 0.55 square to the resistor pattern.

The process for producing thin-film resistors is based on photolithography. The entire surface of the substrate is covered by a layer of resistive material, followed by a layer of conductor material. The conductor pattern is then photoetched. The photoresist used for this process is removed, and a new layer of photoresist is applied. Using a mask containing both the conductor and resistor patterns, the undesired resistor material is removed. There is no special geometry required for resistor terminations in thin-film hybrids since the resistor material is underneath the conductor pattern. The minimum resistor line width is typically 25 μm if a resistor spread of 10% is acceptable. For more precise control, the resistor line width should be increased. The conductor line width is usually greater than 250 μm. If the resistor length is less than 1250 μm, it may be necessary to alter the geometry due to termination effects. Thin-film resistors are usually trimmed using a 25-μm spot diameter from a laser. In many cases, a trimming tab is added to the resistor geometry like those shown in Figure 11-6. The "top hat" trimming tab may be trimmed from

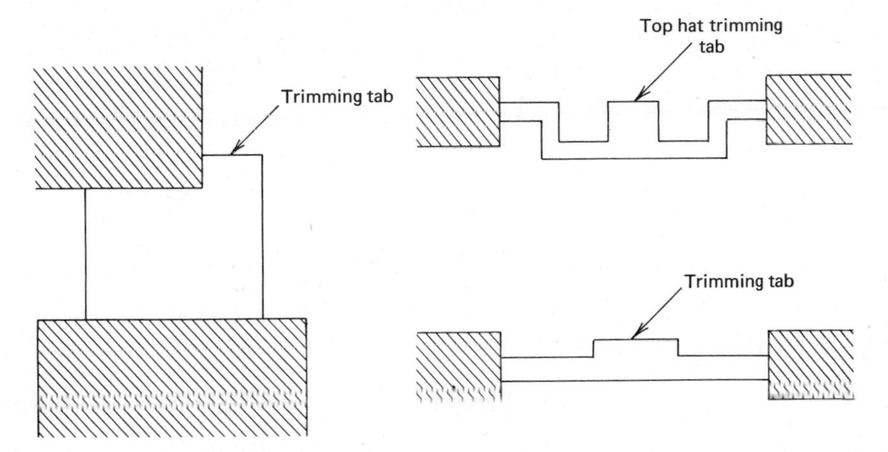

Figure 11-6. Thin-film resistor geometries for laser trimming.

the top or the bottom. It is also possible to use an "L-cut" trim on a "top hat" configuration as shown in Figure 11-7. Note that the second portion of the "L-cut" has a more significant effect on the resistance than the first portion, which is the opposite situation from using an "L-cut" on a rectangular resistor geometry. A serpentine pattern can be created from a rectangular resistor by a series of laser cuts from alternate sides, as shown in Figure 11-8. An alternative form of trimming, which is applicable to tantalum nitride resistor films, is anodization. In this case, the thickness of the film is reduced along the entire length of the resistor, eliminating the possibility of "hot spots," which are encountered in laser-trimmed resistors due to the localized nature of the trim. Special layout rules are imposed by the anodization process so that the trimming of one resistor does not influence the trimming of another. The minimum spacing between resistors is 1250 μm for simultaneous trimming by

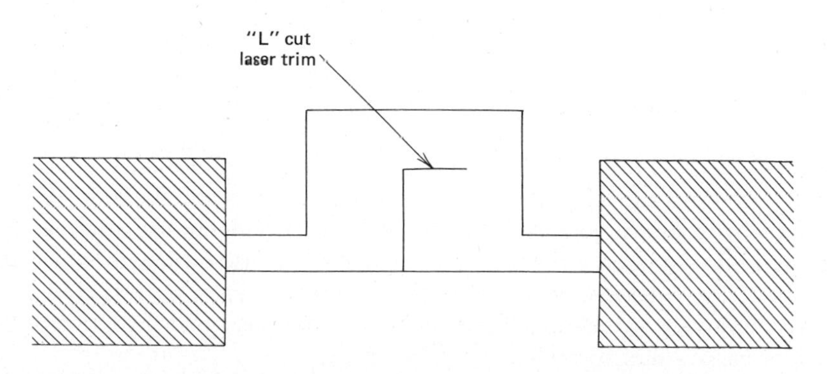

Figure 11-7. Laser trimming a thin-film "top hat" resistor.

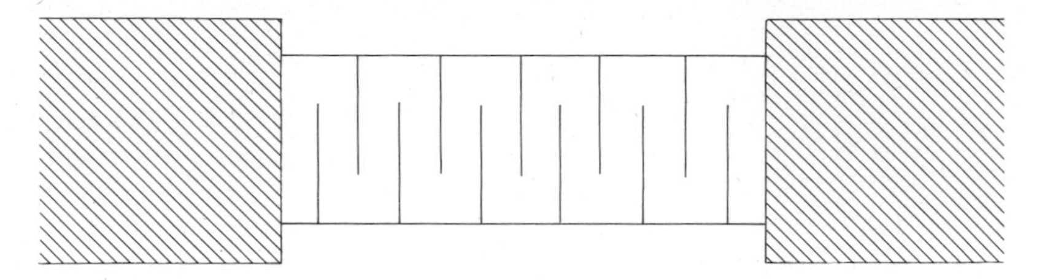

Figure 11-8. A serpentine geometry formed from a rectangular thin-film resistor by laser trimming.

anodization. Typical trimming tolerance for thin-film resistors is ±0.5% by either process.

Thin-Film Capacitor Design

As in the case of thick-film hybrids, it is common practice to use chip add-on capacitors for most applications in thin-film hybrids. It is, however, possible to fabricate capacitors with good characteristics and a reasonably high capacitance density by thin-film techniques. If tantalum nitride is used as the resistor material, it can also serve as the bottom electrode of a capacitor, the dielectric of which is formed by anodization. This is a polarized capacitor, and the bottom plate must be positive with respect to the upper plate in the circuit design. The working voltage must be less than the forming voltage. Typical capacitors of this type have a working voltage of 30 V and a capacitance density of $0.1\ \mu F \cdot cm^{-2}$. The maximum size of these capacitors is usually limited to $1\ cm^2$, because the probability of defects becomes prohibitively high for larger components. The upper electrode is usually a double layer of titanium and gold. Another popular dielectric in thin-film hybrids is evaporated silicon monoxide. A dielectric thickness of 1400 Å results in a capacitance density of $10{,}000\ pF \cdot cm^{-2}$ and a working voltage of 12 V. Since one electrode of a thin-film capacitor is usually made from resistive material, distributed *RC* structures can be readily fabricated in thin-film hybrids.

Thin-Film Crossovers

Crossovers can be fabricated in thin-film hybrids by several techniques. One method is to use a capacitor dielectric and two levels of metalization. In many cases, since wire-bonding is used for the electrical connections to semiconductor devices, wire-bonded crossovers can be used. The use of flat ribbons instead of round wires is also popular. Special structures using electroplated beams over organic layers (which are later removed) have also been used to provide crossovers in thin-film hybrids.

HYBRID MICROWAVE STRUCTURES

Hybrid microcircuits play an important role in the realm of high-frequency electronics. Parasitic effects like inductance, capacitance, and substrate conductivity become important in the frequency range above 10 MHz, and, as the frequency approaches 1 GHz, the skin effect, surface smoothness, and the dielectric properties of the substrate beome important. One distinct advantage of hybrid microcircuits as compared to discrete component circuits at these high frequencies is the reproducibility of the component layout, resulting in relatively fixed parasitic effects.

The parasitic inductance and capacitance in a particular circuit can be reduced by careful layout considerations. In general, to reduce inductance the following guidelines should be considered. Components should be placed as close as possible to one another on the substrate. Meander patterns should be used for resistors. Higher sheet resistance is desirable. The space-width/line-width ratio should be reduced while maintaining a constant length. Capacitance between and within components can be reduced by the following. A thin substrate with a low dielectric constant should be used. Rectangular resistor patterns are preferred over meandering patterns. The patterns should be reduced in size and a higher sheet resistance should be used. The space-width/line-width ratio should be increased. Clearly, some of the guidelines for reducing inductance are in conflict with those for reducing capacitance, resulting in a need for compromise to satisfy the requirements for a particular circuit.

Parasitic resistance occurs in the form of series resistance in the conductor pattern and shunt resistance between components. The conductor resistance can be reduced by using a thicker conductor pattern of gold or copper. The shunt resistance between components depends on the resistivity of the substrate, or on surface leakage. In most cases, surface leakage dominates, and the sheet resistance is on the order of $10^{10}\ \Omega/\square$. This is only significant in closely spaced meander patterns fabricated from high sheet resistance materials.

As operating frequencies progress above 100 MHz, the entire circuit takes the form of a transmission line or waveguide. Three configurations are shown in Figure 11-9, with the microstrip being highly popular. At these frequencies, the skin effect can be important. It is desirable for the conductor pattern to be at least three skin depths thick at the lowest frequency of operation, so that the Q of the conductor is high enough for microwave operation. For gold, the skin depth as a function of frequency is shown in Figure 11-10. Also shown in this figure is the number of skin depths in 7.62-μm-thick gold as a function of frequency. Both thick- and thin-film conductors can be used for this purpose. Thick-film processing is, in general, less expensive for line widths greater than 125 μm. This places an upper limit on the transmission line impedance that can be economically processed by thick-film techniques at 88 Ω. For impedances

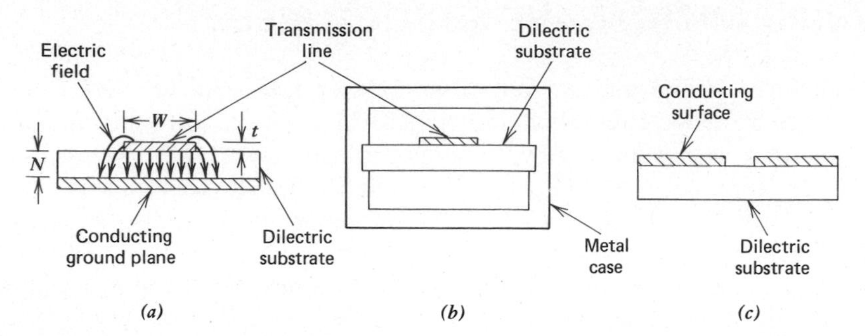

Figure 11-9. Microwave hybrid transmission line structures. (*a*) Microstrip. (*b*) Suspended stripline. (*c*) Slot line.

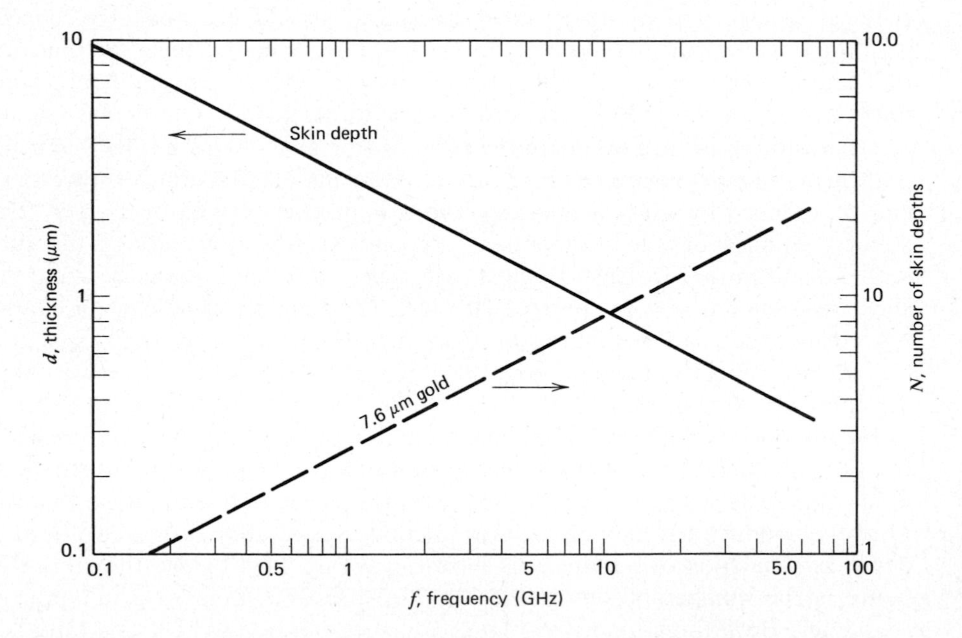

Figure 11-10. Skin depth as a function of frequency for a 7.6-μm layer of gold (12). Reprinted from *Solid State Technology*, "Microwave Integrated Circuits Using Thick- and Thin-Film Technologies," by R. Waugh et al., Vol. 16, April 1973. Copyright © Cowan Publishing Corp.

above this level, thin-film processing is indicated. It is also difficult to print uniform lumped inductors by thick-film techniques. Resistor terminations for microstrip transmission lines can be produced by either thick- or thin-film processing, or by chip resistors.

SUMMARY

The major components fabricated on hybrid substrates are resistors. It is also necessary to produce conductors and crossovers. Thick-film hybrids are capable of dissipating higher powers than the other microcircuit forms. Another prominent application of thick-films is the multilayer interconnection of large-scale integrated circuits. Thin-film hybrids feature precise resistors with long term stability. Both technologies are useful for producing microwave hybrids, but the higher microwave frequencies are limited to thin-film fabrication.

REFERENCES

1. D. W. Hamer and J. V. Biggers, *Thick Film Hybrid Microcircuit Technology* (New York: Wiley-Interscience, 1972).
2. C. A. Harper, ed., *Handbook of Thick Film Hybrid Microelectronics* (New York: McGraw-Hill, 1974).
3. R. A. Rikoski, *Hybrid Microelectronic Circuits, The Thick Film* (New York: Wiley-Interscience, 1973).
4. M. L. Topfer, *Thick Film Microelectronics* (New York: Van Nostrand-Reinhold, 1971).
5. J. E. Sergent and W. G. Dryden, *Solid State Technology* 20, 10 (1977).
6. R. W. Berry et al., *Thin Film Technology* (New York: Van Nostrand-Reinhold, 1968).
7. J. R. Peak, *Solid State Technology* 10, 5 (1967).
8. A. H. Mones and R. M. Rosenberg, *Solid State Technology* 19, 10 (1976).
9. K. Kurzweil and J. Loughran, *Solid State Technology* 16 5 (1973).
10. L. Holland, ed., *Thin Film Microelectronics* (New York: Wiley, 1965).
11. G. R. Madland et al., *Integrated Circuit Engineering* (Cambridge: Boston Technical Publishers, 1966).
12. R. Waugh et al., *Solid State Technology* 16, 4 (1973).
13. E. I. du Pont de Nemours and Co., Data Sheet ML-1, A-66420, 1971.

PROBLEMS

11-1. Design a thick-film rectangular power resistor that must dissipate 3 W and have a resistance of 3750 Ω after trimming.

11-2. Design a thick-film "top-hat" power resistor to dissipate 4 W with a resistance of 9250 Ω. Assume that laser trimming is to be used. How would the design change if air-abrasive trimming is to be used?

11-3. A thick-film capacitor with a value of 50,000 pF is to be screen printed, using a K1200 (relative dielectric constant of 1200) dielectric paste. If a double print of the dielectric is used, with each layer 18.75 μm thick after firing, how large an area is required for this capacitor?

11-4. Tantalum nitride with a sheet resistance of 50 Ω/□ is to be used to form an array of resistors on an alumina thin-film substrate. All of the resistors are to be designed to support 8 V. The resistor values are 80, 500, 2000 and 8000 Ω. Design geometrical layouts for these resistors.

11-5. A tantalum nitride resistor measures 3900.0 Ω at 27°C. What value would be measured at 200°C?

Chapter 12
Microelectronic Circuits

In the preceding chapters, the processes and design techniques associated with the components and devices used in microelectronic circuits have been discussed. There are a number of books devoted to the design of microelectronic circuits, and many more which discuss the principles of operation of electronic circuits. This chapter provides the opportunity for only a brief glimpse at these topics, as well as some relatively recent applications in which the properties of the substrates play an important role on the operation of the systems.

It is interesting to note that essentially none of the processes basic to the microelectronics industry were originated specifically for the purpose of fabricating microcircuits. Of course, each of these processes has been refined and specially adapted by the manufacturers to the point where the results are almost incredible. As an example, consider the growth of single-crystal silicon. It was difficult to obtain boules with 25-mm diameters in the early 1960s. By 1976, 75- and 100-mm wafers were common, and, in 1978, wafers the size of salad plates began to appear. Since the area of a wafer is proportional to the square of the diameter, these large increases in diameter represent an even larger increase in area. It is important to recognize that virtually all of the other processes have had to undergo significant changes to remain compatible with the increasing diameters of the wafers. Diffusion furnace tubes had to be increased in size, the heating elements and the temperature controllers had to be redesigned, and the methods for inserting and removing the wafers from the furnace had to be automated, because of the increased diameter and weight of the wafers. Similarly, changes were necessary in photomasks, photolithography, ion implantation, metalization, etching, and so forth, all of which could lead to a drastic reduction in yield. The overall effect of these changes has been an increase in yield, and a reduction in prices, even in the face of inflation. The engineers in the microcircuits industry are to be commended for the tremendous progress made in three decades after the invention of the transistor.

The optimum size of integrated circuit dice has not changed significantly since the introduction of large-scale integrated circuits in the early 1970s. Dice larger than 1 cm on a side usually suffer from yield problems, while those which are smaller than 0.6 cm are not complex enough, electronically, to compete with the products from other manufacturers. The first prototypes of a particular circuit design are usually made using a mask set with greater than minimum line widths. After the product has been demonstrated to perform satisfactorily, a new set of masks is made which are scaled down from the first set, so that the line widths approach the minimum allowed values. In some cases, it may also be necessary to scale down voltage or current levels when the dimensions are reduced. Since the minimum line widths have steadily decreased over the same time period as the dramatic increase in the wafer diameter, it may be possible to perform more than one down-scaling on a design during the production "life" of a particular product. The die size does not scale directly, because the bonding pads must remain approximately the same size while the rest of the circuit is reduced. The result of circuit down-scaling is that once the die size goes below approximately 0.5 cm on a side, another level of complexity is added to the circuit design so that a new product is developed with a large die size. This cyclic development of the complexity of integrated circuits has led to predictions of multimillions of devices on a single die in the foreseeable future.

The reduction in line width is not the only contributor to increased circuit complexity through the years. Innovative circuit design techniques have significantly altered the space that is required for the performance of a particular electronic function. Examples of this include the one transistor per cell MOS dynamic random access memory, and the nonisolated devices in integrated injection logic.

BIPOLAR INTEGRATED CIRCUITS

In bipolar circuit design, the integrated circuit process imposes certain constraints, but also provides the possibility for structures that can not be realized in discrete component circuits. It is difficult to produce quality *pnp* transistors in a process designed primarily for *npn* transistors. This does not exclude the use of *pnp* transistors from bipolar integrated circuits. They are key elements, serving as active loads, in the most popular analog integrated circuit, the 741 type of operational amplifier. The use of active loads in the input stage of this amplifier increases the gain of this stage, permitting the use of one less intermediate gain-stage. Since each intermediate gain-stage in an operational amplifier must be frequency compensated, reducing the number of these stages from two to one permitted the designer to use internal frequency compensation by including a relatively large MOS capacitor in the amplifier.

Another important application of *pnp* transistors in integrated circuits is as the injector element of integrated injection logic (I^2L). This type of logic is an outstanding example of innovative circuit design, and the use of unusual structures not available to the designer of electronic circuits using discrete components. The basic I^2L gate is shown in Figure 12-1. Note that the *pnp* and *npn* transistors are merged, in that the collector of the *pnp* transistor serves as the base of the *npn* transistor. Since all of the *npn* transistors in a cell have a common emitter, there is no need for isolation between the components within the individual logic cells. This represents a substantial reduction in the area required to perform a particular logic function, compared to the other bipolar logic families. Another unique feature incorporated in I^2L is the "inverted" *npn* transistor with multiple collectors. These transistors use the conventional emitter diffusion to form the collectors. While these transistors have a lower gain than the conventional structures, it is sufficient for the operation of the circuit. I^2L has made it possible to fabricate large-scale

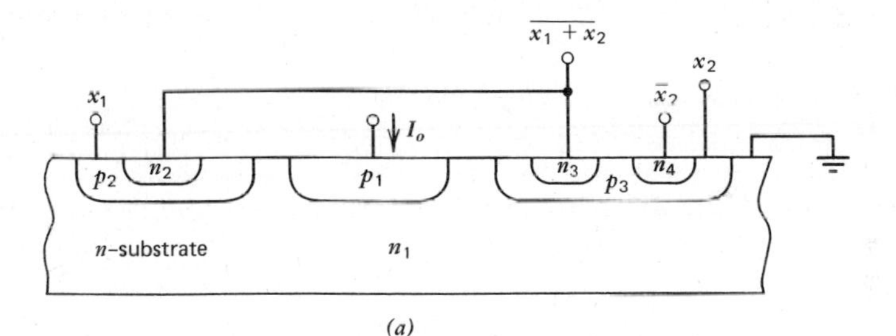

(*a*)

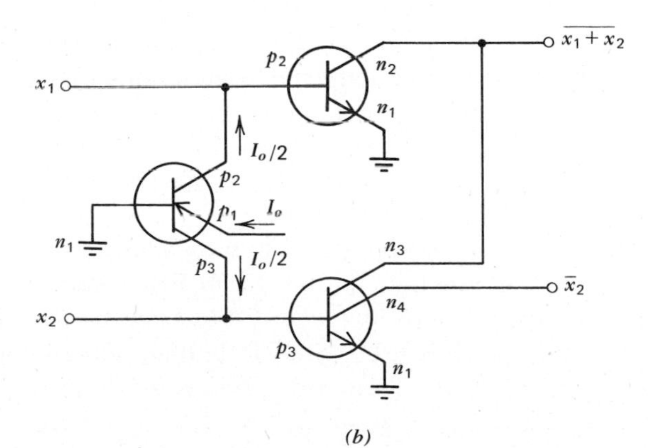

(*b*)

Figure 12-1. An integrated injection logic (I^2L) gate. (*a*) Cross section. (*b*) Schematic. (1). From *Basic Integrated Circuit Engineering*, by D. J. Hamilton and W. G. Howard. Copyright © 1975. Used with permission of McGraw-Hill Book Company.

integrated bipolar circuits that are competitive with MOS circuits in digital watch circuits and microprocessors. Since the basic processes for I^2L are essentially the same as those for conventional bipolar circuits, it is possible to include both types of circuits within a single integrated circuit, to provide desirable interface levels.

A very popular form of bipolar logic is called transistor-transistor logic (TTL), which makes use of multiple emitter transistors. There are a large number of compatible logic circuits in this family in both small-scale and medium-scale forms. In addition to the standard TTL group, most of these circuits are available in low-power, high-speed, Schottky, and low-power Schottky versions. Since the Schottky clamping diodes only require a slight change in the process, these variations are primarily due to circuit design techniques.

The highest speed bipolar logic system is emitter-coupled logic (ECL). These circuits sacrifice power dissipation for speed by operating in the nonsaturated mode. The major processing difference between ECL and the other bipolar logic systems is an emphasis on small size and shallow diffusions. Washed emitter structures are frequently used to reduce the width of the emitter stripe.

Analog bipolar integrated-circuit design makes use of some of the characteristics peculiar to integrated circuits. Transistors fabricated close to one another on the same wafer usually have electrical characteristics that are very similar. This results in the popularity of differential amplifier configurations. The thermal coupling of components indicates that transistor biasing by using diodes fabricated from the emitter-base junction is a viable technique for thermal stability. Diffused resistors within the same circuit also exhibit similar characteristics, even though the diffusion process may not be uniform over the entire area of a 100-mm-diameter wafer. Most of the nonuniformities, which also affect transistor characteristics, are long-range variations that result in a reasonable degree of uniformity within the limited area of an individual circuit. Thermal tracking of electrical characteristics is observed in resistors located close to one another. For these reasons, bipolar circuit design is primarily based on resistor ratios, rather than resistor values.

Some of the most interesting integrated circuits fabricated by bipolar technology are those involving junction and/or MOS field-effect transistors. The addition of these components has made possible operational amplifiers with higher input impedance and larger bandwidth than those made with bipolar transistors alone.

MOS INTEGRATED CIRCUITS

The belated entry of MOS integrated circuits into the market place has not prevented them from seizing a significant portion of the digital applications. The only major drawback to MOS integrated circuits, as compared to bipolar integrated circuits, is their relatively slow operating speed. This is not to imply that MOS circuits are sluggish, but operating frequencies approaching those of ECL have not been obtained. The power dissipation in MOS is generally lower than in comparable bipolar circuits, and n-channel and CMOS on sapphire exhibit excellent speed-power products. The wide variety of processes used for MOS circuit fabrication makes it possible to create many unusual structures including erasable read-only memories and electrically alterable read-only memories. The biggest advantage of MOS integrated circuits is the relatively high yield obtainable from high-density large-scale integrated circuits.

One significant application dominated by MOS technology is the semiconductor memory. Random access memories are available in both static and dynamic forms. Static memories require more components per cell, but they retain their information as long as power is supplied. Dynamic memories consisting of one transistor per cell must be periodically refreshed to retain the stored data. The number of memory cells per circuit has steadily increased through the years, and have reached 65,536 bits, with 262,144 bits under development. The availability of inexpensive memory is the key to the continuing remarkable expansion of the small computer business. Mask programmable read-only memories, in which the gate oxide etch is omitted in desired locations, provide a convenient method for the permanent storage of information and program sequences.

Perhaps the single most important integrated circuit type is the microprocessor. While these circuits are primarily fabricated by MOS processes, they are also available in I^2L versions. The early versions of microprocessors contained the essential circuits of a computer central processing unit, but required a number of supporting circuits to function. The microprocessor has evolved from the relatively simple 4-bit system to 8-bit, 16-bit, and, under development, 32-bit systems. Some of the supporting functions, like clock circuits, scratch-pad memories, read-only memories, input-output ports, and even erasable programmable read-only memories have been included on these remarkable integrated circuits. The microprocessor has brought the computer from the large, expensive systems to the readily affordable small systems, making it possible for the small businessman to have, at his fingertips, much of the same computer power available only to large businesses in the past.

Even more important is the sudden intrusion of electronics and the special purpose microcomputer into the everyday lives of the nontechnical population. Handheld calculators with sophisticated computing power are available at amazingly low prices, and have essentially eliminated the "old standby" of the engineering profession, the slide rule. Sophisticated "toys" have aided the mathematics and spelling education of many children. On-board microcomputers are the projected solutions to the automobile emissions problem, as well as collision avoidance, performance analysis, safety inspection, and so forth. In the home, the ubiquitous microprocessor is found in personal computers, television sets, video tape recorders, sound reproduction systems, video games, nonvideo games, intrusion and fire alarm systems, sewing machines, microwave ovens, and even food blenders.

The integrated circuits based on charge-coupled devices represent another important application of MOS processing. Large shift-registers made from these devices are used in computer systems to fill the gap between fast random access memories and slow mass-storage media. The use of charge-coupled devices in analog shift registers has made possible some unique instrumentation systems. The imaging possibilities of large arrays of these devices are being explored as light-weight broadcast quality color television cameras, which could revolutionize the entire field of photography.

SPECIAL HYBRID STRUCTURES

There are two applications of hybrid technology which deserve special mention. These are magnetic bubble technology and surface acoustic wave devices. These are integrated circuits that make use of thin-film deposition techniques on special substrates in which electric and/or magnetic fields are used to produce effects within the substrate, which perform important electronic functions.

Magnetic bubbles make use of garnet films deposited by liquid-phase epitaxy on (111) oriented gadolinium gallium oxide ($Gd_3Ga_5O_{12}$) substrates. The garnet most frequently used is yttrium iron garnet ($Y_3Fe_5O_{12}$), which exhibits a strong uniaxial anisotropy perpendicular to the surface. In other words, it contains domains that can be easily magnetized perpendicular to the surface, but are very difficult to magnetize parallel to the surface. With the proper magnitude of perpendicular applied magnetic field, cylindrical domains, called "magnetic bubbles," can be created in the garnet film. These bubbles have a magnetic polarization that is opposite in direction to the area surrounding them. The bubbles can be caused to move parallel to the surface by the proper application of fields. A combination of permanent magnets, coils, electric currents, and thin-film magnetic alloy patterns on the surface of the garnet makes it possible to control the motion of the bubbles in shift-register paths to provide memory functions. The permanent magnets are used to maintain the bubble structure when the power is removed, resulting in a nonvolatile memory. Bubble sizes from several micrometers to less than 1 μm have been obtained. In many cases, a single mask level is used to create the magnetic alloy pattern on the surface of the garnet. This eliminates the need for registration between masks on the critical geometries, which may be submicrometer in dimensions. The resulting structures represent the entry of solid state integrated circuits into the field of mass storage previously dominated by magnetic tapes and disk systems.

Surface acoustic wave (SAW) devices are fabricated by delineating thin-film conductor patterns on piezoelectric single-crystal substrates. Typical substrate materials are quartz (SiO_2) and lithium niobate ($LiNbO_3$). In contrast to

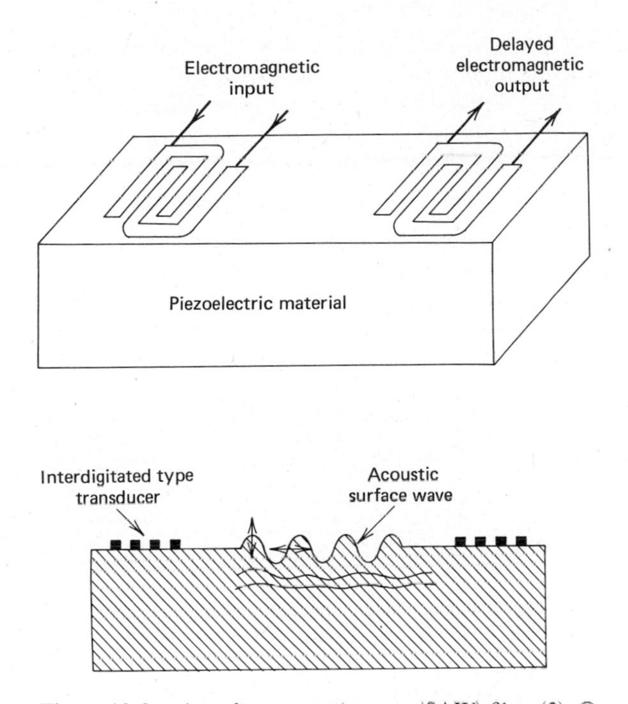

Figure 12-2. A surface acoustic wave (SAW) filter (3). © 1976 IEEE. Reprinted from *Proceedings of the IEEE*, May 1976, p. 581.

the crystals used in crystal controlled oscillators, where a bulk wave is excited, these devices make use of surface acoustic or Rayleigh waves, which propagate at the velocity of sound along the surface of the crystal. A typical SAW device is shown in Figure 12-2. In this type of device, the wave is excited by the input electrical signal applied to one set of electrodes and detected at the other set. The shape and spacing of the electrodes determines the center frequency and band shape of the detected wave. Constructive interference between the waves generated at individual electrodes in the input electrode set is used to determine the amplitude of the propagated wave at a given frequency. These devices are used as filters and delay lines, and special structures have been used to perform complex functions like convolution. The SAW devices provide additional functions to the already impressive collection of electronic capabilities available in the field of microelectronics.

CONCLUSION

The application of a wide variety of processing technologies has resulted in the ability to miniaturize and improve the performance of many aspects of electronics. The invention of the bipolar transistor paved the way for the development of the bipolar integrated circuit, and the experience gained from this development made possible the MOS transistor and integrated circuit. Certain applications requiring high power, precision resistors, high-frequency operation, and high-packing density have assured the position of hybrid technologies in the spectrum of microelectronics. The microelectronics field is dynamic, but, barring any revolutionary changes in processing or the invention of a new device, the continuing progress in this field is based on the application and refinement of the fundamental processes and devices discussed in the preceding chapters. The future of microelectronics is difficult to predict, but I am sure that it will provide fascinating and demanding challenges for the engineering profession for the decades to follow.

REFERENCES

1. D. J. Hamilton and W. G. Howard, *Basic Integrated Circuit Engineering* (New York: McGraw-Hill, 1975).
2. H. Chang, ed., *Magnetic Bubble Technology* (New York: IEEE Press, 1975).
3. A. J. Slobodnik, Jr., *Proceedings of the IEEE* 64, 5 (1976).

Appendix
Semiconductor Device Physics

This appendix contains a review of the semiconductor physics necessary to develop the design equations for the devices discussed in Chapters 9 and 10. This is not an exhaustive treatment of the subject, but should serve to refresh the memories of the readers, while maintaining a notation consistent with the remainder of the book. Many books have been used in the development of this material, but two books, *Introduction to Semiconductor Physics*, by Adler, Smith, and Longini (Wiley, 1964), and *Device Electronics for Integrated Circuits*, by Muller and Kamins (Wiley, 1977), should provide the necessary background for most readers.

BASIC SEMICONDUCTOR PHYSICS

The most important elemental simiconductor is silicon. It is a member of Group IV of the periodic table, which implies that it has four valence electrons per atom. Silicon crystallizes in the diamond lattice, in which each atom is surrounded by four nearest neighbors in a tetrahedral structure. The valence electrons are shared by the atoms so that a complete valence shell is available for each atom.

The electronic structure of pure (intrinsic) single-crystal silicon consists of a valence band, which is full at 0°K, separated by a forbidden gap which is approximately 1.1 eV, as indicated in Figure A-1. In the figure, E_v represents the highest energy level in the valence band, E_c represents the lowest energy level in the conduction band, E_g is the gap energy, given by

$$E_g = E_c - E_v \tag{A-1}$$

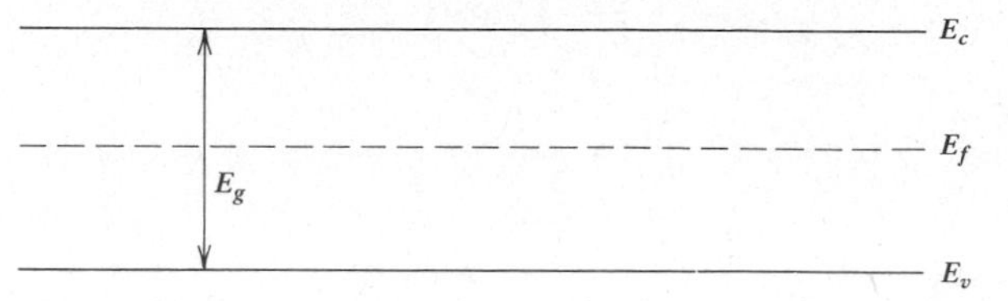

Figure A-1. The band structure of a semiconductor.

and E_f is the Fermi energy. The electrons are distributed into available allowed electron states according to the Fermi-Dirac distribution function, $f_D(E)$, which is given by

$$f_D(E) = \left[1 + \exp\left(\frac{E - E_f}{kT}\right)\right]^{-1} \tag{A-2}$$

where E is the energy level, k is Boltzmann's constant (8.62×10^{-5} eV $\cdot$ °K^{-1}), and T is the absolute temperature. At absolute zero, all electron states below E_f are occupied and all electron states above E_f are empty. For temperatures above absolute zero, $f_D(E_f)$ is one-half. Since an electron state is either occupied or not occupied, the probability of an electron state not being occupied is

$$1 - f_D(E) = \left[1 + \exp\left(\frac{E_f - E}{kT}\right)\right]^{-1} \tag{A-3}$$

This represents the probability of an electron state being occupied by a hole.

The equilibrium electron density in the conduction band, n_o can be determined by the following integral,

$$n_o = \int_{\text{conduction band}} f_D(E) g(E)\, dE \tag{A-4}$$

where $g(E)$ is the density of electron states as a function of energy. Similarly, the hole density in the valence band is

$$p_o = \int_{\text{valence band}} [1 - f_D(E)] g(E)\, dE \tag{A-5}$$

In most cases, for device physics, these equations can be approximated by

$$n_o = N_c \exp\left[\frac{-(E_c - E_f)}{kT}\right] \tag{A-6}$$

and

$$p_o = N_v \exp\left[\frac{-(E_f - E_v)}{kT}\right] \tag{A-7}$$

where N_c and N_v are called the effective densities-of-states at the conduction and valence band edges, and are given by

$$N_c = 2\left(\frac{2\pi m_n^* kT}{h^2}\right)^{3/2} \tag{A-8}$$

and

$$N_v = 2\left(\frac{2\pi m_p^* kT}{h^2}\right)^{3/2} \tag{A-9}$$

The quantities m_n^* and m_p^* are the electron and hole density-of-states effective masses, and h is Planck's constant (4.135×10^{-15} eV $\cdot$ s^{-1}). Equations A-6 and A-7 are called the Boltzmann approximations for the equilibrium carrier concentrations. The Boltzmann approximation does not apply in heavily doped semiconductor regions like the emitter of a bipolar transistor. In this case, the semiconductor is said to be *degenerate*, and n_o is given by

$$n_o = 4\pi\left[2m_n^*\left(\frac{kT}{h^2}\right)\right]^{3/2} F_{1/2}(x_f) \tag{A-10}$$

where $F_{1/2}(x_f)$ is the Fermi integral of order $\frac{1}{2}$, and is given by

$$F_{1/2}(x_f) = \int_0^{\infty} x^{1/2}[1 + \exp(x - x_f)]^{-1}\, dx \tag{A-11}$$

with $x_f = E_f/kT$. The solutions of this integral are tabulated.

In an intrinsic semiconductor, electron-hole pairs are formed by the thermal excitation of an electron from the valence band to the conduction band, resulting in $n_o = p_o = n_i$, where n_i is the intrinsic carrier concentration and is given by

$$n_i = (N_c N_v)^{1/2} \exp\left(-\frac{E_g}{2kT}\right) \tag{A-12}$$

It should be noted, from Equations A-6 and A-7, that

$$n_o p_o = n_i^2 \tag{A-13}$$

is independent of E_f, which varies relative to E_c and E_v when impurities are added to the crystal.

Equations A-6 and A-7 still apply when the conductivity of the semiconductor is dominated by impurities. These materials are called extrinsic semiconductors. Elements from Groups III and V of the periodic table enter the silicon lattice substitutionally. In the case of Group III atoms, like boron, gallium, aluminum, and indium, the three valence electrons are shared in covalent bonds with three of the four silicon nearest neighbors. One electron per impurity atom is lacking to satisfy the crystal bonding structure. The energy associated with supplying the necessary electrons to satisfy these bonds by extracting them from the valence band (leaving holes behind) is a small fraction of the gap energy. These impurities are called *acceptors.* For Group V atoms, like phosphorus, arsenic, and antimony, four of the five valence electrons are shared with the four nearest silicon atoms, leaving the fifth electron loosely bound to the impurity. Again, the energy required to elevate this electron into the conduction band is a small fraction of the gap energy. These impurities are called *donors.*

In most cases, semiconductor devices are operated in the temperature range where essentially all of the donors and acceptors are ionized. A charge balance exists such that

$$n_o - p_o = N_d - N_a \tag{A-14}$$

where N_d is the donor density and N_a is the acceptor density. If N_d is greater than N_a the material is n-type, and, using Equation A-13, n_o is given by

$$n_o = \frac{N_d - N_a}{2} + \left[\left(\frac{N_d - N_a}{2}\right)^2 + n_i^2\right]^{1/2} \tag{A-15}$$

which can be approximated by

$$n_o \simeq N_d - N_a \tag{A-16}$$

when $N_d - N_a \gg n_i$. Similarly, for p-type material,

$$p_o \simeq N_a - N_d \tag{A-17}$$

Current transport in a semiconductor consists of the drift of both electrons and holes due to electric fields, and the diffusion of electrons and holes due to concentration gradients. Electrons and holes traveling in opposite directions contribute to currents in the same direction. For the one-dimensional case, the current density, J, is given by

$$J = J_n + J_p \tag{A-18}$$

where

$$J_n = q\mu_n n\mathscr{E} + qD_n \frac{dn}{dx} \tag{A-19}$$

and

$$J_p = q\mu_p p\mathscr{E} - qD_p \frac{dp}{dx} \tag{A-20}$$

In these equations, $\mathscr{E}$ is the electric field, μ is the carrier mobility, and D is the carrier diffusion coefficient. The mobility and diffusion coefficient for each carrier species are related by the Einstein relation,

$$\frac{D}{\mu} = \frac{kT}{q} \tag{A-21}$$

Under nonequilibrium conditions, there will be either an excess or a deficiency of electrons and holes. The excess electron and hole concentrations are written

$$n' = n - n_o \tag{A-22}$$

and

$$p' = p - p_o \tag{A-23}$$

In general, $n'(x) \simeq p'(x)$, since a large electric field would result if the condition was not met, causing the rapid motion of carriers until the field is substantially reduced. The relationship is not exact because of the differences between diffusion coefficients for holes and electrons. This quasineutral condition applies to all situations in which excess carrier concentrations exist, whether electron-hole pairs are generated locally, or minority carriers are injected into a region, requiring majority carriers to enter from an external contact to provide neutralization. Recombination statistics are characterized by a parameter called the excess-carrier or minority-carrier lifetime. In p-type material, this lifetime is designated τ_n, and in n-type material, τ_p. The recombination process in silicon is dominated by a two-step process in which electrons are captured at recombination centers, usually impurity atoms, followed by the capture of a hole. Thus, lifetime is a strong function of doping density. Particularly effective as a recombination center is gold, which has two energy levels near the middle of the bandgap. Gold is sometimes used as a lifetime "killer" in silicon.

The differential equations for the time rate of change of minority carrier concentrations are called continuity equations. For holes in n-type material,

$$\frac{\partial p}{\partial t} = -\frac{p'}{\tau_p} + G - \left(\frac{1}{q}\right)\frac{\partial J_p}{\partial x} \tag{A-24}$$

where G is the generation rate due to external sources. For electrons in p-type material,

$$\frac{\partial n}{\partial t} = -\frac{n'}{\tau_n} + G + \left(\frac{1}{q}\right)\frac{\partial J_n}{\partial x} \tag{A-25}$$

Under the conditions that $G = 0$, the diffusion term dominating the drift term in the current density expression, homogeneous doping, and steady-state, Equation A-24 can be written

$$0 = -\frac{p'}{\tau_p} + D_p \frac{d^2 p'}{dx^2} \tag{A-26}$$

which has the general solution

$$p' = A \exp\left(\frac{x}{L_p}\right) + B \exp\left(-\frac{x}{L_p}\right) \tag{A-27}$$

where A and B are coefficients determined by the boundary conditions and

$L_p \equiv (\tau_p D_p)^{1/2}$ is called the diffusion length. A similar expression can be written for n' in p-type material.

JUNCTION THEORY

The *pn* junction is a key component of the junction diode, the bipolar junction transistor, the junction field-effect transistor, and the MOS field-effect transistor, essentially all of the devices used in monolithic integrated circuits. This discussion is limited to the abrupt *pn* junction, in which it is assumed that the doping density on the *p*-side is a constant, N_a, and the doping density on the *n*-side is a constant, N_d. The basic geometry is shown in Figure A-2.

When the junction is formed, a space-charge region between $-x_p$ and x_n occurs, to provide an electric field to balance the diffusion components in the current density expressions, since both the electron and hole current densities must be zero in equilibrium. The space-charge region that produces this field is essentially free of mobile charge carriers, and is sometimes called the depletion

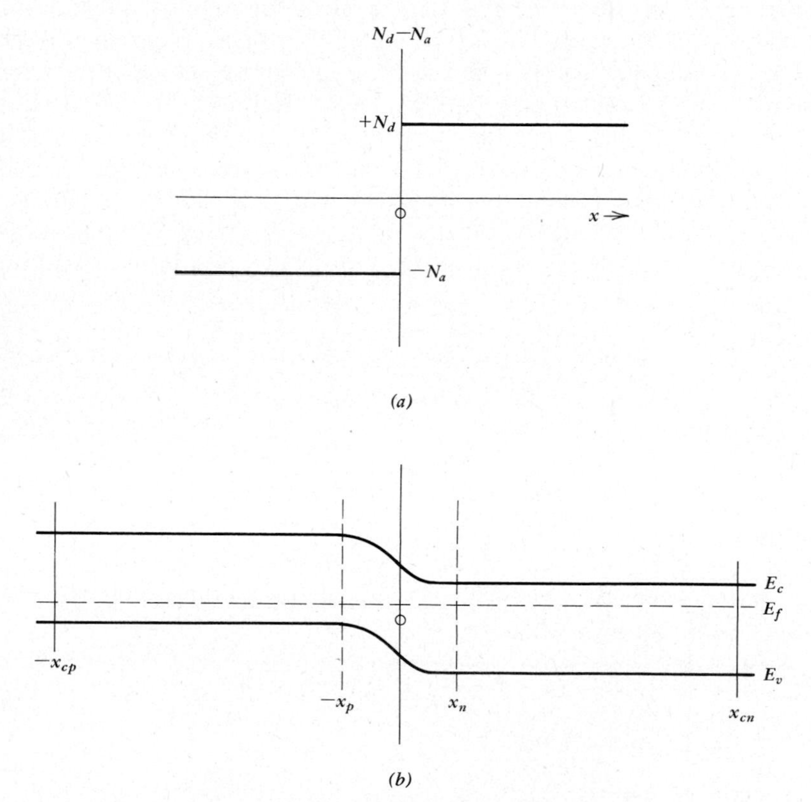

Figure A-2. The geometry of an abrupt *pn* object diode. (*a*) Net doping density. (*b*) The band structure.

or transition region. The electric field in the space-charge region can be found by setting Equation A-19 equal to zero, resulting in

$$\mathscr{E} = -\frac{d\phi}{dx} = -\left(\frac{kT}{qn_o}\right)\frac{dn_o}{dx} \tag{A-28}$$

If this equation is integrated from $-x_p$ to x_n, the result, ϕ_i, is the built-in potential across the junction,

$$\phi_i = \left(\frac{kT}{q}\right)\ln\left(\frac{N_a N_d}{n_i^2}\right) \tag{A-29}$$

which indicates that the n-side is positive with respect to the p-side.

The width of the space-charge region can be determined in the following manner. The charge per unit area on the p-side of the transition region must equal the charge per unit area on the n-side of the transition region, or

$$N_d x_n = N_a x_p \tag{A-30}$$

From Poisson's equation,

$$-\frac{d^2\phi}{dx^2} = \left(\frac{q}{\varepsilon_s}\right)[p(x) - n(x) + N_d(x) - N_a(x)] \tag{A-31}$$

where ε_s is the permittivity of silicon (1.036×10^{-12} F $\cdot$ cm^{-1}). Since both $p(x)$ and $n(x)$ are essentially zero in the space-charge region, an approximate solution for this equation, using Equation A-30 and the continuity of ϕ yields

$$\phi(x_n) - \phi(x_p) = \phi_i = \left(\frac{q}{2\varepsilon_s}\right)(N_d x_n^2 + N_a x_p^2) \tag{A-32}$$

which results in a space-charge region width, x_d, of

$$x_d = x_n + x_p = \left[\left(\frac{2\varepsilon_s\phi_i}{q}\right)\left(\frac{1}{N_a} + \frac{1}{N_d}\right)\right]^{1/2} \tag{A-33}$$

If a reverse bias voltage, $-V_a$, is applied to the junction (V_a is positive for forward bias), the mobile charge density in the space-charge region is even less than that under zero bias conditions, and the transition region width becomes

$$x_d = \left[\left(\frac{2\varepsilon_s}{q}\right)\left(\frac{1}{N_a} + \frac{1}{N_d}\right)(\phi_i - V_a)\right]^{1/2} \tag{A-34}$$

The approximation upon which this is based becomes increasingly weaker with increasing forward bias.

The small-signal capacitance per unit area, C', of an abrupt *pn* junction under reverse bias is determined by differentiating the charge on one side of the capacitor (junction space-charge region) with respect to the applied voltage, or

$$C' = qN_d \frac{dx_n}{dV_a} = qN_a \frac{dx_p}{dV_a} \tag{A-35}$$

resulting in

$$C' = \left[\frac{q\varepsilon_s}{2(1/N_a + 1/N_d)(\phi_i - V_a)} \right]^{1/2} \tag{A-36}$$

The carrier concentrations at the edges of the space-charge region are a function of the applied voltage. It is assumed that the density of injected minority carriers at these points is at least an order of magnitude less than the equilibrium majority carrier densities on both sides of the junction. This condition is called *low injection.* The zero bias situation in a *pn* junction represents a balance between large drift and diffusion components for both carrier types. Thus, the current density for a particular carrier type is approximately zero (the difference of two large numbers) even for moderate applied biases, or

$$J_n \simeq 0 = qn\mu_n \mathscr{E} + qD_n \frac{dn}{dx} \tag{A-37}$$

which leads to

$$\phi(-x_p) - \phi(x_n) = \left(\frac{kT}{q}\right) \ln \left[\frac{n(-x_p)}{n(x_n)}\right] \tag{A-38}$$

This potential difference for a biased junction is $V_a - \phi_i$. The result is

$$n_p'(-x_p) = n_{po}(-x_p)\left[\exp\left(\frac{qV_a}{kT}\right) - 1\right] \tag{A-39}$$

where $n_p'(-x_p)$ represents the excess electron density at the edge of the space-charge region on the p-side, and $n_{po}(-x_p)$ is the equilibrium electron density at the same point. In a similar manner,

$$p_n'(x_n) = p_{no}(x_n)\left[\exp\left(\frac{qV_a}{kT}\right) - 1\right] \tag{A-40}$$

The minority carrier current density in the field-free n-side of the junction (between x_n and the contact, x_{cn}) is found by using Equation A-40 as one boundary condition on Equation A-27, zero as the boundary condition at x_{cn}, and substituting the result in Equation A-20, with $\mathscr{E} = 0$. The result is

$$J_p(x) = \frac{qp_{no}(x_n)D_p \cosh[x_{cn} - x)/L_p]}{L_p \sinh[(x_{cn} - x_n)/L_p]}\left[\exp\left(\frac{qV_a}{kT}\right) - 1\right] \tag{A-41}$$

which yields

$$J_p(x_n) = \frac{qp_{no}(x_n)D_p}{L_p \tanh[(x_{cn} - x_n)/L_p]}\left[\exp\left(\frac{qV_a}{kT}\right) - 1\right] \tag{A-42}$$

Similarly,

$$J_n(-x_p) = \frac{qn_{po}(-x_p)D_n}{L_n \tanh[(x_{cp} - x_p)/L_n]}\left[\exp\left(\frac{qV_a}{kT}\right) - 1\right] \tag{A-43}$$

If there is no recombination within the space-charge region, the number of electrons per unit time entering the space-charge region at x_n is equal to the number of electrons per unit time leaving the space-charge region at $-x_p$, or $J_n(-x_p) = J_n(x_n)$. For this one-dimensional situation, the total current density, J, is everywhere the same, so that

$$J = J_n(x_n) + J_p(x_n)$$

or

$$J = \left\{\frac{qn_{po}(-x_p)D_n}{L_n \tanh[(x_{cp} - x_p)/L_n]} + \frac{qp_{no}(x_n)D_p}{L_p \tanh[(x_{cn} - x_n)/L_p]}\right\}\left[\exp\left(\frac{qV_a}{kT}\right) - 1\right] \tag{A-44}$$

Since, from Equation A-13, $n_{po}(-x_p) = n_i^2/N_a$, and $p_{no}(x_n) = n_i^2/N_d$, for a junction with area A, the current can be written

$$I = JA = qn_i^2A\left\{\frac{D_n}{N_aL_n \tanh[(x_{cp} - x_p)/L_n]} + \frac{D_p}{N_dL_p \tanh[(x_{cn} - x_n)/L_p]}\right\}\left[\exp\left(\frac{qV_a}{kT}\right) - 1\right] \tag{A-45}$$

or

$$I = I_s\left[\exp\left(\frac{qV_a}{kT}\right) - 1\right] \tag{A-46}$$

where I_S is called the *saturation* current and represents everything in front of the exponential factor in Equation A-45.

BIPOLAR JUNCTION TRANSISTORS

In this section, a model is developed for an *npn* bipolar junction transistor with uniform doping in each region. The geometry for this device is shown in Figure A-3. The transistor is biased in the forward-active region of operation, with the

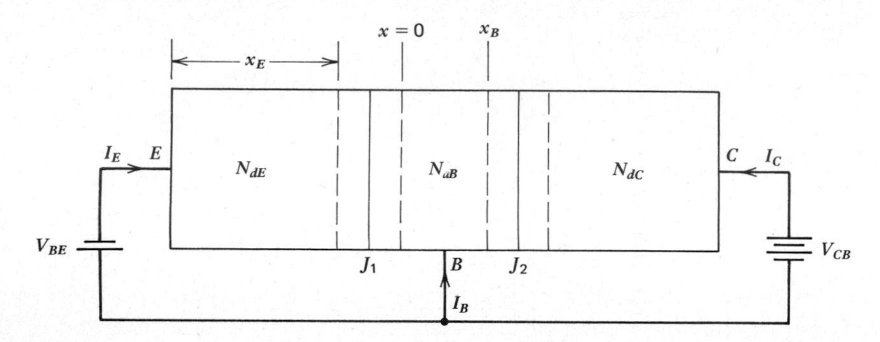

Figure A-3. The geometry of an *npn* transistor with uniform doping in each region.

base-emitter junction forward biased, and the collector-base junction reverse biased.

The dc common-base short-circuit current gain is given by

$$\alpha_F = -\frac{I_C}{I_E} \tag{A-47}$$

and the dc common-emitter short circuit current gain is

$$\beta_F = \frac{I_C}{I_B} = \frac{\alpha_F}{1-\alpha_F} \tag{A-48}$$

It is possible to write α_F as the product of three factors,

$$\alpha_F = \gamma \alpha_T M \tag{A-49}$$

where

$$\gamma \equiv \frac{I_{nE}}{I_E} \tag{A-50}$$

is called the *emitter efficiency*,

$$\alpha_T \equiv -\frac{I_{nC}}{I_{nE}} \tag{A-51}$$

is called the *base transport factor*, and

$$M \equiv \frac{I_C}{I_{nC}} \tag{A-52}$$

is called the *collector multiplication factor.*

The emitter efficiency can be written

$$\gamma = \left(1 + \frac{I_{pE}}{I_{nE}}\right)^{-1} \tag{A-53}$$

which, for uniform doping in each region, becomes

$$\gamma = \frac{1}{1 + \dfrac{D_{pE} L_{nB} N_{aB} \tanh(x_B/L_{nB})}{D_{nB} L_{pE} N_{dE} \tanh(x_E/L_{pE})}} \tag{A-54}$$

The base transport factor is found by using the electron current density expression analogous to Equation A-41, assuming that the reverse-biased collector-base junction forces the excess electron density to zero at x_B. The result is

$$\alpha_T = \operatorname{sech}\left(\frac{x_B}{L_{nB}}\right) \tag{A-55}$$

The collector multiplication factor is given by the empirical relation

$$M = \left[1 - \left(\frac{V_{CB}}{BV}\right)^n\right]^{-1} \tag{A-56}$$

where BV is the avalanche breakdown voltage for the collector-base junction.

The frequency response of the bipolar transistor can be dominated by the base transport factor. This frequency degradation is related to the transit time across the base. This transit time, τ_B, can be shown to be

$$\tau_B = \frac{x_B^2}{2D_{nB}} \tag{A-57}$$

which results in an alpha cutoff frequency of

$$\omega_\alpha = \frac{1}{\tau_B} \tag{A-58}$$

JUNCTION FIELD-EFFECT TRANSISTORS

The junction field-effect transistor (JFET) is used infrequently in integrated circuits. The basic operating principle is that of using a reverse-biased junction, whose space-charge region width depends on the magnitude of the bias voltage, to control the area available for majority carrier current between the drain and source. The geometry of the device under consideration is shown in Figure A-4. For this analysis, it will be assumed that the space-charge region at the substrate extends primarily into the substrate so that $x_w \simeq (t - x_d)$. The resistance of the channel is given by

$$R = \frac{\rho L}{x_w W} \tag{A-59}$$

where $\rho = (q\mu_n N_d)^{-1}$ is the resistivity of the channel region, and L and W are

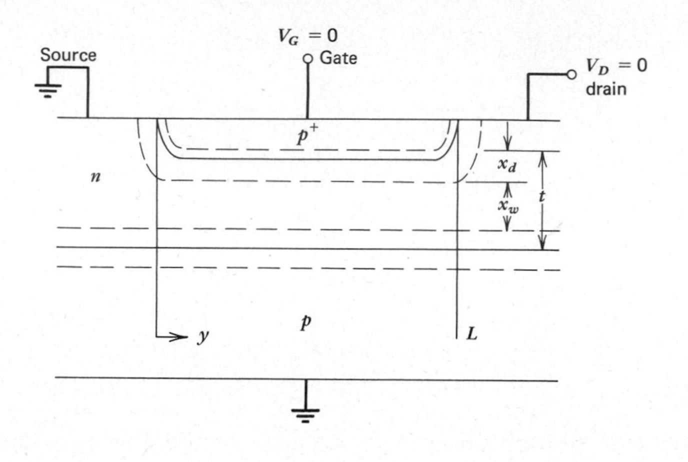

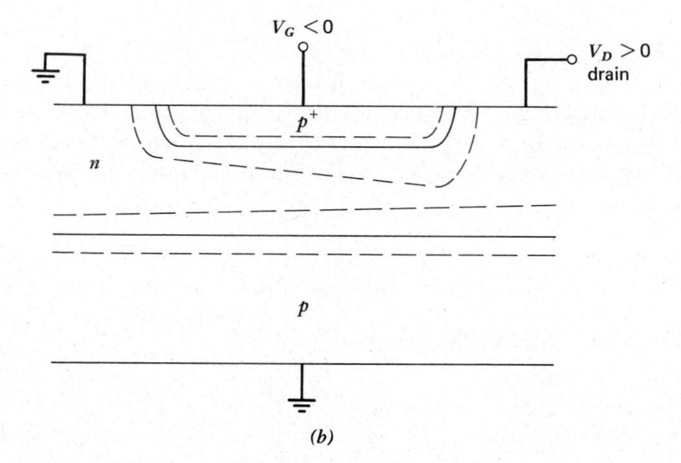

(b)

Figure A-4. The n-channel junction FET. (a) Equilibrium. (b) Under bias.

the length and width of the gate diffusion. The drain current, I_D, is then

$$I_D = \frac{V_D}{R} = \frac{q\mu_n N_d x_w V_D W}{L} \tag{A-60}$$

The space-charge region width, x_d, is, from Equation A-33,

$$x_d = \left[\left(\frac{2\varepsilon_s}{q}\right)\left(\frac{1}{N_d} + \frac{1}{N_a}\right)(\phi_i - V_G)\right]^{1/2}$$

or

$$x_d \simeq \left[\left(\frac{2\varepsilon_s}{qN_d}\right)(\phi_i - V_G)\right]^{1/2} \tag{A-61}$$

since $N_a \gg N_d$. Thus

$$I_D = \left(\frac{Wq\mu_n N_d t}{L}\right)\left\{1-\left[\left(\frac{2\varepsilon_s}{qN_d t^2}\right)(\phi_i - V_G)\right]^{1/2}\right\}V_D \qquad \text{(A-62)}$$

or

$$I_D = G_o\left\{1-\left[\left(\frac{2\varepsilon_s}{qN_d t^2}\right)(\phi_i - V_G)\right]^{1/2}\right\}V_D \qquad \text{(A-63)}$$

for small drain voltages. For larger values of V_D, the space-charge region width becomes a function of position along the channel. If it is assumed that the electric field along the channel is much less than the electric field perpendicular to the channel, the space-charge region width is given by

$$x_d = \left[\left(\frac{2\varepsilon_s}{qN_d}\right)[\phi_i - V_G + \phi(y)]\right]^{1/2} \qquad \text{(A-64)}$$

where $\phi(y)$ is the potential in the channel at point y. This is called the *gradual-channel approximation.* Then the incremental potential drop along the channel is given by

$$d\phi = -I_D\, dR = \frac{-I_D\, dy}{Wq\mu_n N_d(t - x_d)} \qquad \text{(A-65)}$$

Integrating this expression from the source to the drain yields

$$I_D = G_o\left\{V_D - \left(\frac{2}{3}\right)\left(\frac{2\varepsilon_s}{qN_d t^2}\right)^{1/2}[(\phi_i - V_G + V_D)^{3/2} - (\phi_i - V_G)^{3/2}]\right\} \qquad \text{(A-66)}$$

This equation is valid until $x_d(L) = t$, at which point the channel is said to be *pinched-off*, and the drain current saturates. The drain voltage to achieve this condition is, from Equation A-64,

$$V_{D\text{sat}} = \frac{qN_d t^2}{2\varepsilon_s} - (\phi_i - V_G) \qquad \text{(A-67)}$$

and the drain current is

$$I_{D\text{sat}} = G_o\left[\frac{qN_d t^2}{6\varepsilon_s} - (\phi_i - V_G)\left\{1 - \left(\frac{2}{3}\right)\left[\frac{2\varepsilon_s(\phi_i - V_G)}{qN_d t^2}\right]^{1/2}\right\}\right] \qquad \text{(A-68)}$$

It should be noted that if the gate voltage is more negative than

$$V_T = \phi_i - \frac{qN_d t^2}{2\varepsilon_s} \qquad \text{(A-69)}$$

there is no channel, and the device is off. V_T is called the *turn-off* voltage.

The gain parameter for the JFET is called *transconductance*, g_m, and is defined by

$$g_m \equiv \left. \frac{\partial I_D}{\partial V_G} \right|_{V_D = \text{constant}} \tag{A-70}$$

For the nonsaturated region of operation,

$$g_m = G_o \left(\frac{2\varepsilon_s}{qN_d t^2} \right)^{1/2} [(\phi_i - V_G + V_D)^{1/2} - (\phi_i - V_G)^{1/2}] \tag{A-71}$$

and, for saturation

$$g_{m\text{sat}} = G_o \left\{ 1 - \left[\left(\frac{2\varepsilon_s}{qN_d t^2} \right) (\phi_i - V_G) \right]^{1/2} \right\} \tag{A-72}$$

MOS FIELD-EFFECT TRANSISTORS

The charge in the channel of an enhancement mode metal-oxide-semiconductor field-effect transistor (MOSFET) is controlled by the gate voltage. The basic geometry of this device is shown in Figure A-5. The model developed in this section is a relatively simple model based on charge-control analysis.

For gate voltage levels below a value, V_T, called the *threshold voltage*, the drain and source diffused regions are isolated by a space-charge region, the width of which depends on the gate voltage, V_G. For values of V_G greater than V_T, the surface is inverted and a conducting channel exists between the drain and source. Once this channel forms, increasing the gate voltage results in an increase in the charge in the channel, Q_n, and the space-charge region width is primarily controlled by the drain voltage.

For low values of drain voltage, V_D, the drain current, I_D, is related to the channel charge by

$$I_D = -\frac{Q_n}{T_{tr}} \tag{A-73}$$

where T_{tr} is the channel transit time. This transit time is primarily due to drift in the channel, with a velocity, v_d, given by

$$v_d = -\mu_n \mathscr{E} = \frac{\mu_n V_D}{L} \tag{A-74}$$

and the transit time is

$$T_{tr} = \frac{L}{v_d} = \frac{L^2}{\mu_n V_D} \tag{A-75}$$

The channel charge is given by

$$Q_n = -C'_{ox}(V_G - V_T)WL \tag{A-76}$$

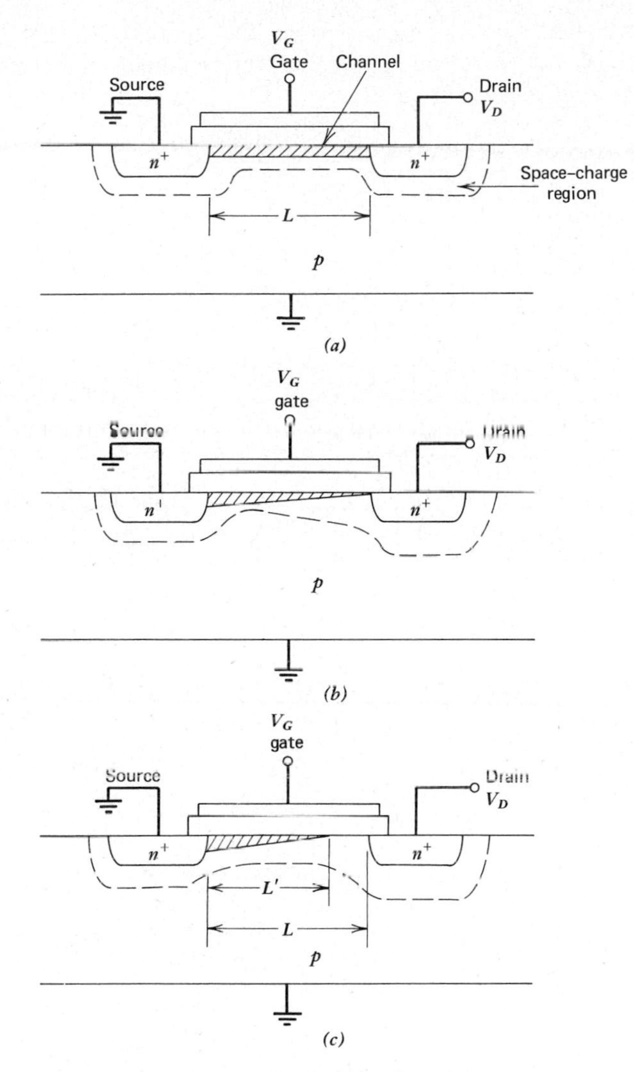

Figure A-5. The n channel MOSFET. (a) $V_G > V_T$, $V_D \simeq 0$. (b) $V_G > V_T$, $V_D < (V_G - V_T)$. (c) $V_G > V_T$, $V_D > (V_G - V_T)$.

where C'_{ox} is the capacitance per unit area of the gate, and W and L are the width and length of the channel. The drain current can be written

$$I_D = \mu_n\left(\frac{W}{L}\right)C'_{ox}(V_G - V_T)V_D \tag{A-77}$$

for low values of drain voltage.

For increasing values of V_D, the channel charge in the vicinity of the drain is reduced. As a first order approximation, the average voltage (above the threshold voltage) between the gate and the channel is assumed to be $(V_G - V_D/2)$ resulting in

$$Q_n = -C'_{ox}\left(V_G - V_T - \frac{V_D}{2}\right)WL \tag{A-78}$$

and

$$I_D = \mu_n\left(\frac{W}{L}\right)C'_{ox}\left(V_G - V_T - \frac{V_D}{2}\right)V_D \tag{A-79}$$

This equation is a good approximation up to the point where $V_D = (V_G - V_T)$, which is called the saturation drain voltage, $V_{D\,\text{sat}}$. For drain voltages higher than this value, the channel is pinched-off, resulting in a reduction of channel length to L', and a constant drain current given by

$$I_{D\text{sat}} = \mu_n\left(\frac{W}{2L}\right)C'_{ox}(V_G - V_T)^2 \tag{A-80}$$

A more exact treatment of MOSFET's is given in Chapter 10.

INDEX